W9-BYU-993

Ecology of small mammals

Ecology of small mammals

Edited by
D. Michael Stoddart
B.Sc, Ph.D.
Lecturer in Zoology
King's College
University of London

London
Chapman and Hall
A Halsted Press Book
John Wiley & Sons, New York

First published 1979
by Chapman and Hall Ltd
11 New Fetter Lane, London EC4P 4EE

Photoset in English Times by
Red Lion Setters, Holborn, London
Printed in Great Britain at the
University Press, Cambridge

ISBN 0 412 14790 4

Distributed in the U.S.A. by Halsted Press
a Division of John Wiley & Sons, Inc., New York

Library of Congress Cataloging in Publication Data
Main entry under title:

Ecology of small mammals.
"A Halsted Press Book."
Includes bibliographical references.
1. Mammals—Ecology. 2. Rodentia—Ecology.
I. Stoddart, David Michael. II. Title: Small mammals.
QL703.E26 1979 599.05 79-8
ISBN 0-470-26676-7

Contents

Preface

From their largely descriptive beginnings about a half century ago, studies on the ecology of small mammals have mushroomed in number, scope, content and complexity. Yet strangely, or perhaps not so strangely if one considers the extent and complexity of ecological interactions, the main problems for which the early workers sought answers still defy complete analysis, and basic hypotheses remain untested if not even untestable. The same holds true for so many branches of animal ecology that it seems to be the complexity of the concepts that frustrates efforts rather than the subject species. Like all branches of science, small mammal ecology has been subject to a series of fashionable approaches, one following another as technology penetrates previously impregnable regions. Doubtless the future development of our science will be punctuated by wave upon wave of new endeavour in whole fields that are perhaps even yet unidentified.

Answers to the complex questions which ecologists ask do not come easily. Increasingly though, they arise in direct proportion to the efforts expended upon their elucidation. Many studies have achieved such a high level of elegance, in terms of manpower and apparatus, that there is a feeling that questions asked when such resources are unavailable are not worth asking. Nothing could be further from the truth. Many a complex model has failed fully to explain the phenomenon for which it was constructed because of a lack of basic field data on the species' natural history. The growing tendency to skimp on the field work in favour of increasing the effort expended on elegant and fashionable analyses is to be deplored; sooner or later the architects of these procedures will yearn for more substantial footings.

This book is, in part, a review of the main pure and applied avenues from which the map of small mammal ecology is drawn. Additionally, it serves as a filter to point the way towards areas where less than satisfactory data are to be found. It should be of value to the pest control worker, conservationist and pure scientist alike and, because it is written with underlying principles very much in mind, it should remain a standard text for some time to come. Most of the book deals with rodent ecology. There are three good reasons for this. Firstly, on account of their numerical importance, both of species and of individuals, much more time and effort has been spent researching their ecology than the ecology of any other mammalian group. Secondly, the often occur in great densities making the practical considerations of trapping and population sampling somewhat easier than if they were sparsely dispersed, and thirdly, they eat huge quantities of man's crops and infect his family and stock with debilitating diseases. It is important, if trite, to point out that all small mammals are not rodents. The second most abundant order of mammals, in terms of number of species, is the bats, yet relatively little effort has been expended upon their behalf. This is perhaps because they are mainly difficult to find, catch, mark and study. Mostly they lead inoffensive lives impinging little on man, but a few species carry disease and a few more are locally serious pests of commercial fruit growers. They are an extremely successful group of small mammals and they may be found over land everywhere from the high northern latitudes to high southern latitudes. The principles underlying their success are well worth knowing.

The small marsupials of Australia are a somewhat isolated and little known group of small mammals. The frailness and vulnerability of the natural habitats of Australia and its associated islands make it imperative that the strongest of conservation measures be taken in order to prevent further extinctions. This seemingly huge and, in European terms, unspoiled continental assemblage has suffered mightily during just 200 years of European colonization with the native mammalian fauna bearing the brunt of forest clearance and stock rearing. Different ecological principles apply to small marsupials than to small placentals, the latter of which can survive sometimes quite severe environmental perturbations. Which features of their evolutionary past and ecological present have led the marsupials to their present predicament? These, and other questions, are not easily answered at the present state of our knowledge, but growing awareness of their existence may act as a further spur to further endeavour.

Most of the world's 3940 or so species of mammal weigh less than 5 kg; medium sized mammals are rare, while there is a fair number of those weighing over 30 kg. In this volume we have restricted our discussion to those species weighing 5 kg or less. We accept the arbitrary nature of this restriction and nowhere has the guillotine separated a large from several small cousins. Much of the discussion relates with equal applicability to larger species, many of which owe their existence to the presence of multitudes of small forms but, in restricting the scope of our coverage, we are accepting two principles: the need for brevity of the book and the disproportionately large amount of data on the smaller forms. Much of what can be said about the evolution of life-history strategies or energetic relationships between small mammals populations and their environments can be said about other groups of animals, and may well have been said. No apology is made for this; in fact it emphasizes the point that there is nothing fundamentally special about small mammals that singles them out from the remainder of the vertebrates. Perhaps the only reason for recognizing their corporate existence at all is their success. If one half of the stored cereals spoiled each year by rodents could be given to the underdeveloped countries of the world, massive inroads into malnutrition could be made. If the money spent on mammalian pest eradication each year could be channelled into these countries' agricultural programmes, human starvation would rapidly become a thing of the past. Such is the force of destiny surrounding small mammals.

During the past decade, the meaning of the word 'ecology' has taken on a series of wider, more vernacular cloaks than the single one which enshrouded it at its birth. By and large, this is desirable because it focuses the attention of the general populace upon matters affecting their natural surroundings. There is a danger though, that ecology comes to mean whether or not a particular development should or should not go ahead, and whether roads, airports and pollution-emitting motor cars are necessary adjuncts to an expected and sought after high quality human life. Few non-scientists see the need to separate autecological studies from synecological studies but the need nevertheless exists, and it is more important now than ever before to build up data on as many species as possible. In writing this book we are trying to define the characteristics of the free lives of small mammals; others may deem our coverage insufficient. In spurring them to extend what we have started here we are, in a very real sense, making a great deal of scientific research available to those upon whom the mantle of environmental decision making must fall. If we have achieved this, our venture will have been worth while.

It has been an honour to have had the opportunity to work with my colleagues on the compilation of this book and I salute their patience in dealing with my requests. I acknowledge the help and forbearance I have had from the publishers, in particular from Denis Ingram and Daphne Richardson.

London 1979 D. MICHAEL STODDART

Contributors

F.E.G. Cox	Department of Zoology, King's College, University of London, Strand, London WC2R 2LS, U.K.
T.H. Fleming	Department of Biology, University of Miami, Coral Gables, Florida 33124, U.S.A.
J. Gaisler	Department of Biology of Animals and Man, Faculty of Science, J.E. Purkyně University, 61137 Brno, Kotlařská 1, Czechoslovakia.
G.F. Hayward	Animal Ecology Research Group, Department of Zoology, University of Oxford, South Parks Road, Oxford OX1 3PS, U.K.
A. Myllymäki	Department of Pest Investigation, Agricultural Research Centre, SF-01300 Vantaa 30, Finland.
J. Phillipson	Animal Ecology Research Group, Department of Zoology, University of Oxford, South Parks Road, Oxford OX1 3PS, U.K.
H.N. Southern	Department of Zoology, University of Oxford, South Parks Road, Oxford OX1 3PS, U.K.
C.H. Tyndale-Biscoe	Division of Wildlife Research, C.S.I.R.O., P.O. Box 84, Lyneham, ACT 2602, Australia.

Life-history strategies

1

THEODORE H. FLEMING

1.1 Introduction

In terms of numbers of species and individuals, small mammals, defined as species weighing about 5 kg or less, dominate the Class Mammalia. As pointed out by Bourliere (1975), 10 of the 16 contemporary mammalian orders contain mostly small species, and 90 per cent of the 3900 recognized recent species weigh <5 kg. Small mammals are represented in each of the three major lines of mammalian evolution (monotremes, marsupials, and placentals), and the diversity of their trophic and ecological adaptations is impressive, ranging from fossorial grazers (e.g. pocket gophers and mole rats) to aerial carnivores (e.g. several species of phyllostomatid bats). Given such a diverse array of lifestyles, it should not be surprising to learn that small mammals have evolved numerous life-history strategies in adapting to a wide range of environments. In this chapter, I propose to discuss the basic concepts underlying the evolution of the life-history strategies of small mammals and will illustrate these concepts with data from the recent mammalian literature. Before outlining what I believe to be the major components of mammalian life-histories and how such life-histories evolve, I wish to briefly review the advantages and disadvantages of (1) being small and (2) being a small mammal. As we will see, size *per se* can place major constraints on many aspects of an organism's general adaptive strategy

A wide array of biological traits are either positively or inversely correlated with body size. Traits that are often positively correlated with size include (1) generation time, (2) competitive ability (both within and between species), (3) social dominance, (4) mobility, (5) absolute brain size, and (6) absolute metabolic rate. In addition, habitat stability, in a temporal sense, tends to increase with body size (Southwood *et al.*, 1974). Traits that are often inversely correlated with size include: (1) rate of population increase, (2) vulnerability to predation, (3) mass specific metabolic rate, and (4) cost of locomotion. Compared with large species, therefore, small species are more vulnerable to the vicissitudes of their environments and tend to have higher population turnover rates. Their relative energetic demands may be quite high, but their absolute energetic demands can be quite low.

Specific advantages and disadvantages of being a small mammal have been outlined by Bourliere (1975). The disadvantages of being small include the following: (1) a high cost of homeothermy, (2) faster metabolic rates that lead to shorter lifespans, (3) the high rate of daily food intake per unit size, (4) the high cost of locomotion, and (5) the constraints of a short lifespan on the evolution of social behaviour. Being small places obvious energetic constraints on many species of mammals. On the positive side, the advantages of being small include: (1) the ease of concealment from predators and hence a lower energetic commitment towards escape behaviour, (2) a wider range of potential food types, (3) a wider range of potential microhabitats, and (4) a higher rate of population increase that favours a rapid response to environmental changes. To judge from both the taxonomic and numerical dominance of small mammals, the advantages of small size clearly outweigh the disadvantages. On an evolutionary timescale, small mammals have a 33 per cent higher rate of formation of new genera than do large mammals, and the extinction rate of small-sized genera is less than that of large-sized genera. These size-related differences suggest that a kind of group selection at the species level is operating in the evolution of mammals (Van Valen, 1975).

Throughout this chapter, I will be using the term 'strategy' in a non-teleological sense to refer to a suite of adaptive responses that a species accumulates over evolutionary time (Wilbur *et al.*, 1974). Life-history strategies have traditionally included demographically important traits such as juvenile and adult mortality schedules, age at sexual maturity, age-specific fecundity, degree of parental care, and so forth. In this review, however, I wish to expand the number of traits considered as a part of a life-history strategy to include energetic traits (i.e. food habits and metabolic strategies), and behavioural traits (i.e strategies of habitat selection and social organization). I do this because there exists a close relationship between a species' habitat(s), diet, energetic characteristics (i.e. homeothermy v. heterothermy), social organization and its demographic characteristics. To discuss the evolution of demographic strategies without consideration of a species' physical, energetic, and social environment would be a futile exercise indeed.

1.2 Evolution and life-history strategies

My thesis in this chapter is that the demographic, energetic, and behavioural characteristics of small mammals are the products of organic evolution operating via natural selection acting at the individual level. Although this is the conventional view of the origin of adaptations held by most biologists, there exists enough confusion in the literature as to the functions of various adaptations, and hence the evolutionary mode by which they arose, to warrant a brief examination of how natural selection operates. Much more detailed treatments of the evolutionary process are to be found in Dobzhansky (1970) and Stebbins (1977).

As noted by Lewontin (1970), natural selection can operate on any level of biological organization that shows variation, reproduction, and heritability; these levels can range from molecules on up to ecosystems. For our purposes, however, three levels (individuals, kin groups, and non-related groups) are of especial importance in the evolution of life-history strategies. Each of these levels of selection will be discussed below.

1.2.1 Individual selection

This is 'the force on the genetic composition of a population that arises directly from differences in age-specific mortality and fertility schedules of different genotypes, independent of the genotypes of their ancestors, their descendents, or their contemporaneous relatives' (Lewontin, 1970). Evolution at the individual level operates via differential reproductive success, and fitness is measured in the currency of surviving, reproducing offspring. Selection acts to maximize mean reproductive performance, regardless of its effect on long-term population survival. While individual selection is often viewed as operating via competition for limited environmental resources, Williams (1966) points out that it can operate in the complete absence of conventional ecological competition but always involves reproductive competition. The outcome of individual selection is selective gene substitution which produces *organic adaptations*, defined by Williams (1966) as mechanisms designed to promote the success of individual organisms. From a behavioural standpoint, individual selection favours selfish, rather than altruistic, behaviour – behaviour that is aimed at promoting an individual's own genetic self-interests at the expense of other (unrelated) individuals.

1.2.2 Kin selection

This is a form of natural selection that deals with gene frequency changes 'owing to effects on the reproduction of relatives of the individual(s) in which a character (allele) is expressed ... ' (West-Eberhard, 1975). Whereas individual selection promotes selfish behaviour, kin selection will favour altruistic behaviour when it is directed towards an individual's close relatives that share a large fraction of its genes. Lewontin (1970) views kin selection as arising as an extension of maternal (parental) behaviour. Kin selection doesn't necessarily require a group-like social structure but can operate 'whenever relatives live close to one another ... ' and hence can influence each other's survival and reproductive success (Maynard-Smith, 1976). In his original formulation of the theory of kin selection, Hamilton (1964) noted that an altruistic gene can increase in frequency in a population if the following inequality holds: $K > 1/r$, where K is the ratio of the advantage to the disadvantage of an altruist's act and r is the genetical correlation between the altruist and the beneficiary of the act (e.g. $r = 0.5$ for diploid parents and offspring, $r = 0.25$ for parents and grandchildren, etc.).

Related to kin selection is the concept of *inclusive fitness*. As discussed at length by West-Eberhard (1975), inclusive fitness has two basic components: (1) an individual's personal fitness (fitness in the classic sense that results from individual selection) and (2) his effect on the fitness of neighbours, times his relatedness to them. In algebraic terms, inclusive fitness (a_i) $= a + a_k$, where a refers to personal fitness and a_k refers to the lifetime sum of the effects on the fitness of relatives, each effect being weighted by the degree of relatedness of the individuals affected (West-Eberhard, 1975). In the evolution of many life-history traits, particularly those dealing with parental and other social behaviours, it is likely that natural selection is maximizing inclusive fitness rather than simply individual fitness.

1.2.3 Group selection

When evolution operates at this level, gene frequency changes occur as a result of differential survival of groups. Group selection favours survival of the group and can therefore favour adaptations that generally oppose an individual's self-interests, such as reproductive restraint aimed at preventing the overexploitation of a population's resources. This evolutionary mechanism favours altruistic behaviour (directed towards non-relatives), and, according to Williams (1966), produces *biotic adaptations* which are traits designed to promote the success (i.e. survival) of biotas (e.g. populations, communities, biomes, etc.). Group selection is often explicitly or implicitly invoked in discussions of population regulation and evolution of co-operative behaviour (e.g. Wynne-Edwards, 1962, 1965). According to a number of recent models (reviewed by Maynard-Smith, 1976), three conditions must exist in order for group selection to operate: (1) average group size must be small so that genetic drift can affect gene frequencies, (2) migration rates between groups must be low, and (3) groups containing selfish alleles must have higher extinction rates than groups containing altruistic alleles. It remains to be seen how often these conditions exist in nature. If group advantageous effects can evolve by both selfish and altruistic methods, the former method will prevail. Most superficially altruistic behaviours such as alarm calls have probably evolved via individual selection (Wilson, 1977) or kin selection (Sherman, 1977).

Although most biologists subscribe to the view that adaptations are the product of individual selection, there is ample evidence in the mammalian literature that subtle, or not so subtle, forms of 'group selection thinking' exists in the minds of some ecologists or behaviourists. Two examples will suffice to illustrate my point. Fisler (1971) has noted that the sex ratio in three populations of two species of harvest mice (*Reithrodontomys*) are skewed in favour of males in the youngest and oldest of six age classes. This skew is puzzling because, according to Fisler, fewer males are needed by populations to inseminate females so that males ought to be the 'expendable' sex; females should be at a premium because they bear young. Fisler explains this apparent anomaly by suggesting that in small, secretive species such as harvest mice 'there may be a selective advantage in having many reproductively active males moving about to assure insemination of any females that come into oestrus ... '. Fisler clearly interprets the excess of males as being advantageous to the population, apparently because it maximizes reproductive efficiency. If his thinking was centered on selection operating at the individual level, he would not have stated that males are expendable in populations. Although it is often true that sex ratios in small mammal populations favour females, it is incorrect to interpret this as being a population adaptation (for reducing population pressure on limited resources?). A deficiency of males in a population probably reflects their higher mortality rates that can arise, in part, from the risks they take to maximize their fitness by inseminating as many females as possible.

A second example of group selection thinking comes from Christian's (1970) discussion of the role of socially subordinate individuals in the evolution of species of small mammals. He states that the ' ... inhibition of maturation, whether density-dependent or density-independent, of young born late in the breeding season may be a mechanism that has evolved that insures an adequate breeding population an optimum habitat for the succeeding breeding season.' Again, this example seems to ascribe a population advantage to an observed ecological phenomenon. An alternative explanation that invokes individual rather than group selection is that selection favours those individuals born late in the breeding season that do not undergo sexual maturation until the following spring because their overwinter survival will be higher and they will achieve greater reproductive success under more favourable breeding conditions in the spring.

1.3 Demographic components of life-history strategies

My general approach in this and the following sections will be to state the general evolutionary problem that species face in a given portion of their life-history strategy, to discuss the theoretical aspects of the subject, and to briefly review pertinent empirical data, covering as wide a range of small mammals as possible. My overall aim here is to provide a conceptual framework for an understanding of the ecological and behavioural characteristics of small mammals.

Because fitness is measured by the number of surviving, reproducing offspring an individual produces during its lifetime, demographic characteristics have come to represent the major focus of interest in the evolution of life-history strategies. The basic evolutionary problem is this: given a particular environmental setting, how should an individual distribute its reproductive commitment to maximize its contribution to future generations? (Cody, 1971). There are, of course, many parameters that could be varied to arrive at an optimal demographic strategy for a given environment. These include traits such as age at first reproduction, number of reproductions per season or lifetime, number and size of offspring at birth, the amount of parental care invested in each litter, and the total amount of energy or other resources allocated to reproduction as opposed to growth, maintenance, competitive ability, and predator escape (Cody, 1966; Wilbur *et al.*, 1974). It is highly unlikely that these parameters evolve independently of each other, but instead evolve as a group of interdependent traits so that a change in one parameter affects a variety of other parameters. Demographic strategies thus represent suites of co-evolved characteristics that reflect an organism's response to its environment (Cody, 1971).

1.3.1 Theoretical considerations

Beginning with the work of Cole (1954), a large body of theoretical literature dealing with the evolution of demographic strategies has appeared. Much of this material has been excellently reviewed by Stearns (1976, 1977) and only a small portion of the results of these studies will be dealt with here. Before proceeding with this material, I need to briefly review two topics; (1) the basic ingredients of any demographic model and (2) the demographically important features of environments.

Basic demographic ingredients

These include a set of age-specific life-history parameters that deal with rates of mortality and natality (reproduction) within a population and are best illustrated in the form of a *life and fecundity* table. Table 1.1 represents such a table for the American grey squirrel (*Sciurus carolinensis*) using data reported by Barkalow *et al.* (1970). The various symbols found in that table are defined as follows: l_x = the probability of a newborn surviving x time units into the future (alternatively, it can represent the proportion of a cohort of newborns born at the same time surviving x time units into the future); d_x = the proportion of individuals dying during each time interval; q_x = the mortality rate during interval x, calculated as the quotient of the number dying divided by the number alive at the beginning of x; e_x = the future life expectancy of an x-year old (e_0 = mean life expectancy in the population); m_x = age-specific fecundity as determined by the product of the age-specific litter size, times the proportion of females breeding at age x; $l_x m_x$ = the product of age-specific survivorship and fecundity values, which when summed over all age intervals, results in the parameter R_o, the net reproductive rate ($R_o = \Sigma l_x m_x$). The final parameter, T, represents mean generation time or the length of time that elapses between the birth of a female and the birth of its offspring. Details of these calculations are more fully discussed by Mertz (1970) and Caughley (1977).

Life and fecundity tables are usually constructed for only the females of populations (because it is usually easier to assess female fecundity), and when such a convention is followed, the magnitude of R_o automatically tells us whether the population from which the data were drawn is increasing, stable, or decreasing in size as follows: when $R_o > 1.0$, the population is increasing; when $R_o = 1.0$, it is stable; and when $R_o < 1.0$, it is decreasing. The *instantaneous* rate of population increase, symbolized by r, can be approximated by the formula $r = R_o/T$. r is maximal when a population has attained a stable age structure, which requires that its l_x and m_x schedule remain constant for several generations (Caughley and Birch, 1971).

The information summarized in a life and fecundity table can also be presented graphically (Fig. 1.1). Here, the mortality data are presented in the form of a *survivorship curve*, which in the case of *Sciurus carolinensis* takes the form of a type IV curve of Slobodkin (1961) in which the mortality

Table 1.1 *A life and fecundity table of the American gray squirrel,* Sciurus carolinensis. *Data are from Barkalow* et al. *(1970)*

x (years)	l_x	d_x	q_x	e_x	m_x	$l_x m_x$
0-1	1.000	0.753	0.753	0.99	0.05	0.050
1-2	0.247	0.135	0.545	1.82	1.28	0.316
2-3	0.112	0.030	0.266	2.41	2.28	0.255
3-4	0.083	0.026	0.316	2.10	2.28	0.189
4-5	0.056	0.015	0.216	1.84	2.28	0.128
5-6	0.042	0.025	0.588	1.32	2.28	0.096
6-7	0.017	0.000	0.017	1.48	2.28	0.039
7-8	0.017	0.017	1.000	0.50	2.28	0.039
						R_o = 1.112
						T = 2.601

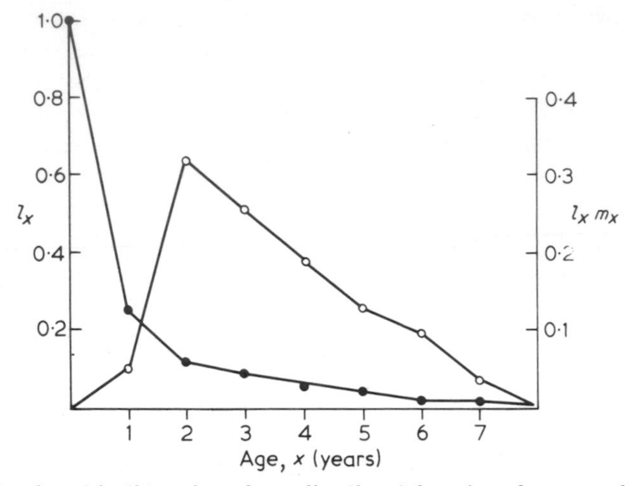

Fig. 1.1 Survivorship (l_x) and net fecundity ($l_x m_x$) functions for a population of the American grey squirrel, *Sciurus carolinensis*, based on data presented in Table 1.1.

rate is high early in life, tapers off during the middle years of life, and then increases again at the end of life. Such a 'U'-shaped distribution of mortality is characteristic of many species of mammals (Caughley, 1966). The net fecundity function of the grey squirrel is triangle-shaped, as it is in many species of animals. Note that sexual maturity is attained at one year of age in some, but not all, females and that two-year-olds make the largest reproductive contribution to the population. The area under the $l_x m_x$ curve represents the net reproductive rate or R_0, which is 1.11 in this example. Although usually expressed in algebraic form, most demographic models deal with optimizing the location (along the age axis) and shape of the net fecundity function so that r, which represents the difference between instantaneous birth (b) and death (d) rates, is maximal under a given set of environmental conditions.

Another important demographic concept is that of *reproductive value*, symbolized as V_x. Reproductive value, which can be defined as the age-specific expectation of all present and future offspring, has two basic components, the number of progeny produced by a female in the current year plus expected future progeny, each discounted by the probability that the female will live long enough to produce them. Following Pianka and Parker (1975), the reproductive value of an x-year old female can be symbolized as:

$$V_x = m_x + \sum_{t=x+1}^{\infty} (l_t/l_x)\, m_t.$$

The far right-hand term of this expression is known as the *residual reproductive value*, i.e. future expected progeny. Most demographers believe that an inverse relationship exists between the amount of energy devoted to the production of current offspring and the residual reproductive value; a large reproductive effort (defined more formally below) in the current breeding season reduces adult survivorship to the next breeding season, thereby reducing the value of future offspring.

The final demographic ingredient I wish to mention is dispersal. The movement of animals into and out of a population can obviously have an important influence on population size and composition. Rather than discuss dispersal strategies at length here, I propose to postpone this discussion until I have dealt with the other demographic parameters in more detail.

Demographically important aspects of environments

This aspect of demographic theory has been of special interest to Southwood (1976, 1977), who recognizes three important features of habitats and environments. The first feature is the *durational stability* of a habitat, or the time that an area remains inhabitable. The concept of ecological succession becomes important here in that early successional stages have low durational stability whereas late successional stages have high durational stability. Other things being equal, species living in early successional habitats should have very different demographic strategies, especially dispersal strategies, than those living in late successional habitats.

The second habitat feature is *temporal variability*, which Southwood defines as the extent to which the carrying capacity (K) of a habitat varies during the time the site is inhabitable. Temporal variability measures the degree to which resources and living conditions vary seasonally. As we will see when we discuss the concepts of r and K selection, habitat seasonality can have a profound influence on the evolution of demographic strategies.

The final habitat feature is *spatial heterogeneity* or the degree of habitat patchiness. Spatial variation in the distribution of food, shelter, and potential mates figures importantly in a species' total life-history strategy, especially in its choice of diet, preferred habitat, and mating system. By affecting local population density, spatial heterogeneity can influence the intensity of density-dependent factors such as competition, disease, and predation and can thus affect various demographic parameters. Wiens (1976) discussed in detail the ecological implications of spatial heterogeneity.

A species' demographic response to a particular environment will depend on two factors, the length of time a habitat remains favourable for survival and reproduction (designated H by Southwood) and the generation time (T) of the species (which is itself an evolved response to a particular set of environmental conditions). A ratio of $T/H > 1.0$ means that a species must constantly recolonize that habitat. This situation will favour a very different set of demographic characteristics compared to stable habitats in which the ratio of T/H is low.

The evolution of reproductive effort

At the core of a species' demographic strategy is the age-specific distribution of *reproductive effort*, which is defined by Stearns (1976) as 'the rate at which resources in excess of maintenance requirements are diverted into reproduction rather than growth.' In the case of mammals, Millar (1977) defined reproductive effort in terms of the feeding intensity needed to support a litter of offspring rather than the cumulative amount of energy devoted to offspring production. However it is defined, reproductive effort and its distribution over the lifetime of individuals figures importantly in models of life-history evolution, which deal with the following kinds of questions. What is the optimal age of first reproduction? Will semelparity (single reproduction) or iteroparity (repeated reproductions) result in highest fitness? What is the optimal balance between offspring size and number? Should reproductive effort increase, decrease, or remain constant with age? What is the importance of the relative magnitude of juvenile v. adult mortality rates on the evolution of reproductive effort?

Also prominent in discussions of life-history evolution are the concepts of *r* and *K* selection, first popularized by MacArthur and Wilson (1967). *r* selection is a form of individual selection which operates in environments that are periodically or chronically unsaturated with organisms in relation to their carrying capacity (*K*); this unsaturation can result from periodic fluctuations in *K* or from periodic bouts of density-independent mortality. As a result of low demand:supply ratios of potentially limiting resources, selection will favour genotypes that maximize their reproductive effort at every breeding opportunity. Early maturity and frequent large litters of small offspring will be favoured under *r* selection. By contrast, *K* selection is a form of individual selection which operates in environments that are usually saturated with organisms – ones in which mortality is primarily density-dependent and in which the demand:supply ratio for limiting resources is high. In such situations, selection will favour genotypes that do not make a maximum reproductive effort at each breeding opportunity. Instead, it favours genotypes in which maturity is delayed and which produce smaller litters of larger, more fit offspring. Whereas *r* selection favours maximum production of offspring each with relatively low fitness, *K* selection favours maximum efficiency in the use of resources and the production of a few highly competitive offspring. Other characteristics of *r* and *K* selected genotypes (or phenotypes) are listed in Table 1.2. It should be noted that these characteristics are stated in relative terms and that *r* and

Table 1.2 *Characteristics of the extremes of the* r-K *selection continuum. Modified from Southwood (1977).*

r species	*K* species
Short generation time	Long generation time
Small size	Large size
High level of dispersal	Low level of dispersal
Much density independent mortality	High survival rate, especially of reproductive stages
High fecundity	Low fecundity with high parental investment or iteroparity
Panmictic	Territorial
Intraspecific competition via exploitation	Intraspecific competition via interference
Low investment in 'defence' and other interspecific competitive mechanisms	High investment in 'defence' and other interspecific competitive mechanisms
Time efficient	Food and space resource efficient
Populations often overshoot *K*	Populations seldom overshoot *K*
Population density variable – often 'boom or bust'	Population density relatively constant at *K* from generation to generation
T/H large	T/H small

K 'syndromes' should be viewed as endpoints of a continuous demographic spectrum.

Figure 1.2 schematically illustrates the major demographic differences between *r* and *K* strategists. At any population density below *K*, the instantaneous birth and death rates of the *r* strategist are usually greater than those of the *K* strategist. As a result, the maximum instantaneous rate of increase (r_{max}) of the former is greater than that of the latter. In contrast, because of its better competitive ability and its more efficient use of resources, the *K* strategist has a higher equilibrium density (*K*) than does the *r* strategist. Pianka (1972) discusses these differences in great detail.

While the concepts of *r* and *K* selection have sparked considerable theoretical and empirical interest (*see* Pianka, 1972), various workers have criticised these concepts as being too simplistic and too inadequate to explain the variation in life-history traits seen in nature (Wilbur *et al.*, 1974). For example, a given array of '*r*-selected' demographic traits can result from a variety of different selective pressures, not simply from low demand:supply ratios and/or density-independent mortality. Nicols *et al.* (1976) further criticise the concepts because they do not take into account

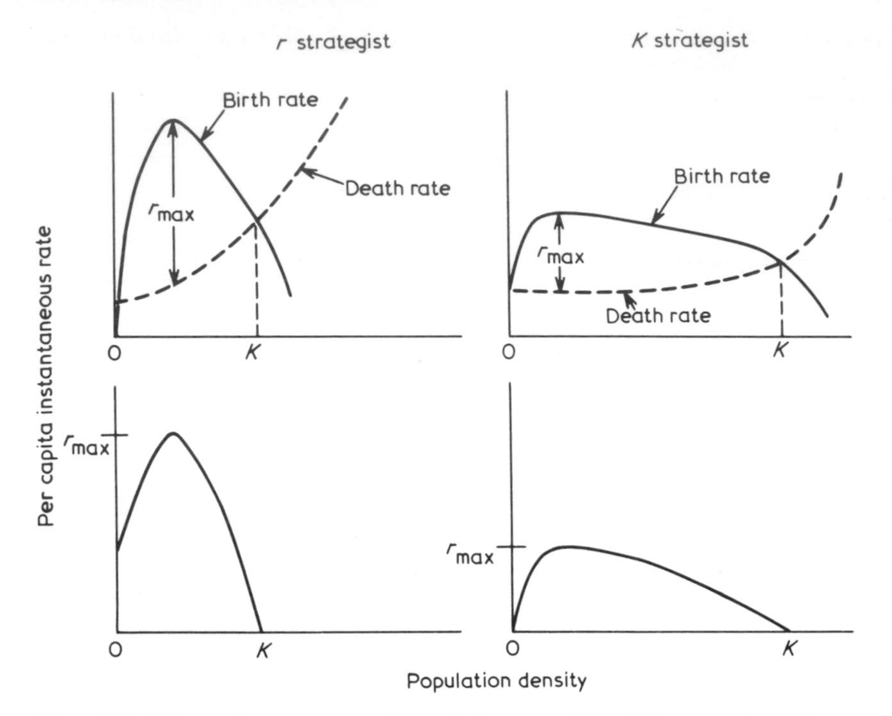

Fig. 1.2 Schematic representation of demographic differences between an r strategist (left) and a K strategist (left). r is calculated as the difference between instantaneous birth and death rates. K represents the environmental carrying capacity above which populations have negative growth rates. (After Pianka, 1972).

the flexible nature of the reproductive strategies of many species, especially those living in highly variable environments such as low rainfall deserts. In the end, to ascribe the evolution of high reproductive effort per breeding attempt to 'r selection' or its converse, low reproductive effort, to 'K selection' is to gloss over the many ways in which a given evolutionary result can be attained.

From a review of the major papers dealing with the evolution of reproductive effort and its attendant tactics (e.g. age at maturity, frequency of breeding, litter size, etc.), it appears that the following kinds of information are necessary (and sufficient?) for a full understanding of demographic evolution: (1) stability and predictability of the environment; (2) the trophic and successional position of a species; (3) the demand:supply ratio of critical resources; and (4) the predictability of juvenile and adult mortality patterns. Given this information about a species and its environment, various predictions about its demographic characteristics can be made. These predictions are summarized in Table 1.3, where the end results of either an r or K selection regime or a 'bet-hedging' strategy, in which reproductive risks are minimized rather than offspring production being

Table 1.3 *Demographic predictions resulting from r and K selection and bet-hedging strategies. From Stearns (1976).*

r and *k* selection and bet-hedging with adult mortality variable	
Stable environments	Fluctuating environments
Slow development and late maturity	Rapid development and early maturity
Iteroparity	Semelparity
Smaller reproductive effort	Larger reproductive effort
Fewer young	More young
Long life	Short life
Bet-hedging with juvenile mortality variable	
Early maturity	Late maturity
Iteroparity	Iteroparity
Larger reproductive effort	Smaller reproductive effort
Shorter life	Longer life
More young per brood	Fewer young per brood
Fewer broods	More broods

maximized, are contrasted. Stearns (1976) provides a full discussion of the bases for these predictions.

Two basic suites of demographic characteristics can be recognized in this table: (1) Early maturation, large reproductive effort (in terms of litter size) per reproduction, and short lifespan, and (2) delayed maturity, small reproductive effort per reproduction, and long lifespan. The former strategy is selectively favoured in fluctuating environments when adult mortality is variable on a seasonal or yearly basis and in stable environments when juvenile mortality is variable from one breeding season to the next. The latter strategy is selectively favoured in stable environments when adult mortality is variable and in fluctuating environments when juvenile mortality is variable. Semelparity is advantageous only in fluctuating environments in which adult mortality is variable. In the other three situations, iteroparity is favoured. While many of the same predictions emerge from different theoretical studies, Stearns (1976) and Bell (1976) warn us that they result from relatively simple mathematical models which ignore important real-world complexities, such as sexual reproduction. The predictions should thus be viewed as hypotheses on how the demographic characteristics of animals living in different environments should differ and not as rigid expectations about the real world.

Theoretical studies indicate that there should exist significant differences in the magnitude of reproductive effort as a function of environmental stability and the variability of mortality. Reproductive effort should also vary with potential lifespan and age within a lifespan. In species with a short potential lifespan, such as those living in habitats of low durational stability, a high reproductive effort early in life will obviously be advantageous. Conversley, if potential lifespan is long, a large reproductive effort at any breeding attempt will probably incur a greater cost in terms of reduced residual reproductive value than the immediate gain in fecundity. Therefore, low reproductive effort will be advantageous in long-lived species (Goodman, 1974). Reproductive effort should increase with age whenever immediate benefits outweigh costs (Stearns, 1976), or if the efficiency of reproduction increases with age (Hirshfield and Tinkle, 1975). Although in an evolutionary sense current reproductive effort (i.e. the amount of energy devoted to reproduction in the current breeding season) is weighed against potential losses of fitness in the future, it should also be sensitive to current environmental conditions and should temporarily increase when conditions are good and decrease when conditions are bad (Hirshfield and Tinkle, 1975; Pianka and Parker, 1975; Ricklefs, 1977).

The evolution of dispersal strategies

As previously mentioned, dispersal, defined as any movement of an individual or its propagules from its home area (Lidicker, 1975), is an important demographic process. The demographic consequences of dispersal include potential changes in birth and death rates, population density, age structure and sex ratio, social organization, and patterns of habitat utilization. In addition, dispersal can also have important evolutionary consequences, including the differential movement of genes among populations, reduced inbreeding and opposition to the improvement of local adaptations, and an increased probability of the occurrence of novel gene combinations (Lidicker, 1975). Dispersal is sometimes considered to be a biotic adaptation (*sensu* Williams, 1966; Howard, 1960), and Van Valen (1971) has stated that individual selection opposes dispersal whereas group selection favours it. However, I will adopt the view that dispersal strategies are products of individual selection.

Regardless of whether it occurs in juveniles or breeding adults, dispersal movements away from natal sites or current home ranges must result in a net gain in fitness if they are to be selectively advantageous. Otherwise, assuming that dispersal tendencies have a genetic basis, they will be selected against. The risks incurred as a result of dispersal include increased exposure to biological and physical hazards during the move from one home site to another, especially if unfavourable habitats must be traversed, and the possibility that the new home site is located in sub-optimal habitat where probabilities of survival and successful reproduction (owing to a closed social system, etc.) are low. Although the risks involved in dispersal may be high, the potential gain in fitness upon dispersal can also be high owing to: (1) increased access to critical resources and mates, (2) decreased exposure to predators and disease, and (3) an opportunity to produce more fit offspring via heterosis.

As in the case of other aspects of demographic strategies, the evolution of dispersal strategies will be especially sensitive to patterns of environmental variability or stability (Gagdil, 1971). Well-developed dispersal tendencies will be selectively favoured under the following environmental conditions: (1) when habitats have low durational stability; (2) when suitable habitat patches are relatively close together; and (3) when catastrophic mortality creates temporarily vacant (or markedly unsaturated) habitat patches. The opposite conditions will generally favour reduced dispersal tendencies. Even in stable environments, however, selection may favour genotypes with dispersal ability (Hamilton and May, 1977).

1.3.2 Empirical data

Several decades of work on the population ecology of a wide variety of species inhabiting many different environments have produced an enormous amount of information on the basic demographic properties of small mammals. Most of these data must be considered 'baseline' as the studies producing them were not designed to test general theories of life-history evolution. The number of studies that provide sufficient amounts of quantitative data for assessing mean values and estimates of variances of critical life-history parameters are relatively few; studies summarizing the demographic data in life and fecundity table form are still fewer. Thus, although we know a great deal about the general demographic features of numerous species, there still is considerable need for life table studies of many species, and more important, there is a need for studies specifically designed to test general theory using an experimental and/or comparative intraspecific approach wherever possible. Hirshfield and Tinkle (1975) and Stearns (1976) discuss the kinds of studies that are needed for testing demographic theory.

Out of the wealth of available data, I have chosen to discuss the following topics: (1) intraspecific variation in the demographic parameters of small mammals; (2) interspecific variation in life-history patterns; and (3) general trends in dispersal strategies.

Intraspecific variation in demographic parameters

Small mammals display considerable intraspecific variation in their demographic traits. If it has a genetic basis (and this question has rarely been investigated in small mammals), this variation forms the raw material with which evolution works in optimizing a species' demographic response to its environment. If this variation has a purely phenotypic basis, then its existence indicates that selection has probably placed a premium on flexibility of an individual's response to a fluctuating environment. Variation has been reported in the following life-history traits: timing and duration of the breeding season, age at sexual maturity, litter size and number per season, and juvenile and adult mortality. As discussed below, this variation can arise as a result of factors that are either extrinsic or intrinsic to the population or species.

Since many of the habitats occupied by small mammals undergo significant seasonal fluctuations regarding climatic conditions and food availability, the timing of reproduction can have a critical effect on an individual's fitness, and selection should favour those individuals producing young at the most favourable time(s) of the year. Some species, such as the pika (*Ochotona princeps*) show little yearly variation in the onset of the breeding season; first litters in Alberta, Canada, are conceived in May, at the beginning of the vegetation growth season, and second litters are produced under mild climatic conditions (Millar, 1972). In contrast, the beginning of the breeding season in the red squirrel (*Tamiastriatus hudsonicus*) in southern British Columbia varies from February to May, depending on the severity of the previous winter (Millar, 1970a). Voles (*Microtus*) tend to have breeding seasons that vary in their onset, which has been experimentally shown to be affected by chemical signals in their plant resources (Negus and Berger, 1977), and their duration; breeding sometimes occurs in winter during the increase phase of vole cycles (Krebs and Myers, 1974). The presence of a competing species, *Microtus townsendii*, has been shown to delay the onset of breeding in field-dwelling populations of *Peromyscus maniculatus* in British Columbia. The mechanism behind this suppression is unknown (Redfield *et al.*, 1977).

Although many studies have documented variation in the onset and duration of breeding in various species, few studies have addressed the question of the adaptive significance of such variation. In some environments, particularly those in which the availability of critical resources fluctuates widely from year to year, it is easy to understand the significance of reproductive variation. In low rainfall deserts of the southwestern United States, for example, heteromyid rodents do not reproduce unless rainfall sufficient to cause the germination of winter annual plants occurred in the previous autumn (Beatly, 1969). Winter annuals are a critical resource for the rodents because they provide a source of seeds, dietary water, and perhaps vitamins that are necessary for successful reproduction. In other situations, however, the advantage of variation in the timing of breeding is not as clear as the following example will demonstrate. Female *Peromyscus maniculatus* in British Columbia can be roughly divided into two classes, 'early' and 'late' breeders (Fairbairn, 1977a). The early breeders, which represent less than 50 per cent of the population, breed in February or March, before environmental conditions are optimal for successful reproduction, and suffer relatively heavy mortality compared to females that wait until conditions for breeding are improved. Early breeders that do survive until conditions are better can produce a second litter, and daughters from their first litter may reproduce once before the breeding season has ended. Fairbairn calculates that the seasonal net productivity of early and late breeders is nearly the same, which accounts for this reproductive 'polymorphism' in the population.

Reproduction is generally considered to involve a risk to parental survival and future fecundity. The magnitude of this risk to adults and young will depend on the timing of gestation, lactation, and weaning in relation to environmental conditions. Those species in which periods of gestation coincide with peaks in food (or other critical resources) are minimizing the risk to pregnant females relative to that of lactating females and dispersing young. When food peaks coincide with lactation periods, pregnant females are exposed to greater risk (Bradbury and Vehrencamp, 1977b). In general, species living in relatively stable environments should minimize adult reproductive risk by timing gestation periods to coincide with the most favourable conditions and have more than one reproduction per year, whereas species living in more seasonal environments should minimize juvenile mortality by timing lactation and weaning periods to coincide with the most favourable period(s).

The breeding patterns of four species of tropical emballonurid bats conform closely to these expectations (Bradbury and Vehrencamp, 1977b). Two species, *Rhynchonycteris naso* and *Saccopteryx leptura*, experience relatively small seasonal fluctuations in their insect food supplies and time the first of two yearly pregnancies to coincide with food peaks; the second pregnancy is timed so that the lactation and weaning periods coincide with high food levels (Fig. 1.3). Births tend to be asynchronous in these species. In contrast, two other species, *Saccopteryx bilineata* and *Balantiopteryx plicata*, experience greater food seasonality, and they have only one highly synchronous birth per year whose period of lactation and dispersal coincides with food peaks (Fig. 1.3). The annual female mortality (disappearance rates of *R. naso* and *S. leptura* are lower than those of *S. bilineata* and *B. plicata*, suggesting that females of the latter two species are taking greater reproductive risks than females of the former two species.

In addition to the timing of reproduction, parameters such as age at maturity and litter size and number per season can vary intraspecifically. In many species of rodents, young born early in the breeding season will mature fast enough to breed in their season of birth whereas late-born young often do not breed until the following season. In microtine rodents,

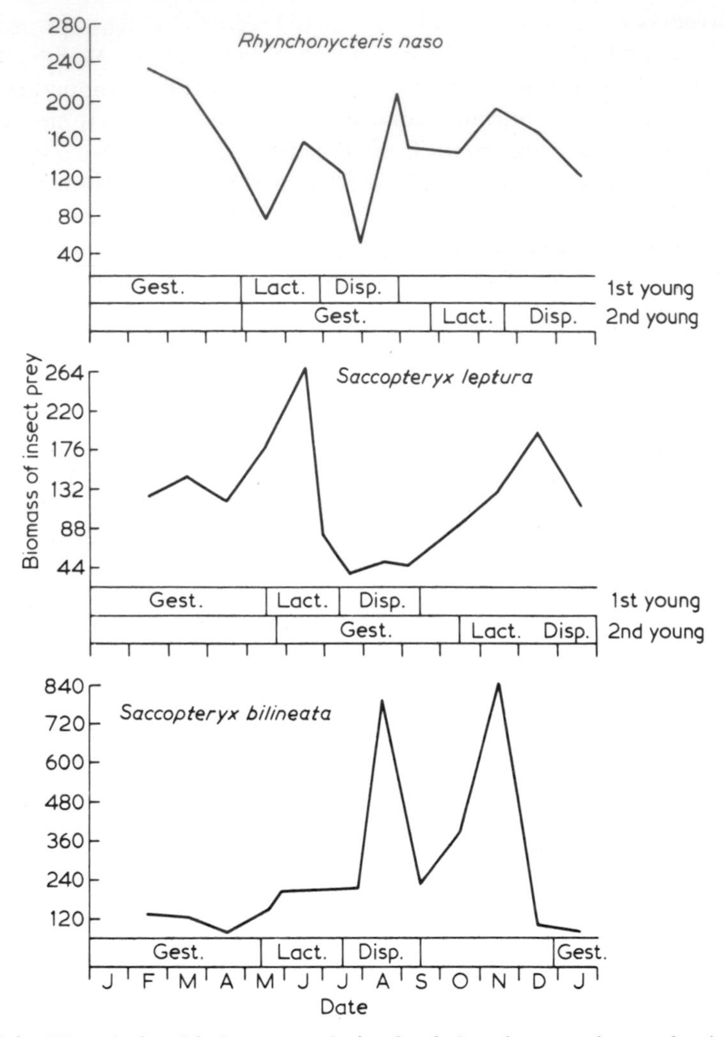

Fig. 1.3 The relationship between relative food abundance and reproductive events in three species of tropical emballonurid bats. The estimates of food abundance are bat species-specific, not habitat-specific, and refer to total prey biomass available in a species' foraging area at a given time of the year. (After Bradbury & Vehrencamp, 1977b).

size (age?) at maturity varies with the cycle of population density; size at maturity is lowest during the increase phase and highest during the peak and decline phases (Krebs and Myers, 1974). In the heteromyid rodent *Perognathus parvus*, young of the year breed only in exceptionally favourable years (O'Farrell *et al.*, 1975). As another indication of intraspecific variation in maturation, females in many species mature at an earlier age than males. Delayed maturity in males would be individually advantageous in situations in which males compete among themselves for sexually receptive females, and size and/or experience gives individuals a competitive advantage. Delayed maturity in females, which sometimes occurs in the red squirrel *Tamiasciurus hudsonicus*, for example, (Millar, 1970b), might be favoured if a considerable store of food (either in the form of fat or a food cache) is necessary in order for reproduction to be successful.

Intraspecific variation in litter size occurs as a result of a number of factors, including female age and/or parity, time of the year, population density, and geographical or altitudinal location. Age and parity-related variation has been observed in some, but not all, species of *Peromyscus*; litter size increases with parity during the first half of a female's reproductive lifespan and then declines (Drickamer and Vestal, 1973). In captive females of *Peromyscus leucopus*, the ability to successfully wean large litters (seven or eight young) increases with parity so that the most productive litter size of multiparous females is at least one young larger than that of primiparous females (Fleming and Rauscher, 1978). Seasonal differences in average litter size have been reported in *Microtus californicus* (Lidicker, 1973), *Tamiasciurus hudsonicus* (Smith, 1968), *Sciurus carolinensis* (Barkalow *et al.*, 1970), and the snowshoe hare, *Lepus americanus* (Dolbeer and Clark, 1975). This variation could result from changes in food availability (as affected by density or fluctuations in *K*) or physiological condition resulting from previous reproduction. A positive correlation between latitude or altitude and litter size has been documented in a variety of small mammals by Fleming (1973), Lord (1960), Smith and McGinnis (1968), and Dunmire (1960). Spencer and Steinhoff (1968) postulate that the positive correlation between litter size and altitude in *Peromyscus maniculatus* results from selection favouring larger litters in environments with reduced growing seasons, but the possibility that it results from an age structure skewed towards older, more parous females at higher elevations needs to be investigated (Fleming and Rauscher, 1978).

An *absence* of significant litter-size variation occurs in many small mammals, including most species of bats in which females produce one young per birth. Pikas (*Ochotona princeps*) show no variation in litter size as a function of age, habitat, or location where available habitat is constantly saturated as in Alberta, Canada (Millar, 1973), but litter size is significantly larger at low elevations near Bodie, California, where available habitat is not saturated and juveniles stand a better chance of becoming established in the population (Smith, 1978). Although population density can vary tremendously during microtine cycles, there are no striking trends in litter size with the cycle in most species (Krebs and Myers, 1974).

Intraspecific variation in the number of litters produced per female per season has been reported in a number of species. The most striking

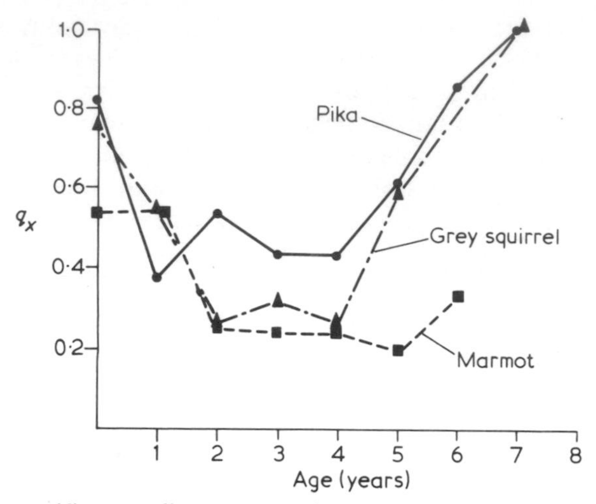

Fig. 1.4 Age-specific mortality rates (q_x) in three species of mammals. (Data are based on Armitage & Downhower, 1974, Barkalow *et al.*, 1970, and Millar & Zwickel, 1972b.)

examples are perhaps found in desert – or arid grassland – dwelling heteromyid rodents. In *Perognathus parvus,* for example, females will produce two litters in 'good' years (i.e. years following a relatively wet winter); in an 'average' year they will produce 1.1; and in a 'poor' year only one out of three adult females will attempt to breed (O'Farrell *et al.*, 1975). These observations conform with theoretical expectations regarding the relationship between current reproductive effort and environmental favourability (Hishfield and Tinkle, 1975). Similar but less dramatic variability occurs in the chipmunk, *Tamias striatus*, in eastern North America. All adult females breed in the spring, but the number breeding in summer is variable, ranging from no females to most females in various years (Pidduck and Falls, 1973). The proximate and ultimate causes of this variation have not been elucidated.

As in the case of fecundity and natality rates, mortality rates vary intraspecifically as a result of factors such as age, reproductive condition, season, and social status. Higher mortality rates in juveniles than in adults have been reported in a wide variety of small mammals. Fig. 1.4 displays some of these data in the form of 'q_x' curves for several species living in widely different environments. Age-specific mortality rates tend to vary inversely with reproductive value, which is highest at or just after the onset of sexual maturity.

Most models of life-history evolution assume that reproduction feeds back negatively on future adult survivorship and hence current reproductive effort has a 'cost' in terms of reduced future expected fecundity. There is both circumstantial and direct evidence that reproduction does indeed entail

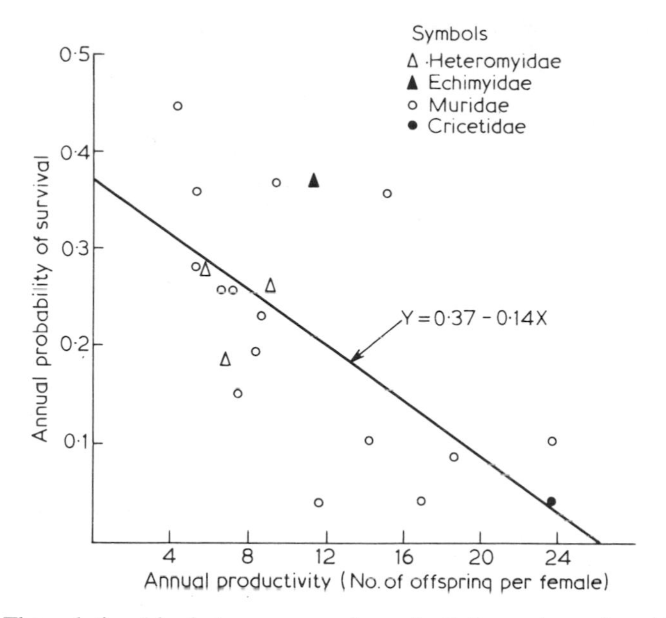

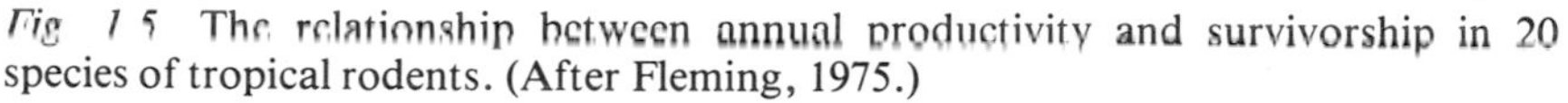
Fig. 1.5 The relationship between annual productivity and survivorship in 20 species of tropical rodents. (After Fleming, 1975.)

a mortality risk in small mammals. The circumstantial evidence is in the form of a significant negative correlation between annual productivity and annual probability of survival in a number of species of tropical rodents (Fig. 1.5). More direct evidence is as follows. Fairbairn (1977b) has noted that early breeding female *Peromyscus maniculatus* have higher mortality rates than nonbreeding females. Similarly, females of *Clethrionomys glareolus* in Poland survive better in winter than in the breeding season (Bujalska, 1975). The poor overwinter survival of female yellow-pine chipmunks, *Eutamias amoenus*, may result from the energy drain they experience from previous breeding (States, 1976). Cameron (1973) reported that female desert woodrats (*Neotoma lepida*) producing exceptionally large litters of five died as a result of emaciation. Finally, studies have suggested that adult males have higher mortality rates than females, but the causes of this mortality difference have rarely been documented. It seems reasonable to postulate that at least a portion of this difference can be attributed to movements by males, especially activity associated with locating suitable mates.

Additional sources of variation in mortality include season of the year and social status. Young born early in the breeding season have lower survival rates than those born at the end of the season in *Peromyscus maniculatus* (Healy, 1967) and *Clethrionomys gapperi* (Fuller, 1977). Winter survival is higher than summer survival in *Clethrionomys gapperi*,

C. rutilus, Microtus ochrogaster and *Peromyscus maniculatus* (Fuller, 1977; Gaines and Rose, 1976; Redfield *et al.*, 1977; Whitney, 1976). In showshoe hares, *Lepus americanus*, low overwinter survival of adults is correlated with extended periods of cold and snow, and low survival of juveniles correlates with periods of cold, wet weather (Meslow and Keith, 1971). Subordinate social status results in lower survival in species such as the arctic ground squirrel (*Spermophilus undulatus*), Richardson's ground squirrel (*Spermophilus richardsonii*), yellow-bellied marmots (*Marmota flaviventris*), and *Peromyscus maniculatus* (Armitage and Downhower, 1974; Carl, 1971; Healy, 1967; Michener and Michener, 1977).

All of the above sources of mortality must be taken into account before realistic models of life-history evolution can be constructed. Unfortunately, most current models take a very 'gross' approach to sources of mortality. In his model of reproductive effort in birds, Ricklefs (1977) separates the effects of mortality resulting from reproductive and non-reproductive causes and finds that reproductive effort is more sensitive to the latter factor than the former. More modelling efforts along this line are needed.

Interspecific variations in life-history patterns

Owing to the enormous ecological and taxonomic diversity of small mammals, one should expect to find, and does find, considerable diversity in their life-history strategies. Given this diversity, it is natural to ask what kinds of patterns emerge regarding the relationship between reproductive effort and the following factors: taxonomic, trophic, and successional status, body size, and geographic location? The two basic suites of demographic characteristics described on p.13 have been identified in lizards and birds (Tinkle, Wilbur, and Tilley, 1970). Can small mammals be similarly divided into two demographic groups? To my knowledge, no-one has yet attempted to synthesise all of the available data on small mammals into one overview of their demographic strategies. A start towards identifying trends and patterns has been made, however, by Millar (1977) and French, Stoddart, and Bobek (1975).

An obvious first step in any attempt to identify general trends in the reproductive adaptations of small mammals is to ask how many parameters are size-dependent. That is, to what extent is size a good predictor of a species' reproductive characteristics? Millar (1977) has analysed data from 100 studies of mammalian reproduction, growth, and development and finds the following parameters to be significantly positively correlated with adult weight: relative birth weight, relative weaning weight, and absolute growth rates. The following parameters are poorly correlated with adult weight: age at weaning, litter size, and reproductive effort, which Millar defines as the ratio of the metabolic needs of a litter of young at weaning to the metabolic needs of the female (in a non-reproductive state). Much of the variability in reproductive characteristics cuts across taxonomic and trophic lines with the exception of reproductive effort per litter, which is highest in squirrels and lowest in lagomorphs. Millar concludes that of the various parameters he has analysed, only two, litter size and time to weaning, are non-conservative traits that vary independently of body size. The other parameters appear to have important size-related constraints.

In the most extensive analysis of mammalian reproductive and demographic attributes, French *et al.*, (1975) have attempted to organize small mammals into groups according to demographic characteristics utilizing 168 studies from the recent literature. They identify three general groups of species characterized by the following features: (1) High reproductive rates, low survival rates, and high density tolerance; this group includes microtine and murid rodents; (2) moderate reproductive rates, medium survival rates (*see* Fig. 1.6), and moderate population densities; this group includes cricetine rodents and soricine shrews; and (3) low reproductive rates, high survival rates, and rather low population densities; this group includes sciurid, heteromyid, zapodid, and fossorial rodents.

Ranges of variation of various demographic characteristics of these groups are as follows. Regarding the number of litters per female per year, values are high (> 3.0) in microtines, murids, insectivores, and cricetines, and low (< 2.3) in other taxonomic groups. Reproduction by young-of-the-year typically occurs only in the 'high' group. Litter size is large (> 5.3) in murids,

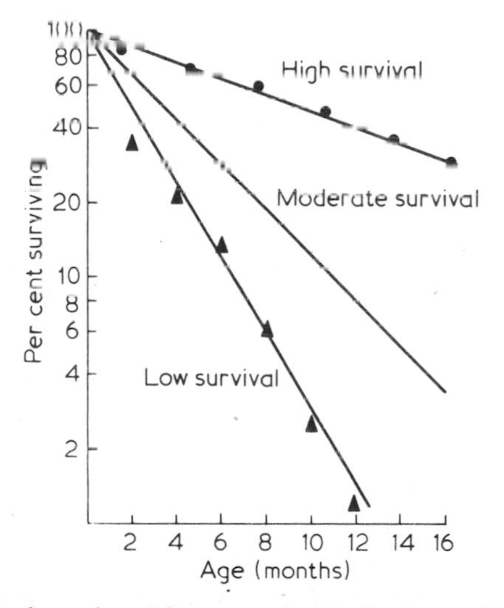

Fig. 1.6 Three typical survivorship curves in small mammals. Species represented include the pocket mouse *Perognathus formosus* (high survival), the white-footed mouse *Peromyscus leucopus* (moderate survival), and the bank vole *Clethrionomys glareolus* (low survival). (After French *et al.*, 1975.)

insectivores, microtines, sciurids, and zapodids, and is low (< 4.5) in cricetines, fossorial rodents, and heteromyids. The prevalence of pregnancy, expressed as the proportion of the adult female population that is pregnant per month of breeding, is high (> 80 per cent) in sciurids, heteromyids, and microtines, intermediate in insectivores, and relatively low (< 70 per cent) in murids, cricetines, zapodids, and fossorial forms. Life expectancy at birth (e_0) is high (7.4–12.5 months) in sciurids, heteromyids, zapodids, insectivores, and fossorial species; it is intermediate (3.1–3.6 months) in cricetines and some microtines; and it is low (≈ 1.8 months) in other microtines and murids. Population densities tend to be high and volatile (66–118 per hectare) in microtines and murids, intermediate (< 31 per ha) in fossorial species, low (7–15 per ha) in sciurids, heteromyids, cricetines, and insectivores, and very low (~0.5 per ha) in zapodids.

Further generalizations that emerge from this study include the following. (1) With the inevitable exceptions (which would be interesting to study further), various taxonomic groups form natural demographic groups. That is, demographic similarities tend to occur along taxonomic rather than geographic lines. This trend is well-illustrated by the demographic differences that exist in three sympatric Panamanian rodents, *Liomys adspersus* (Heteromyidae), *Oryzomys capito* (Cricetidae), and *Proechimys semispinosus* (Echimyidae) (Fleming, 1971). Although living in the same habitat, these species have very different reproductive strategies (Fig. 1.7). *Oryzomys capito* has the attributes of an *r*-selected species – early maturity and high reproductive effort during a short lifespan. *Proechimys semispinosus* has the opposite attributes and appears to be *K*-selected. *Liomys adspersus* differs from both species in that it is a relatively long-lived seasonal breeder in which yearlings make the largest reproductive contribution to the population. (2) Despite considerable variation in the energetic strategies and social organization within heteromyid and sciurid rodents, most species have similar demographic characteristics. (3) As illustrated in Figure 1.5, annual reproductive productivity and survival are inversely related. Furthermore, daily or seasonal periods of reduced metabolic activity markedly increase expected longevities. (4) As found in other vertebrates, two general demographic strategies exist in small mammals: (i) High reproductive effort per season and low survival rates, and (ii) low reproductive effort per season and high survival rates.

Neither Millar's nor French *et al.*'s study considered bats, which represent the second largest mammalian order, in any detail. Demographically, bats clearly fall in the 'low reproductive effort – high survival' group of small mammals. Age at maturity is usually one year in females and one or two years in males (Carter, 1970). Fecundity is low; most species produce one young per birth. Temperate Zone bats have only one birth per year, but females of tropical species may give birth two or three times a year

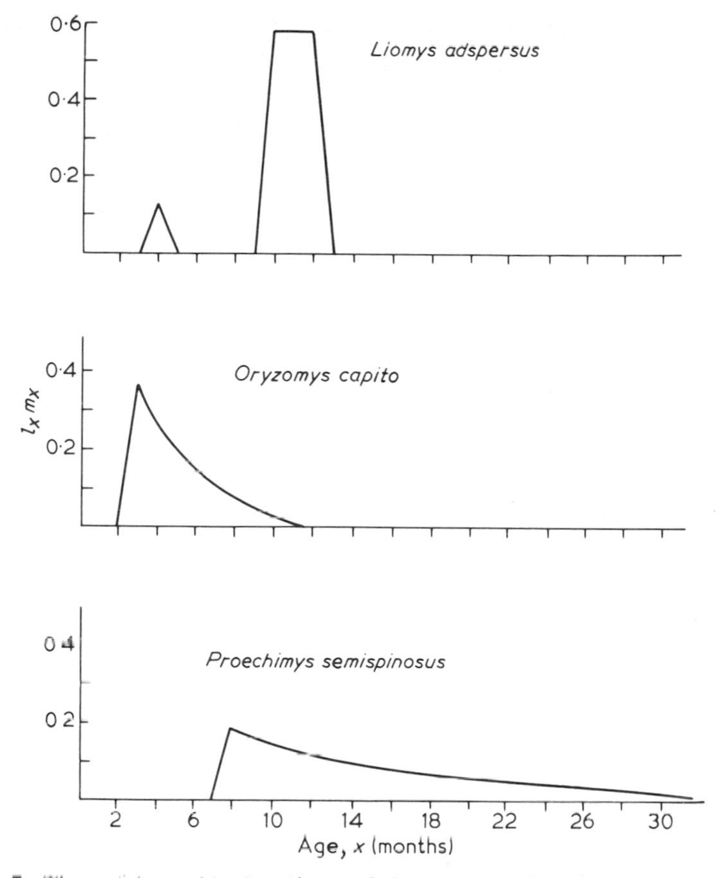

Fig. 1.7 The net fecundity functions of three sympatric species of Panamanian rodents (based on data in Fleming, 1971.)

(Fleming, Hooper, and Wilson, 1972; Wilson and Findley, 1970). Annual mortality rates are correspondingly low. Lifespans of the hibernating *Myotis sodalis* can be as long as 14.8 years in females and 13.5 years in males (Humphrey and Cope, 1977). A 28 year-old male *Myotis lucifugus* has been reported from Ontario, Canada (M.B. Fenton, pers. comm.). Since many species of bats, including the above *Myotis* species, are among the smallest mammals, their demographic performances are remarkable. Reduced metabolic expenditures (in Temperate Zone species) and relative freedom from predation owing to safe roosting sites appear to play a large role in moulding their reproductive strategies (Humphrey, 1975).

The two major demographic groups of small mammals roughly correspond to an '*r*-selected' and a '*K*-selected' dichotomy. However, as previously discussed, applying these labels to any species or group of species tells us relatively little about which of the many possible selective pressures have actually produced these suites of demographic traits. A much more fine-grained analysis of selected species is needed to elucidate the basic features of demographic evolution. One such attempt is that of Fleming (1974), who reported that of two species of tropical heteromyid rodents, the one inhabiting the more seasonal environment (*Liomys salvini*) appeared to be more *r*-selected than the one inhabiting a less seasonal habitat (*Heteromys desmarestianus*). More studies comparing the demographic traits of closely related species living in different kinds of habitats are needed before the generality of the above results is known. Studies of intraspecific demographic variation such as Tinkle and Ballinger's (1972) study of the iguanid lizard *Sceloporus undulatus* are needed in small mammals before the selective advantage of various suites of demographic characteristics are understood. As Stearns (1976) points out, we presently have a plethora of theoretical studies of life history evolution, but we lack well-designed empirical tests of these models. (For an *a posteriori* test, *see* Schaffer and Tamarin, 1973.)

General trends in dispersal strategies

Dispersal strategies of small mammals have been reviewed by Bekoff (1977), Christian (1970), and Lidicker (1975). According to these authors, dispersal is perhaps more common in juveniles than in adults and results from a variety of causes, including social pressures that may or may not be density-related and individual (genetic?) dispersal tendencies. Males tend to disperse more frequently than females in species such as microtine rodents (Krebs and Myers, 1974), *Peromyscus maniculatus* (Fairbairn, 1977b), and various species of ground squirrels (e.g., *Spermophilus richardsonii*; Michener and Michener, 1977). Females are more likely to disperse in water voles, *Arvicola terrestris* (Stoddart, 1970). No sexual differences in dispersal tendency have been found in cotton rats (*Sigmodon hispidus*) and harvest mice (*Reithrodontomys megalotis*) (Joule and Cameron, 1975). Although mortality rates tend to be high in dispersers, moving away from a natal site may be the only way young animals can gain an opportunity to reproduce. This appears to be especially true in ground squirrels (e.g., *Spermophilus armatus*; Slade and Balph, 1974). From a review of dispersal tendencies in canids and marmots, Bekoff (1977) suggests that "knowledge of the behavioral interactions that occur before dispersal may provide a key to understanding both interspecific and intraspecific differences in social organization and dispersal patterns." He concludes that the opposing selective forces of potentially higher reproductive output but higher rates of mortality in dispersing individuals can result in behavioural polymorphism for dispersal strategies.

1.4 Energetic components of life-history strategies

Although fitness is measured in the currency of surviving, reproducing offspring, a more fundamental, biological problem – the acquisition and allocation of energy – must be solved before reproductive activities can be undertaken. The over-riding importance of the actual and potential availability of energy in determining reproductive strategies of small mammals is demonstrated by the seasonal occurrence of reproductive activity in most species and by the flexible reproductive responses possessed by many species inhabiting widely-fluctuating environments. In this section two aspects of energetic strategies will be considered: The evolution of food habits and food niche breadth, and the evolution of homeothermic v. heterothermic metabolic strategies.

1.4.1 The evolution of food habits and food niche breadth

The basic problem in this area of life-history evolution is this: given an array of potential food items differing in abundance, size, nutritional characteristics, and spatial and temporal distribution patterns, which items should individuals of a species select to fulfill two basic requirements? (1) The diet must satisfy quantitative and qualitative nutritional needs, and (2) it must be harvested in an energy- and/or time-efficient manner. As in the previous section, the theoretical aspects of this problem will be discussed before reviewing the pertinent empirical data.

Theoretical considerations

Beginning with the work of MacArthur and Pianka (1966) and continuing through the work of Estabrook and Dunham (1976), many authors have modelled the evolution of optimal diets. Much of the relevant literature is reviewed by Pyke, Pulliam, and Charnov (1977), Rapport and Turner (1977), and Schoener (1971), and only the major conclusions of these studies will be mentioned here. At the onset of this discussion, I need to define what is meant by an 'optimal diet.' According to Estabrook and Dunham (1976), an optimal diet is one which includes 'the set of kinds of prey which if eaten wherever encountered will maximize the intake of food value per unit time.' In most models 'food value' means net energy content (which takes into account the energetic costs of searching for, pursuing, and handling the prey) of the prey item, but for certain kinds of consumers (e.g., seed predators and herbivores) other measures of food value may be more appropriate. The presence of toxic or noxious compounds in seeds and the vegetative parts of plants sets important constraints on the diets of seed predators and herbivores. Janzen (1971) and Levin (1976) discuss the co-evolution of plants and their consumers, and Freeland and Janzen (1974) and Westoby (1974) treat the evolution of herbivorous strategies in detail.

Predicting optimal diet breadth has been the objective of most models of the evolution of food habits. As in the case of demographic strategies, two basic strategies can be visualized: food generalists and food specialists. Whether or not a generalized or specialized diet will be optimal depends on a number of factors, including the absolute abundance of potential food, the relative value of potential food, and the relative abundance of potential food types (Estabrook and Dunham, 1976).

The results of a number of theoretical studies indicate that a generalized diet containing a relatively high diversity of food items will be selectively favoured under the following conditions: (1) During a decrease in the absolute abundance of food; (2) during a period of increased energetic requirements (e.g., during a breeding season); (3) during a decrease in the diversity of food types; (4) when foraging occurs via pure searching (e.g., in herbivores and frugivores); (5) when selection favours maximizing the rate of energy intake per unit time; and (6) when food is uniformly distributed in space. Furthermore, relatively mobile animals that encounter potential food items in the same proportions as they occur in nature are likely to be food generalists; such animals are said to live in *fine-grained* environments (Pianka, 1974).

In contrast, a specialized diet containing a relatively low diversity of food items will be selectively favoured under the following conditions: (1) During an increase in the absolute abundance and/or diversity of food; (2) during periods of reduced energetic requirements; (3) when foraging occurs via pure pursuit (e.g., in insectivorous bats); (4) when selection favours minimizing the amount of time spent foraging; and (5) when food is clumped in space. Relatively sedentary animals that encounter some food items more frequently and others less frequently than expected by chance, are likely to be food specialists; such animals are said to live in *coarse-grained* environments. As a final prediction, food generalists will be selected for in highly seasonal environments whereas specialists will be favoured in aseasonal environments.

Empirical data

The mammalian literature abounds in food-habit studies, but, as in the case of demographic studies, relatively few of these studies have been conducted to test optimal diet theory. Specific questions that need to be answered about the diets of small mammals include: (1) to what extent does diet choice depend upon the absolute or relative abundance of potential food items; (2) how does food quality and handling characteristics affect diet choice; and (3) how extensively do the diets of sympatric species of similar food habits overlap and how important is interspecific competition in shaping diet choice? I will first discuss the evidence for food specialization v. generalization before addressing the above questions.

Most species whose diets have been examined in detail are food generalists rather than specialists. After reviewing the food habits of a wide variety of species, Landry (1970) concluded that rodents as a group are omnivores. This generalization applies to granivorous species, such as desert-dwelling heteromyid rodents (Reichman, 1975) and herbivorous species, such as various microtines (e.g., *Microtus chrotorrhinus*; Whitaker and Martin, 1977). An important reason why broad diets are so common is that the availability of food changes seasonally in most environments. Seasonal dietary changes have been documented in many species, including the common hamster, *Cricetus cricetus* (Górecki and Grygielska, 1975), the prairie vole, *Microtus ochrogaster* (Meserve, 1971), Merriam's kangaroo rat, *Dipodomys merriami* (Bradley and Mauer, 1971), and several species of Costa Rican phyllostomatid bats (Heithaus *et al.*, 1975).

Despite the widespread occurrence of food generalists, food specialization is known to occur in a variety of different species. Most of the 17 species of insectivorous bats studied by Black (1974) in New Mexico could be classified as either moth or beetle specialists. Species whose diets included mostly moths included *Lasiurus cinereus*, *Pipistrellus hesperus*, and *Idionycteris phyllotis;* beetle specialists included *Eptesicus fuscus* and *Antrozous pallidus*. The little brown bat, *Myotis lucifugus*, feeds heavily on mayflies, which usually represent a small fraction of the insect biomass in areas where this bat forages (Buchler, 1976). Several species of desert-dwelling rodents have specialized diets. These species include the sand rat, *Psammomys obesus*, which often feeds on the leaves of a single species of Chenopodiaceae in the Algerian Sahara (Daly and Daly, 1973), the chisel-toothed kangaroo rat, *Dipodomys microps*, which has specialized teeth and feeding behaviour for feeding on leaves of saltbush (*Atriplex*) in the south-western United States (Kenagy, 1972, 1973a), and the desert woodrat, *Neotoma lepida*, which specializes by feeding on toxic or noxious plant parts, including the leaves of creosotebush (*Larrea*) and *Salvia apiana*, and the succulent stems of *Opuntia* cacti (Meserve, 1974). Finally, the beach vole, *Microtus breweri*, which inhabits Muskeget Island, Massachusetts, feeds on the vegetative parts of only two species of plants, beach grass (*Ammophila*) and bayberry (*Myrica*) (Rothstein and Tamarin, 1977). Interestingly, this vole treats its major food species, beach grass, as if it were several resources: It feeds on the roots in August, on the culms in November, and on the leaf blades in May.

A number of studies support the notion that diet choice in small mammals is sensitive to the relative abundance of food, its qualitative and handling characteristics, and the presence of competitors. Many species appear to treat their potential food species in a fine-grained fashion, consuming food items in proportion to their relative abundance. This is true for certain herbivorous rodents (e.g. *Microtus californicus* and *Sigmodon hispidus*; Batzli and Pitelka, 1971; Fleharty and Olsen, 1969), granivorous rodents (e.g. *Dipodomys merriami, Perognathus baileyi,* and *P. intermedius*; Reichman, 1975), and the weasel *Mustela nivalis* (Erlinge, 1975). Other species, especially insectivorous bats, appear to be much more selective in their feeding and do not consume species in relation to their relative abundance (Black, 1974; Buchler, 1976).

Food quality, which includes caloric, nutrient, and toxin content, also influences diet choice. California voles (*Microtus californicus*), for example, prefer palatable, abundant plant species and feed selectively on the stems and seed heads of grasses but are less selective when feeding on forbs. A strong positive correlation exists between the preferences of lab- and field-born animals which implies that the food preferences in this species are strongly selected for and are not simply the results of early experience (Gill, 1977). Pikas (*Ochotona princeps*) are selective in the plants they harvest for their hay piles; they prefer shrubs and clumped herbs (for ease of collection) and plants with high protein content (Millar and Zwickel, 1972). Of the seeds they collect in their cheek pouches, desert-dwelling heteromyid rodents immediately consume those species richest in energy and generally select seeds of higher energy content than are randomly available in the soil (Reichman, 1977).

Handling efficiency influences diet choice in certain species. Of three species of Chenopodiaceae available in their Saharan habitat, sand rats (*Psammomys obesus*) feed most efficiently, in terms of the amount of leaf tissue harvested per second, on *Suaeda mollis*, and they prefer this species in the field and laboratory, even if they have had no prior experience with it (Daly and Daly, 1973). The speed with which individuals can ingest food energy, which depends on handling and husking efficiency, plays an important role in the food preferences of the squirrels *Sciurus carolinensis* and *S. niger* (Smith and Follmer, 1972).

Competitors can also influence diet breadth by either causing it to expand or contract as indicated in the following two examples. Meserve (1976) attributes the generalized diets of *Peromyscus maniculatus* and *Reithrodontomys megalotis* to their small size and socially subordinate position in a California coastal sage scrub rodent community that contains several potential competitors, including two specialists on various species of seeds, flowers, and fruits (*Peromyscus californicus* and *P. eremicus*) and three specialists on the seeds of annuals and grasses (*Dipodomys agilis, Perognathus fallax*, and *Perognathus californicus*). Husar (1976) has documented a significant shift in the diet of the insectivorous bat *Myotis evotis* when it is sympatric with a morphologically similar congener, *M. auriculus*. When allopatric, both species feed preferentially on moths, but when sympatric *M. evotis* consumes more beetles than moths whereas *M. auriculus* remains a moth specialist. This dietary shift has presumably resulted from past interspecific competition between these two species or one or more of the other 12 species of insectivorous bats that have been recorded at the same locality in New Mexico (*see* Black, 1974).

1.4.2 The evolution of energy metabolism: homeothermy v. heterothermy

Birds and mammals are unique among vertebrates in possessing an endothermic mode of energy metabolism. Using endogenously-produced heat, these animals can maintain high and constant body temperatures over a wide range of environmental temperatures. The activity patterns of endotherms are thus much less temperature-dependent than those of ectotherms, and consequently, endotherms have a wider range of habitats available to them. One major disadvantage of endothermy compared to ectothermy is its high 'cost of living.' At a given body temperature, the energetic requirements of an endothermic bird or mammal are about an order of magnitude greater than those of a similar-sized reptile or fish. Since mass-specific energetic requirements vary inversely and exponentially with body size, small endotherms, particularly those weighing less than 10 grams, are in an energetically-precarious position, especially if their food supplies fluctuate on a daily or seasonal basis. The basic problem in this area of life-history evolution thus becomes one in which small mammals must match their energetic expenditures with expected levels of food availability. Species faced with periodically unfavourable environmental conditions in terms of climatic conditions and/or food availability should be under strong selective pressure to utilize periods of hypothermia (on a daily and/or seasonal basis) to reduce their 'cost of living.'

As Bartholomew (1977) notes, the literature on energy metabolism and temperature regulation in vertebrates is voluminous. Most of this literature, however, deals with physiological mechanisms rather than with evolutionary questions. Heinrich (1977) takes an evolutionary approach to temperature regulation and discusses two major questions: (1) Why have animals evolved to regulate stable body temperatures; and (2) why are the temperature set points of homeotherms generally greater than ambient temperatures? In this section I will first review the major relationships between size, energy metabolism, and temperature regulation before examining the basic features of hypothermia (heterothermy) and its adaptive significance.

Size, energy metabolism, and temperature regulation

An animal's size, as expressed in units of mass (weight), has important physiological implications. Metabolic rate and the cost of locomotion vary inversely with body mass. Compared to large species, small mammals have high mass-specific rates of heat loss below their zone of thermoneutrality and high rates of heat gain above this zone; they have high rates of water loss during evaporative cooling; and they can carry only a limited amount of insulative pelage on their bodies (Hill, 1976a). Since the lower critical temperature of an animal's thermoneutral zone is inversely related to its mass, small mammals accrue a metabolic saving by undergoing controlled hypothermia (Morrison, 1960). Rates of cooling and heating when going into or coming out of torpor are inversely related to mass, and the cost of 'heating up' after leaving a torpid state represents only a small fraction of a small species' daily energy budget compared to that of a 'large' species (Bartholomew, 1977). The ability to store fat is an important pre-requisite for hibernation, and the maximum amount of fat that can be stored is not related to mass; this maximum is 50 per cent of the total mass over a wide range of body sizes (Morrison, 1960). Owing to their higher metabolic rate, however, small species can survive on a 50 per cent fat store for a much shorter period of time than large species. A 200 gram squirrel, for example, can survive for about two months of hibernation if it doubles its weight by storing fat whereas a three gram shrew can only survive about 20 days on the same relative amount of fat (Morrison, 1960).

In addition to size, body shape and social tolerance can importantly influence metabolism and temperature regulation. Long, thin animals such as weasels have a higher cost of living at low temperatures than do 'normal-shaped' animals of the same mass. Brown and Lasiewski (1972) have shown that the metabolic rate of the long-tailed weasel (*Mustela frenata*) at 5°C is 50 to 100 per cent higher than woodrats (*Neotoma*) of the same mass but 'normal' shape. The metabolic disadvantage of being long and thin is apparently offset by the advantages of being able to capture rodents in their burrows. A high degree of social tolerance, as seen in clustering or huddling behaviour, can result in significant individual metabolic savings by reducing the rate of heat and/or evaporative water loss. Studies of a number of species of bats have shown that rates of metabolism, weight loss, and evaporative water loss are reduced when bats sleep in clusters in their day roosts (Howell, 1976; Studier *et al.*, 1970; Trune and Slobodchikoff, 1976).

The ontogeny of temperature regulation and adaptive hypothermia

Although as adults small mammals are endotherms, this is not the case in neonates, which are essentially ectothermic during their early stages of development. Because they obtain their body heat from their mothers and are often reared in well-insulated nests (bats are an obvious exception to this), neonates can have relatively constant body temperatures often considerably higher than ambient temperatures (Gebczynski, 1975; Hill, 1972). Shortly after birth, therefore, neonates are effectively homeothermic. The gradual development of endothermy and the ability to thermoregulate probably represents an adaptation for channelling most of the maternally-derived energy and foodstuffs into early growth and development of young rather than into energetically-expensive thermoregulation. During early life, the mother assumes the metabolic cost of maintaining the relatively high body temperatures of her nestlings while they are rapidly growing (Hill, 1976a). The ontogeny of temperature regulation has been most carefully studied in a variety of rodent species, and results of these studies indicate that young mice become fully homeothermic at the ages of eye-opening, rapid pelage development, and when they begin wandering from the nest (Gebczynski, 1975; Hill, 1976b; McManus, 1971; Soholt, 1976).

Once they become fully homeothermic, many species of mammals, both large and small, never deviate from this thermoregulatory state. A small but significant fraction of mammals, however, are 'imperfect' thermoregulators and allow their body temperatures to decline to near-ambient levels on a daily and/or seasonal basis. These species are generally considered to be 'heterotherms' rather than homeotherms, but their 'imperfect' thermoregulation should never cause them to be considered physiologically inferior. Rather, heterothermic species should be viewed as using hypothermia as an adaptation to reduce energy demands during times of actual or potential food (and water) shortage.

There are four basic kinds of adaptive hypothermia: hibernation, estivation, daily torpor, and 'winter sleep.' The first three kinds are different manifestations of similar physiological processes that are well-described in most physiology texts (e.g. Gordon, 1977; Hill, 1976a). These processes include body temperatures that are at or near low ambient temperatures, reduced heartrates and rates of oxygen consumption, long periods of suspended respiration, dormancy more profound than deep sleep, and the ability to spontaneously arouse and re-establish a high body temperature. Winter sleep differs from true hibernation in that animals (e.g. bears, skunks) experience only a modest drop in body temperature that nevertheless represents a significant reduction in heat loss so that fat reserves last longer than if the animal had remained normothermic. The first three kinds of adaptive hypothermia occur only in small species of mammals (and birds). No species larger than 50 grams are known to undergo daily or seasonal bouts of torpor. Marmots (*Marmota*), which weigh from 3–7.5 kg, are the largest true hibernators.

Adaptive hypothermia occurs in a taxonomically and ecologically wide range of small mammals. Hibernation occurs in the echidna, several species of marsupials, most species of Temperate Zone bats, the common hedgehog (*Erinaeceus europaeus*), various species of ground squirrels, prairie dogs, woodchucks and marmots, various dipodoid rodents, hamsters, dormice and jumping mice. Details of the hibernation process, which clearly has polyphyletic origins, differ from group to group. Known aestivators, which become inactive during hot and dry portions of the year, include the pygmy possum (*Cercartetus nanus*), desert ground squirrels, and *Peromyscus eremicus*. Daily torpor occurs in many species of Temperate Zone insectivorous bats, the birch mouse (*Sicista betulina*), pygmy mice (*Baiomys*) and heteromyid rodents of the genera *Perognathus* and *Microdipodops*. Two groups of mammals that have been relatively well-studied regarding trends in the occurrence of adaptive hypothermia are bats [*see* reviews by Dwyer (1971) and McNab (1969)] and ground squirrels [*see* review by Davis (1976)].

The environments inhabited by species utilizing hypothermia as an adaptive strategy generally exhibit one or both of the following features: (1) Food supplies that fluctuate daily and/or seasonally and (2) certain times of the year that are physiologically inimical to the maintenance of constant, high body temperatures and/or a positive water balance. Temperate Zone insectivorous bats, which usually weigh less than 15 g, are good examples of species that experience considerable daily and seasonal fluctuations in food availability. Daily torpor during their active season and hibernation during the winter obviously reduces the energetic stress of fluctuating (or absent) food supplies and harsh climatic conditions. Tropical insectivorous bats generally do not undergo daily torpor nor do they hibernate; their basal metabolic rates, however, tend to be lower than expected, which McNab (1969) interprets to mean that they also experience fluctuations in food availability. It should be noted that not all small mammals living in strongly seasonal environments are heterothermic. Microtine rodents and lagomorphs are obvious examples of nonhibernators. Natural selection has produced a variety of ways for dealing with unfavourable climatic conditions, and hypothermia is just one of those ways.

Regardless of the mechanisms used by different species to achieve adaptive hypothermia, the end results of these physiological strategies are the same – considerable energetic savings that accrue as a result of daily or seasonal bouts of reduced metabolic activity. Quantitative estimates of the magnitude of these savings are available for several species of desert heteromyid rodents. At 5°C, torpid kangaroo mice (*Microdipodops*) have a metabolic rate one-sixtieth that of normothermic mice at that temperature; at 25°C, their metabolic rates are one-fourth those of normothermic mice. The length of their bouts of torpor is inversely proportional to ambient temperatures and levels of available food (Brown and Bartholomew, 1969). From field experiments, Kenagy (1973b) estimates that the little pocket mouse (*Perognathus longimembris*) consumes only about 28 per cent of the seeds it would need if normothermic by undergoing about five months of torpor each year. In addition to providing energetic savings, periods of reduced metabolic activity can significantly increase life expectancies. Life-spans of four and eight years have been reported in the field for jumping mice (*Zapus princeps*) and pocket mice (*Perognathus formosus*) respectively (Brown, 1970; French *et al.*, 1974). For their size, Temperate Zone insectivorous bats are remarkably long-lived.

One group of small mammals conspicuously missing from the list of species utilizing some form of hypothermia to minimize energetic expenditures is the shrews (e.g. *Sorex, Blarina*, etc.). These animals are proverbial examples of species whose metabolic rates are so high that they need to consume their weight or more in food each day. Yet none of them, including arctic species, undergoes torpor or hibernates. Why not? Randolph (1973) has conducted an extremely detailed analysis of the physiological energetics of *Blarina brevicauda* and concludes that although their energetic requirements increase by 43 per cent in the winter, there is sufficient insect food available to support 'normal' shrew densities. Whereas they consume only about 3 per cent of their available food in the summer, these shrews consume about 37 per cent of available prey energy in the winter. Morrison (1960) has, perhaps facetiously, suggested that arctic shrews (*Sorex arcticus*) in Alaska may sustain themselves during the long winter by consuming hibernating arctic ground squirrels; he calculates that one squirrel carcass could support six three gram shrews all winter!

1.5 Behavioural components of life-history strategies

The final broad topic that I wish to discuss in this chapter is behaviour as it affects life-history strategies. Although it rarely enters into models or discussions of demographic evolution, behaviour clearly plays an important role in determining individual fitness. A species' social environment may be more important in determining birth and death rates than its physical environment. Dispersal and mate selection strategies and parent-offspring relationships are three aspects of behaviour that play crucial roles in determining the probabilities of survivorship and reproductive success of individuals. Thus, in order to be at all complete in my survey of the important components of small mammal life-history strategies, I need to devote considerable attention to behaviour. In this section I will discuss two aspects of behavioural evolution: habitat selection and the basic features of social organization.

1.5.1 The evolution of habitat selection

Small mammals are faced with many 'choices' in their lives, and one of the most critical of these is where to live. This choice can be of crucial importance for an animal's survival and reproductive success because not all habitats are equally favourable regarding availability of food, mates, and shelter and pressures from competitors and predators. The basic evolutionary problem here becomes one of selecting a habitat(s) that best satisfies an individual's basic ecological needs as reflected by its morphology and physiological and psychological tolerances. Compared with other areas of life-history evolution, relatively little theoretical work has been done on habitat selection. The following treatment of the theory behind habitat selection comes primarily from two sources, Levins (1968) and Rosenzweig (1974).

Theoretical considerations

Levins' (1968) treatment of the evolution of habitat selection begins with the concept of *fitness sets*, which takes into account differences in the fitnesses of genotypes or phenotypes in different habitats. These fitness

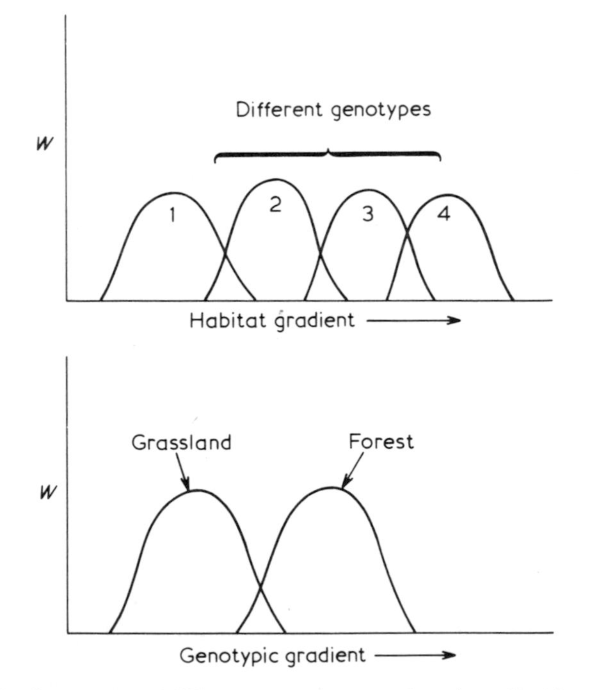

Fig. 1.8 The fitness (W) of different genotypes as a function of habitat.

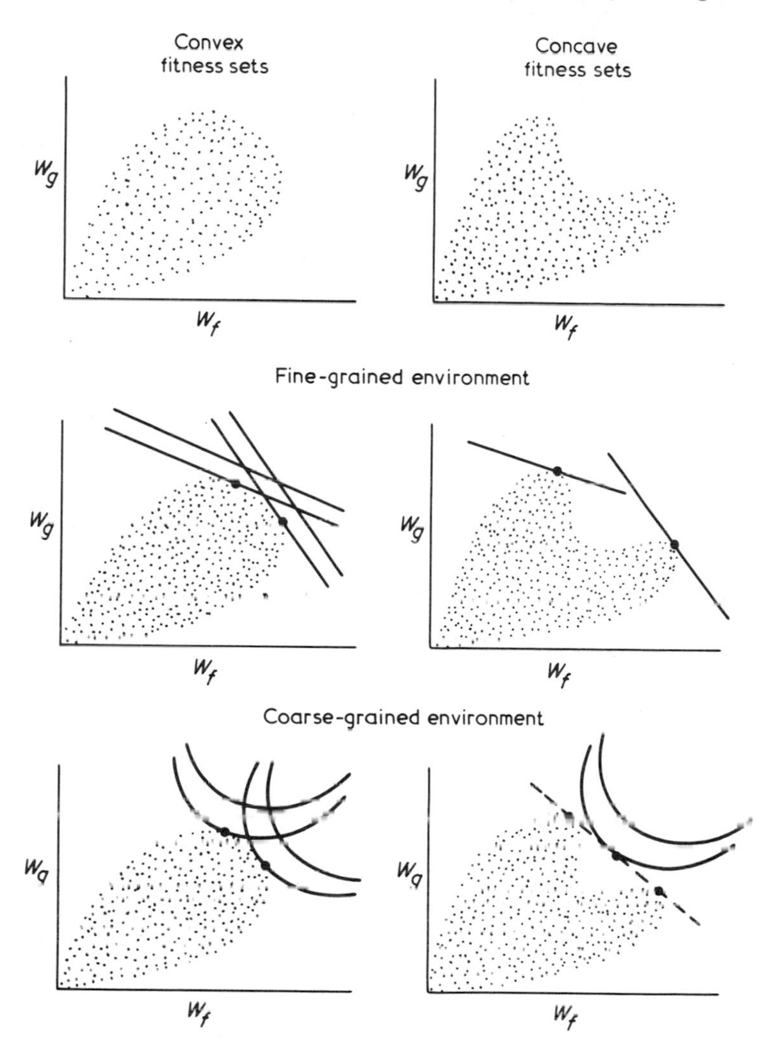

Fig. 1.9 Fitness sets in fine- and coarse-grained environments. W_f and W_g refer to fitness in forest and grassland habitats, respectively. Each dot represents a single genotype. The straight and curved lines in the lower two panels represent adaptive functions or lines of equal fitness. See text for further explanation.

differences can be expressed in two different ways: (1) As changes in the fitness of a particular genotype along an environmental gradient, or (2) as fitness differences between different genotypes in the same environment (Fig. 1.8). Note that for each genotype there will be an optimal position (with greatest W) along the gradient. The breadth of the distribution curves indicates the range of environmental tolerance of each genotype. Fitness sets can be constructed by plotting the fitnesses of all genotypes in a population on a graph whose axes represent different habitats. In the following example only two habitats, forest and grassland, are considered, but the method can be extended to any number of habitats. The cluster of points that results from our plotting represents a fitness set, which can take one of two general shapes depending on the relative magnitude of the differences between the habitats compared to the tolerances of different genotypes (Fig. 1.9). Fitness sets will be *convex* if genotypic tolerances are greater than habitat differences, and they will be *concave* if habitat differences are greater than genotypic tolerances.

The adaptive strategy of a species, as indicated by the existence of individuals that are habitat specialists (e.g. occurring only in grassland or forest) or generalists (e.g. occurring equally frequently in both grassland and forest), will depend on the shape of its fitness set and whether the alternate habitats are potentially encountered in a fine-grained or coarse-grained fashion. As mentioned previously, organisms are said to live in a fine-grained environment if they encounter different habitat patches or food in the same proportion as they occur in nature. If they spend disproportionately large amounts of time living in one patch or feeding on one type of food, they are said to be inhabiting a coarse-grained environment.

Optimal genotypes will be determined by the point of tangency of an *adaptive function* or *fitness isocline* on the fitness set. Adaptive functions in fine-grained environments will be straight lines with a negative slope whose magnitude is determined by the relative frequency of the alternate habitats (Fig. 1.9). In contrast, adaptive functions in coarse-grained environments will be curved lines whose slope is again determined by the relative frequency of alternate habitats (Fig. 1.9).

In both fine- and coarse-grained environments, species with convex fitness sets will contain generalist individuals and a single genotype will be optimal. Exactly which genotype this will be depends on the slope of the adaptive function. In fine-grained environments, species with concave fitness sets will contain one of two possible specialist genotypes (Fig. 1.9), depending on the relative proportions of the alternate habitats. In coarse-grained environments, species with concave fitness sets will be polymorphic and will contain two specialist genotypes whose relative frequency will reflect the proportions of the alternative habitats. Methods used to obtain these results and predictions are treated more fully in Levins (1968).

In addition to factors such as environmental grain size and average genotypic tolerances, other factors, including intra- and interspecific competition, can affect a species' pattern of habitat selection. General competition theory tells us that high levels of intraspecific competition, as might occur when population densities are high, should increase the number of habitats a species utilizes. On the other hand, high levels of interspecific competition might be expected to have just the opposite results: contraction in the use of habitats to just those in which a species' mean fitness is highest. Morse (1974) has reviewed research dealing with the effect of interspecific social dominance, which he defines as 'priority of access to resources that results from successful attacks, fights, chases, or supplanting actions, past or present ... ', on niche breadth, and concludes that subordinate species usually decrease their niche breadths (i.e. number of habitats utilized) in the presence of dominant species.

Empirical data

As in the case of theoretical work in this area, there have been relatively few extensive studies of the proximate and ultimate factors behind the evolution of habitat selection in small mammals, though habitat use and the role of habitat selection in determining the distribution of various species have been discussed in a large number of studies. To date, most studies of habitat selection have dealt with the mechanisms and proximate factors behind habitat choices and address such questions as (1) to what extent is habitat selection innate or learned and (2) what environmental factors do mammals use to select among habitats. There are relatively little data on fitness differences among individuals of a species living in different habitats (either voluntarily or involuntarily). Such data are sorely needed if we are to fully understand why species and individuals choose to live where they do. In addition to mechanisms of habitat selection, I will discuss the following topics below: habitat specialization v. generalization, fitness differences in different habitats, and the effects of population density and interspecific competition on habitat choice.

If one were to survey a wide array of small mammals, I predict one would find that most species are habitat specialists, occurring in only one or a few habitats, rather than habitat generalists, occurring in a wide range of habitats. This prediction parallels similar trends in the structure of communities and faunas: Communities usually contain few species with many individuals and many species with few individuals; faunas tend to contain a few wide-ranging species and many species with restricted ranges. The major cause for each of these trends is probably the same: evolution favours specialization more frequently than it does generalization, even though specialists may run a greater risk of extinction over a given span of geological time.

To check on the validity of my prediction that habitat specialists outnumber generalists in small mammals, I arbitrarily picked four North American genera (pocket mice, *Perognathus;* white-footed mice, *Peromyscus*; chipmunks, *Eutamias*; and rabbits, *Sylvilagus*) and tallied the number of different habitats utilized by each species in these genera as

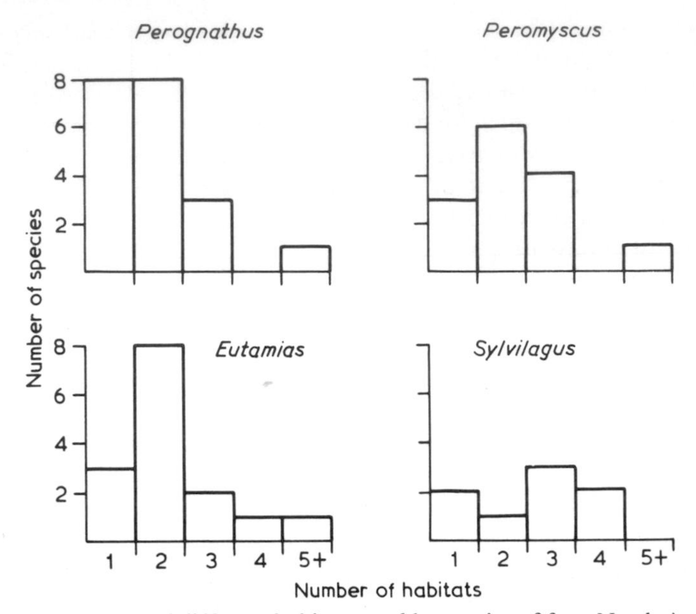

Fig. 1.10 The number of different habitats used by species of four North American general of small mammals.

reported in Burt and Grossenheider (1964). Results of this rough survey are illustrated in Fig. 1.10. Except for the larger rabbits, the modal number of habitats used per species is one or two, in agreement with general expectations.

As previously indicated, there are few data on fitness differences as a function of habitat in small mammals. Grant's (1975) work with *Microtus pennsylvanicus* in large outdoor enclosures containing both grassland and forest habitat indicates that this grassland vole can survive in forest but that adult mortality rates are higher and pregnancy rates lower there than in grassland; two demographic parameters, juvenile survival and maximum longevity, were as high or higher in forest-dwelling voles. Anderson's (1970) discussion of the population structure of small mammals emphasizes the existence of two kinds of habitats in many species, survival habitats and colonizing habitats. He postulates that survival rates are much lower in the latter kind of habitat, which are occupied primarily during times of high population density. Except in Russia, little work has been done to substantiate these ideas. As a final example, I can mention differences in the breeding success of great tits (*Parus major*) studied by Krebs (1971). Birds living in hedgerows, a suboptimal habitat in terms of nestsite availability, suffer higher rates of nest mortality and lower fledgling success than woodland birds.

As expected, a number of studies have demonstrated that various species occupy a wider range of habitats when population densities are high. These species include *Clethrionomys glareolus* in Poland (Bock, 1972), *Microtus pennsylvanicus* in Canada (Grant, 1971), and *M. pennsylvanicus* in the western United States (Stoecker, 1972). One species whose pattern of habitat use is apparently not density-dependent is the common opposum, *Didelphis virginiana*, in the eastern United States (Stout and Sonenshine, 1974).

Habitat expansion in the absence of known or potential competitors has been documented in the pine vole, *Microtus pinetorum*, in the central United States (Goertz, 1971), and in *Peromyscus maniculatus* on Vancouver Island, British Columbia (Redfield *et al.*, 1977). The opposite phenomenon, habitat contraction resulting from the presence of an actual or potential competitor, occurs seasonally in *Clethrionomys gapperi*, which is aggressively restricted from grassland in Manitoba by *Microtus pennsylvanicus* during its breeding season; the two species co-exist in grassland during winter (Iverson and Turner, 1972). Laboratory and field observations suggest that *Apodemus flavicollis*, a woodland rodent, may restrict its sympatric congener, *A. sylvaticus*, to grassland in northern Europe (Hoffmeyer, 1973). Although a similar explanation has been offered to account for the contiguous distributions of several pairs of chipmunks (*Eutamias*) in mountainous parts of the western United States (e.g. Brown, 1971), States (1976) and Meredith (1976) both feel that different habitat preferences, rather than aggressive encounters, determine the distributional limits of these species.

Several studies dealing with small mammals have demonstrated that patterns of habitat selection have an innate component but that early experience (learning) is also important. The most extensive study to date is Wecker's (1963) well-known analysis of habitat selection in a grassland-inhabiting subspecies of *Peromyscus maniculatus*. In this study the response of mice of different hereditary backgrounds and experience in forest, old field, or laboratory enclosures was monitored in a 1600 ft^2 enclosure placed at the interface between forest and field. His results indicated that field-caught animals and their young spent more time in the field half of the enclosure irrespective of previous experience whereas laboratory-reared animals and their young showed no significant preference for field or forest. When reared in the field, however, laboratory-born mice selected the field side. Wecker concluded that the choice of field by this species is predetermined by heredity and that early experience in the field reinforces this innate preference. Early experience in either the lab or the forest cannot reverse the affinity for field habitats. Twelve to 20 generations of experience in the lab, however, was sufficient to reduce the hereditary control of habitat selection such that the response of these mice in the outdoor enclosure was much more variable than that of wild-caught mice. Similar suggestions that habitat selection has a strong genetic component have come from studies of *Microtus pennsylvanicus* (Grant, 1971), *Eutamias minimus* (but not *E. amoenus*?) (Meredith, 1976), and *Neotoma albigula* (Olsen, 1973). That learning can also influence habitat choice is suggested by a study of *Peromyscus maniculatus* (Grant, 1970) in which lab-born mice exposed daily to a pile of grass in their cage chose a simulated grassland habitat rather than their 'normal' habitat of simulated forest floor, even in the presence of a potential competitor, *Microtus pennsylvanicus*.

In choosing a habitat in which to live, animals presumably are more responsive to certain habitat features than others. Various studies have shown that these features include soil type in kangaroo mice (Ghiselin, 1970), the diversity of branching patterns and density of logs in *Peromyscus leucopus*, (M'Closkey, 1975), lack of vegetative cover in *Dipodomys merriami* (Rosenzweig, 1973), presence of low vegetative cover in *Microtus pennsylvanicus* (numerous authors) and *Perognathus penicillatus* (Rosenzweig, 1973), the presence of a specific kind of plant, cholla cactus in the case of *Neotoma lepida* (Brown *et al.*, 1972), and the presence of specific food types in *Peromyscus maniculatus bairdii* (Drickamer, 1976). In each of these cases, the important cues appear to be associated with food or foraging areas and/or shelter, two prime requisites for survival.

1.5.2 The evolution of social organization

As previously mentioned, a species' social environment strongly influences individual fitness. The basic problem in this area of life-history evolution is this: Given a species' trophic position and habitat choice, what social system maximizes an individual's inclusive fitness? Earlier, I introduced the concept of inclusive fitness in the discussion of kin selection (p.4). Whereas we generally believe that individual fitness is being maximized during demographic evolution, it is likely that social evolution operates by maximizing inclusive fitness, which includes individual fitness and the fitness of close genetic relatives. Questions that arise concerning social evolution include the following. When is it individually advantageous to be solitary (except for mating), to be tolerant of contact with and the close proximity of other conspecifics? Under what conditions is it advantageous to be monogamous rather than polygamous? How much parental care should be given to young (especially by males) and how long should parent-offspring contact be maintained once the young are weaned?

The evolution of social behaviour has recently attracted considerable attention from theorists and empiricsts. Wilson's (1975) compendium of the processes and patterns of social evolution is an excellent entrance into the basic literature. Additional reviews and theoretical discussions of this subject are found in Alexander (1974); Bradbury and Vehrencamp (1977a); Crook *et al.*, (1976); Eisenberg (1966, 1975, 1977); Emlen and Oring (1977); King (1973); Kleiman (1977); Maynard-Smith (1977); Orians (1969); Ralls (1977); Trivers (1971, 1972, 1974); and West-Eberhard (1975).

Theoretical considerations

A first question that must be addressed in any discussion of the evolution of social behaviour and social organization is: Why be social? Why should natural selection ever favour group living over a solitary existence characterized by brief contact between members of the opposite sex for mating purposes and, in the case of mammals, the only prolonged contact between individuals being the mother-infant bond? Certainly there are disadvantages to group living. Alexander (1974) lists three of these: (1) Increased competition for resources and mates, (2) increased disease and parasite transmission, and (3) increased conspicuousness. On the other hand, there is at least one important advantage to group living – reduced susceptibility to predators – and two environmental conditions also favour it: (1) food that requires a group effort to locate and/or capture, and (2) the extreme localization of some resource such as a sleeping site (Alexander, 1974). If the advantages of group living outweigh the disadvantages, then social behaviour within groups (e.g. increased individual tolerances, co-operation, and perhaps a division of labour) will evolve for at lest three reasons: (1) It may enhance the original advantage(s) of group living; (2) it may reduce the likelihood of parasite or disease transmission; and (3) it may arise as a result of reproductive competition (e.g. social hierarchies in which subordinates are biding time until they can become dominant and gain mating rights) (Alexander, 1974; West-Eberhard, 1975).

A diverse array of social systems exists in nature. These systems can be characterized by parameters such as average group size, individual and group dispersion patterns, and inter-individual behavioural relationships. Eisenberg (1966) recognizes three general classes of social systems: (1) Solitary or asocial systems; (2) simple social systems such as aggregations, flocks, and unstructured herds; and (3) reproductive social systems with pronounced division of labour and individual interindependence. Several subcomponents can be identified in reproductive social systems, including mating systems (i.e. the number of females (males)/male (female) and the length of the male-female bond), offspring rearing systems (i.e. do both mates stay with the young during rearing and do the young stay with the parent(s) past rearing?), resource exploitation systems (i.e., do individuals co-operate in locating and/or capturing food?), and predator-avoidance systems (i.e. who watches for predators and are alarm signals given once a predator is sighted?) (Crook *et al.*, 1976).

The evolution of these systems involves both selfish and altruistic kinds of behaviour. Selfish behaviour, which enhances individual fitness, includes aggressive, territorial, and co-operative kinds of behaviour as well as reciprocal altruism – the giving of potentially self-sacrificing help with the expectation that such help will be later returned to the donor or its kin. Because the likelihood of 'cheating' (i.e. receiving aid but not returning it) is relatively high under a system of reciprocal altruism, the conditions for its evolution are quite restrictive. These conditions include: (1) Low dispersal rates between populations, (2) a social structure involving small, stable groups, (3) prolonged periods of parental care, and (4) a high benefit/cost ratio for the altruistic act(s) (Trivers, 1971; West-Eberhard, 1975). Nonreciprocal altruism, which enhances the fitness of close relatives, is likely to evolve when the costs of the act or behaviour are low relative to the benefits and/or when it can be exclusively directed towards known kin. Under conditions of intense reproductive and/or ecological competition, the incidence of altruistic behaviour should be low.

Two extreme cases of altruistic behaviour are socially-subordinate behaviour and the reproductive sacrifice that offspring sometimes make by aiding their parents rear young without reproducing themselves. One way in which subordinate behaviour will be selected for is if it increases an individual's inclusive fitness by allowing a reproductively superior relative to raise more offspring, especially if resources necessary for reproduction are limited. By similar reasoning, parents that impose a celibate existence on some of their offspring by forcing them to care for their brothers and sisters will be selectively favoured if they leave more descendents than do parents with 'selfish' broods; the fitness of celibate young depends on how successful they are in rearing their sibs. If their siblings have lower fitness than the helpers would have if they produced their own young, then young should resist parental manipulation and behave selfishly towards parents and sibs (West-Eberhard, 1975).

Ultimately, the evolution of social systems revolves around the question of who mates with whom. The evolution of monogamous and polygamous mating systems has been discussed in detail by Bradbury and Vehrencamp (1977a) and Emlen and Oring (1977). According to these authors, whose ideas are remarkably similar, polygamy will evolve when either resources needed by mates or the mates themselves can be defended; resource or mate defence will be easier when resources are predictable in time and space. Since in most species males make a lower energetic commitment towards the production of offspring than females, selection will favour male polygamy (polygyny) more often than female polygamy (polyandry). Polygamy will evolve under the following three conditions: (1) When one sex assumes most of the parental care, (2) when parental care requirements are minimal, and (3) when food is superabundant so that one parent can care for young.

Monogamy will evolve under conditions that are inimical to the evolution of polygamy. These conditions include: (1) Environments that have low polygamy potential such as distribution patterns that prevent resources and/or mates from being monopolized and defended, and (2) an absence of the opportunity to take advantage of an environment's 'polygamy potential' (e.g., when breeding is highly synchronous among females so that males lose mating opportunities by courting and mating with one female at a time).

Three basic kinds of polygamous systems have been identified by Bradbury and Vehrencamp (1977a) and Emlen and Oring (1977). The first type is *resource defence polygyny* in which males monopolize a resource (food or shelter) required by females. This is basically a territorial system, and the greater the variance in the quality of male territories, the greater will be the degree of polygamy. This is because females that mate with males having high quality territories are likely to raise more young even though they receive little or no help from their mates than do females that mate monogamously with males holding low quality territories (Orians, 1969). Ralls (1977) questions whether this situation occurs in a majority of mammalian species. The second type of polygamous system is *female (harem) defence polygyny* in which females have a clumped distribution irregardless of their reproductive state, and males monopolize them. Females can occur in clumps for a variety of reasons, including protection against predators, exchange of information about food, increased foraging efficiency by observational learning or by the cultural transmission of learned habits. The third type of polygamous system is *male dominance polygyny* in which males form a dominance hierarchy and females choose mates on the basis of social rank. In this system, which includes lek behaviour, males play no role in parental care and there is little potential for the defence of resources or mates. Conditions favouring the evolution of this kind of mating system include: (1) When critical resources are superabundant or widespread, (2) when critical resources are unpredictable in time or space, and (3) when the cost of successful resource or mate defence is too high.

Patterns of parental investment of time and energy in their offspring are closely related to mating systems and may have an important bearing on an individual's inclusive fitness. Trivers (1972, 1974) and Maynard-Smith (1977) discuss the theoretical aspects of this subject. Trivers (1972) defines parental investment as "any investment by the parent in an individual offspring that increases the offspring's chance of surviving (and hence reproductive success) at the cost of the parent's ability to invest in other offspring." Natural selection will favour the parental investment/offspring ratio which maximizes the parent's net reproductive success. This ratio may not be maximally beneficial to the offspring in which case a conflict of interest will arise between parent and offspring. In general, offspring will desire a larger investment than parents should be willing to give. Parents should be especially unwilling to make a larger energetic commitment to current offspring if the ratio of current fecundity (m_x):residual reproductive value is low.

Maynard-Smith (1977) has produced three relatively simple mathematical models of parental investment patterns and finds that in each, two evolutionarily stable strategies are possible – male desertion of the young or female desertion. One parent will always desert its offspring to remate with another partner if two parents are not substantially more effective at rearing a brood than a single parent. Because of sexual differences in the amount of energy devoted to producing offspring, it always pays males to desert and remate as many times as possible in species that provide minimal care for their offspring. In species with parental care, a male will care for its brood if the female has invested so much energy in producing the brood that she cannot care for it or if the population sex ration is skewed towards males. Females will care for the brood if males have increased opportunities for mating by deserting and/or if the population sex ratio is skewed towards females. Males should always desert or kill a brood if its paternity is uncertain.

Empirical data

An increasing number of workers is turning its attention to the social organization of mammals. Because of their conspicuousness and relative ease of study, many data are available on the behaviour of primates and ungulates. Fewer data are available for small mammals, which tend to be much more difficult to study. Not surprisingly, the best-studied small mammals – various species of ground squirrels – tend to inhabit open environments and are diurnal. As in the case of other areas of life-history evolution, relatively few studies of either large or small mammals have been directed towards testing theories of the evolution of sociality, most of which are relatively new. In this section I will discuss general trends in the social systems of small mammals and will illustrate these trends with some selected examples.

The most common pattern of social organization in small mammals is a dispersed population of solitary individuals. The mother-offspring group is the universal nuclear unit of mammal societies and is the basis of more complex mammal societies which are organized along matrilineal lines. Since females are committed to expending considerable amounts of time and energy to reproduction, they are often the limiting sex in the evolution of mating systems, and polygynous systems are far more common than monogamous systems. Kleiman (1977) estimates that less than three per cent of the mammal species are monogamous. Examples among small mammals include the honey possum (*Tarsipes spencerae*), several insectivores (e.g., *Solenodon paradoxus, Microgale talazaci*), bats (e.g., *Vampyrum spectrum, Lavia frons*), and rodents (e.g., *Ondatra zibethica, Notomys alexis*). Monogamy tends to occur in species with a low annual reproductive output and delayed sexual maturation (i.e., *K*-selected species).

Wilson (1975) states that because the data base is fragmentary and because social traits are very labile in mammals, it is difficult to synthesize the available data into a single evolutionary scheme. Some tentative generalizations can be offered, however. Evolutionary stem groups, such as the marsupials and insectivores, tend to have solitary social organizations. Species foraging nocturnally or below ground also tend to be solitary. The most complex social systems tend to occur in the largest members of various orders, including marsupials, rodents, ungulates, carnivores, and primates, for two possible reasons: (1) Larger species forage above ground and are diurnal, and (2) these species tend to have larger, more complex brains and are more intelligent. Species living in open environments tend to be more social than those living in closed habitats. This latter trend suggests that predator detection and avoidance is an important driving force in the evolution of mammalian sociality.

Several authors, including Eisenberg (1966), Fisler (1969), and Crook *et al.* (1976) have attempted to classify mammalian social systems using a variety of different criteria. Rather than review these schemes, I have chosen to illustrate the diversity of mammalian social systems by concentrating on a single order of small mammals – order Chiroptera. Bats belong to the second largest order of mammals, and although details of the social systems of only a few of the 850 current species are known, available data, which have been excellently reviewed by Bradbury (1977), indicate that an amazing diversity of behavioural strategies exists in this group.

In discussing the social organization of bats (as well as other kinds of mammals), it is useful to divide the year into seasonal periods that correspond to different phases of the reproductive cycle. Three periods are recognizable in most species of tropical and temperate mammals: the non-breeding period, the period of mate selection and copulation, and the period of parturition and lactation. Superimposed on this cycle are three basic patterns of sexual dispersion, which may be seasonally variable or invariant: (1) Mixed sex groups containing roughly equal numbers of adult males and females; (2) harem systems in which several females are monopolized by a single male; other males are solitary or occur in bachelor groups; and (3) sexually segregated groups in which females live in groups and males live solitarily or in groups.

All of these patterns of sexual dispersion occur in bats, which, unlike other mammals, are much more likely to be found living in groups rather than as solitary, dispersed individuals. Solitary species are the exception in bats but are known to occur in the tropical suborder Megachiroptera (e.g., *Epomops franqueti*) and temperate and tropical members of suborder Microchiroptera (e.g., *Diclidurus alba, Lasiurus borealis*). A large number of species, including virtually all temperate species, have seasonally variable sexual dispersion patterns. These patterns include unisexual groups during the nonbreeding season (e.g., *Pteronotus parnellii, Myotis adversus,* and various species of *Pteropus*) and the typical 'temperate cycle' in which the sexes occur in mixed groups in the winter, segregate during parturition and lactation, and re-establish contacts in various ways (e.g., in transient harems, at mating caves, and in mating swarms) in the fall. In temperate species, physiological constraints associated with hibernation prevent males from maintaining harems during the winter. Hence, unlike most other kinds of mammals, competition among males for mates may be quite low. Since any copulation that occurs between fall and the end of winter may result in a conception at ovulation in the spring, the optimal mating strategy for males is to inseminate as many females as possible before going into hibernation.

Additional social patterns in tropical bats include the following: (1) Seasonally invariant social structures consisting of either year-round harems (e.g., in *Saccopteryx bilineata* and *Phyllostomus hastatus*) and year-round groups of several to many males and females (e.g., in *Saccopteryx leptura* and *Pteropus giganteus*); (2) monogamous families (e.g., in *Nycteris hispida* and *Vampyrum spectrum*); and (3) leks or communal male display grounds (e.g. in *Hypsignathus monstrous*).

Our knowledge of the ecology of most species of bats is quite fragmentary so that attempts to relate social organization to ecological factors are premature. According to Bradbury (1977), diets and roost sites are probably the most important ecological determinants of bat social structure. Roosts, which provide protection against predators and optimal physiological conditions for energy conservation, can determine colony size and the amount of social behaviour that occurs when the group is maximally concentrated. Foraging patterns may play a role in shaping social organization through several presently poorly-known routes: through information exchange about the location of good feeding areas, through co-operative food capture, by foraging in groups to monitor the renewal rates of food in various habitat patches, and by colony defense of common foraging areas. Where food resources are defensible, males should attempt to control them and gain access to several females. Such behaviour is known to occur in colonies of *Saccopteryx bilineata* (Bradbury and Vehrencamp, 1976).

Ecological factors bearing on the evolution of mammalian social organization are much better known for marmots (*Marmota*). The ecology and social behaviour of the yellow-bellied marmot (*M. flaviventris*) has been particularly well-studied at middle elevations in the Rocky Mountains by Armitage and his students (e.g., Armitage 1973, 1974, 1975; Downhower and Armitage, 1971). The social structure of this herbivore includes colonies of harems centred around rocky den sites, which provide protection against predators and serve as hibernation sites, and satellite colonies whose members suffer higher mortality and lower reproductive success than do members of the main colonies. Harems contain several females that behave amicably towards each other; harem males, which are at least three years old, do not force females to remain in their harem. Only about 50 per cent of the females born into a harem disperse, but nearly 100 per cent of the males disperse as yearlings. There is a basic conflict of reproductive interest between a harem master and his females because male fitness increases with harem size (it is maximum in harems of two or three females) whereas female fitness decreases with harem size; solitary females have higher net fecundity than do harem females. Since a female stands to suffer a loss of fitness whenever additional females join a harem, one wonders why females are not more aggressive towards each other! Elliot (1975) has suggested that the answer to this apparent paradox is that the lifetime reproductive potential of harem females is greater than that of solitary or satellite females because of their higher survival rates. Thus the disadvantage of the reduced reproductive success of harem females in any single year is offset by their greater lifetime production.

In addition to *Marmota flaviventris*, two other species, *M. monax* and *M. olympus*, have been extensively studied, and Barash (1974) has used these data to propose a general theory of the evolution of marmot sociality. These three species differ greatly in the environments they inhabit and in their demographics and social structure. *M. monax* inhabits low elevation fields in the eastern United States where the growing season is about 150 days long. This species displays low social tolerance among adults which live solitarily. Females breed annually, and the young disperse in the year of their birth at about 33 per cent adult weight. As already discussed, *M. flaviventris* inhabits fields at various elevations in the Rocky Mountains. At mid-elevation sites, where the growing season is 70 – 100 days long, this species displays higher social tolerance, and adults and young live in colonies in which individuals have distinct home ranges. Females usually breed every year, and the young disperse as yearlings, at about 55 per cent adult weight. *M. olympus* inhabits high-elevation fields in Washington, where the growing season is 40 – 70 days long. This species displays the highest social tolerance: individuals live in amiable colonies and feed in social groups rather than in individual home ranges. Females breed every other year, and young disperse at three years of age, at about 70 – 80 per cent adult weight.

The following correlations emerge from this three-species comparison:

(1) A negative correlation between length of the growing season and age at dispersal, (2) a negative correlation between length of the growing season and social tolerance, and (3) a positive correlation between length of the growing season and frequency of reproduction. From these correlations, Barash postulates that greater maturity may be required for success in more severe environments and hence selection favours prolonging social tolerance of young until they have nearly attained adult size at higher elevations. An alternate explanation for these trends has been offered by Anderson, Armitage, and Hoffmann (1976). They suggest that the density of hibernacula may determine the degree of sociality in marmots. It is likely that the density of these critical resources varies inversely with environmental severity, as reflected by the length of the growing season. Also the degree of saturation of these resources is probably greatest at high elevations, especially if adult survival rates vary inversely with reproductive rates. A low density and high saturation of good hibernacula would selectively favour the production of fewer young of higher quality (± larger size at dispersal) and would favour a greater degree of social tolerance between adults and young until an optimal dispersal size had been attained. Demographic differences between the three species suggest that the degree of environmental saturation varies directly with elevation and lends credence to this alternate explanation of marmot sociality.

As a final example of the evolution of sociality in small mammals I would like to address the following question posed by Eisenberg (1967): "Do organisms that evolve independently in different geographical locations toward the same ecological niche also evolve similar social organizations?" Eisenberg attempted to answer this question by comparing various kinds of individual and social behaviour of desert or semi-desert rodents of the following four families: Cricetidae (*Peromyscus* spp.), Heteromyidae (*Perognathus, Microdipodops*, and *Dipodomys*), Dipodidae (*Allactaga*, and *Jaculus*), and Muridae, Gerbillinae (*Tatera, Pachyuromys, Dipodillus, Gerbillus,* and *Meriones*). Although living in desert-like habitats in the Old and New Worlds, these species differ markedly in numerous morphological, physiological, and ecological attributes so it should not be surprising that they definitely have not converged towards the same type of social organization. From detailed laboratory observations and available field data, Eisenberg recognized three types of sociality in these species: (1): Tolerant species, which includes the communal *Tatera indica* and pair-tolerant *Peromyscus californicus, P. maniculatus*, and *P. eremicus*; (2) species of intermediate tolerance, including two species of *Meriones, Gerbillus gerbillus, Peromyscus crinitus, Jaculus orientalis,* and *Allactaga elator*; and (3) intolerant species, which includes all of the heteromyid species examined. Eisenberg concludes that "Adaptation to xeric habitats with dispersed food supplies is not necessarily in and of itself conducive to selection for a dispersed, solitary existence. Equally important are other aspects of the species ecology including its mode of assembly of foodstuffs, its shelter construction, and its reproductive rate." As a final point he states that where defence of a food cache is possible, selection may or may not favour a solitary existence.

This point is similar to ones I have tried to make at various places in this chapter: different species have evolved different demographic, physiological, and behavioural strategies for coping with similar kinds of environments. It can be argued that no two species ever live in exactly the same environment (competitive exclusion and historical events would not allow this to happen) so that biological differences between species living in the 'same' environment really represent adaptations to somewhat different environmental conditions. Identifying exactly what environmental conditions favour the diversity of adaptive strategies seen among small mammals represents an exciting scientific challenge for present and future generations of ecologists.

1.6 Summary

In this chapter I have attempted to review both the theoretical and empirical aspects of the major components of the life-histories of small mammals. My basic premise has been that the demographic, physiological, and behavioural adaptations of small mammals are the result of individual, or in some cases, kin selection. Any population benefits that might accrue from these adaptations are the result of individual benefits.

As in certain other groups of vertebrates, two general demographic strategies exist in small mammals: (1) High reproductive effort per breeding season and low annual survival rates, and (2) low reproductive effort per season and high survival rates. Examples of the first group include microtine and murid rodents; examples of the second group include heteromyid rodents and bats. Demographic similarities tend to occur along taxonomic rather than geographic lines. Although these two strategies correspond to the expected results of *r* and *K* selection, the selective pressures that have produced them are probably diverse and are presently poorly known. Dispersal appears to be an important demographic process in many species, but its adaptive significance at the individual level has received little attention.

The energetic strategies of small mammals include diet choice and homeothermic v. heterothermic metabolic strategies. Most species are food generalists and include a relatively diverse array of food items in their diets. Seasonal changes in food availability is a major cause of this pattern. Food specialists include certain species of desert rodents and insectivorous bats. Within a few weeks after birth most small mammals remain homeothermic on a daily and seasonal basis. A taxonomically diverse array of species, including certain marsupials, insectivores, bats, and rodents, utilize controlled hypothermia on a daily and/or seasonal basis to reduce their energetic expenditures in environments that are periodically unfavourable regarding food availability and/or climate. Heterothermic and homeothermic species often co-exist under the same climatic conditions which indicates that hypothermia is just one way of dealing with unfavourable conditions.

The behavioural strategies of small mammals include patterns of habitat selection and social organization. Most species are habitat specialists, and intra- and interspecific interactions can strongly influence habitat choice. Habitat selection appears to have a strong genetic component, but learning, which gives individuals flexibility in their response to local conditions, is probably also important. Most species of small mammals exist in populations of dispersed, solitary individuals. Male polygamy in which critical resources or mates are often monopolized is the most common mating system; monogamy is an uncommon social system. Bats differ from other small mammals in that they are usually gregarious; they display a diverse array of mating systems. The environmental conditions responsible for various knds of social systems are poorly known. As in the case of demographic and physiological strategies, a variety of different social systems can be found among small mammals inhabiting a given type of environment (e.g., desert). A recurring theme in the evolution of small mammal life-history strategies has been multiple solutions to common environmental problems.

1.7 Acknowledgements

I thank J.S. Millar for his careful review of this chapter, which was prepared under partial support from the U.S. National Science Foundation.

References

Alexander, R.D. (1974) The evolution of social behaviour. *Annual Review of Ecology and Systematics,* **5**, 325-383.

Anderson, D.C., Armitage, K.B., & Hoffman, R.S. (1976) Socioecology of marmots: female reproductive strategies. *Ecology,* **57**, 552-560.

Anderson, P.K. (1970) Ecological structure and gene flow in small mammals. *Symposium of the Zoological Society of London,* **26**, 299-325.

Armitage, K.B. (1973) Population changes and social behavior following colonization by the yellow-bellied marmot. *Journal of Mammalogy,* **54**, 842-854.

Armitage, K.B. (1974) Male behaviour and territoriality in the yellow-bellied marmot. *Journal of Zoology,* London, **172**, 233-265.

Armitage, K.B. (1975) Social behavior and population dynamics of marmots. *Oikos,* **26**, 341-354.

Armitage, K.B., & Downhower, J.F. (1974). Demography of yellow-bellied marmot populations. *Ecology,* **55**, 1233-1245.

Barash, D.P. (1974) The evolution of marmot societies: a general theory. *Science,* **185**, 415-420.
Barkalow, F.S., Jr., Hamilton, R.B. & Soots, R.F., Jr. (1970) The vital statistics of an unexploited gray squirrel population. *Journal of Wildlife Management,* **34**, 489-500.
Bartholomew, G.A. (1977) Body temperature and energy metabolism. In: *Animal physiology: principles and adaptations.* M.S. Gordon (ed). Third edition. pp.364-449. MacMillen Publishing Co., New York.
Batzli, G.D., & Pitelka, F.A. (1971) Conditions and diet of cycling populations of the California vole, *Microtus californicus. Journal of Mammalogy,* **52**, 141-163.
Beatley, J.C. (1969) Dependence of desert rodents on winter annuals and precipitation. *Ecology,* **50**, 721-724.
Bekoff, M. (1977) Mammalian dispersal and the ontogeny of individual behavioural phenotypes. *American Naturalist,* **111**, 715-732.
Bell, G. (1976) On breeding more than once. *American Naturalist,* **110**, 57-77.
Black, H.L. (1974) A north temperate bat community: structure and prey populations. *Journal of Mammalogy,* **55**, 138-157.
Bock, E. (1972) Use of forest associations by bank vole populations. *Acta Theriologica,* **17**, 203-219.
Bourliere, F. (1975) Mammals, small and large: the ecological implications of size. In: *Small mammals: their productivity and population dynamics.* F. Golley, & K. Petrusewicz, and L. Ryszkowski (Eds). pp.1-8. Cambridge University Press, Cambridge.
Bradbury, J.W. (1977) Social organization and communication. In: *The biology of bats.* W. Wimsatt, (ed). pp.1-72. Volume 3. Academic Press. New York.
Bradbury, J.W., & Vehrencamp, S.L. (1976) Social organization and foraging in emballonurid bats. I. Field studies. *Behavioral Ecology and Sociobiology,* **1**, 337-381.
Bradbury, J.W., & Vehrencamp, S.L. (1977a) Social organization and foraging in emballonurid bats. III. Mating systems. *Behavioral Ecology and Sociobiology,* **2**, 1-17.
Bradbury, J.W. & Vehrencamp, S.L. (1977b) Social organization and foraging in emballonurid bats. IV. Parental investment patterns. *Behavioral Ecology and Sociobiology,* **2**, 19-29.
Bradley, W.G., & Mauer, R.A. (1971) Reproduction and food habits of Merriam's kangaroo rat, *Dipodomys merriami. Journal of Mammalogy*, **52**, 497-507.
Brown, J.H. (1971) Mechanisms of competitive exclusion between two species of chipmunks. *Ecology,* **52**, 305-311.
Brown, J.H., & Bartholomew, G.A. (1969) Periodicity and energetics of torpor in the kangaroo mouse, *Microdipodops pallidus. Ecology,* **50**, 705-709.
Brown, J.H. & Lasiewski, R.C. (1972) Metabolism of weasels: the cost of being long and thin. *Ecology,* **53**, 939-943.
Brown, J.H., Lieberman, G.A. & Dengler, W.F. (1972) Woodrats and cholla: dependence of a small mammal population on the density of cacti. *Ecology,* **53**, 310-313.
Brown, L.N. (1970) Population dynamics of the western jumping mouse (*Zapus princeps*) during a four-year study. *Journal of Mammalogy,* **51**, 651-658.
Buchler, E.R. (1976) Prey selection by *Myotis lucifugus* (Chiroptera: Vespertilionidae). *American Naturalist*, **110**, 619-628.
Bujalska, G. (1975) Reproduction and mortality of bank voles and the changes in the size of an island population. *Acta Theriologica,* **20**, 41-56.

Burt, W.H., & Grossenheider, R.P. (1964) *A Field Guide to The Mammals*. Houghton Mifflin Company, Boston.

Cameron, G.N. (1973) Effect of litter size on postnatal growth and survival in the desert woodrat. *Journal of Mammalogy,* **54**, 489-493.

Carl, E.A. (1971) Population control in arctic ground squirrels. *Ecology,* **52**, 395-413.

Carter, D.C. (1970) Chiropteran reproduction. In: *About bats*. Slaughter, B.H., & Walton, D.W. (Eds.) pp.233-246 Southern Methodist University Press, Dallas.

Caughley, G. (1966) Mortality patterns in mammals. *Ecology,* **47**, 906-918.

Caughley, G. (1977) *Analysis of Vertebrate Populations*. John Wiley and Sons, New York.

Caughley, G., & Birch, L.C. (1971) Rate of increase. *Journal of Wildlife Management,* **35**, 658-663.

Christian, J.J. (1970) Social subordination, population density, and mammalian evolution. *Science,* **168**, 84-90.

Cody, M.L. (1971). Ecological aspects of reproduction. In: *Avian Biology*. Farner, D., & King, J. (Eds.) pp.461-512 Volume 1. Academic Press, New York.

Cole, L.C. (1954) The population consequences of life-history phenomena. *Quarterly Review of Biology,* **29**, 103-137.

Crook, J.H., Ellis, J.E., & Goss-Custard, J.D. (1976) Mammalian social systems: structure and function. *Animal Behaviour,* **24**, 261-274.

Daly, M., & Daly, S. (1973) On the feeding ecology of *Psammomys obesus* (Rodentia, Gerbillidae) in the Wadi Saoura, Algeria. *Mammalia,* **37**, 545-561.

Davis, D.E. (1976) Hibernation and circannual rhythms of food consumption in marmots and ground squirrels. *Quarterly Review of Biology,* **51**, 477-514.

Dobzhansky, T. (1970) *Genetics of the evolutionary process*. Columbia University Press, New York.

Dolbeer, R.A. & Clark, W.R. (1975) Population ecology of snowshoe hares in the central Rocky Mountains. *Journal of Wildlife Management,* **39**, 535-549.

Downhower, J.F., & Armitage, K.B. (1971) The yellow-bellied marmot and the evolution of polygamy. *American Naturalist,* **105**, 355-370.

Drickamer, L.C. (1976) Hypotheses linking food habits and habitat selection in *Peromyscus*. *Journal of Mammalogy,* **57**, 763-766.

Drickamer, L.C., & Vestal, B.M. (1973) Patterns of reproduction in a laboratory colony of *Peromyscus*. *Journal of Mammalogy,* **54**, 523-528.

Dunmire, W.W. (1960) An altitudinal survey of reproduction in *Peromyscus maniculatus*. *Ecology,* **41**, 174-182.

Dwyer, P.D. (1971) Temperature regulation and cave-dwelling in bats: an evolutionary perspective. *Mammalia,* **35**, 424-455.

Eisenberg, J.F. (1966). The social organization of mammals. *Handbuch der Zoologie, VIII,* **10**, 1-92.

Eisenberg, J.F. (1967) A comparative study in rodent ethology with emphasis on evolution of social behavior, I. *Proceedings of the United States National Museum,* **122**, 1-51.

Eisenberg, J.F. (1975) Phylogeny, behavior, and ecology in the Mammalia. In: *Phylogeny of the primates*. Luckett, W.P. & Szalay, F.S. (Eds.) pp.47-68. Plenum Publishing Corporation, New York.

Eisenberg, J.F. (1977) The evolution of the reproductive unit in the Class Mammalia. In: Rosenblatt, J. & Komisaruk, B. (Eds.) Lehrman Memorial Symposium, No. 1.

Elliot, P.F. (1975) Longevity and the evolution of polygamy. *American Naturalist,* **109**, 281-287.

Emlen, S.T., & Oring, L.W. (1977) Ecology, sexual selection, and the evolution of mating systems. *Science,* **197**, 215-223.

Erlinge, S. (1975) Feeding habits of the weasel *Mustela nivalis* in relation to prey abundance. *Oikos*, **26**, 378-384.

Estabrook, G.F. & Dunham, A.E. (1976) Optimal diet as a function of absolute abundance, relative abundance, and relative value of available prey. *American Naturalist,* **110**, 401-413.

Fairbairn, D.J. (1977a) Why breed early? A study of reproductive tactics in *Peromyscus. Canadian Journal of Zoology,* **55**, 862-871.

Fairbairn, D.J. (1977b) The spring decline in deer mice: death or dispersal? *Canadian Journal of Zoology,* **55**, 84-92.

Fisler, G.F. (1969) Mammalian organizational systems. *Los Angeles County Museum Contributions in Science,* **167**, 1-32.

Fisler, G.F. (1971) Age structure and sex ratio in populations of *Reithrodontomys. Journal of Mammalogy,* **52**, 653-662.

Fleharty, E.D. & Olson, L.E. (1969) Summer food habits of *Microtus ochrogaster* and *Sigmodon hispidus. Journal of Mammalogy,* **50**, 475-486.

Fleming, T.H. (1971) Population ecology of three species of neotropical rodents. Miscellaneous Publications of the University of Michigan Museum of Zoology, **143**, 1-77.

Fleming, T.H. (1973) The reproductive cycles of three species of opossums and other mammals in the Panama Canal Zone. *Journal of Mammalogy,* **54**, 439-455.

Fleming, T.H. (1974) Population ecology of two species of tropical heteromyid rodents. *Ecology,* **55**, 493-510.

Fleming, T.H. (1975) The role of small mammals in tropical ecosystems. In: *Small mammals: their productivity and population dynamics.* Golley, F., Petrusewicz, K, & Ryszkowski, L. (Eds.) pp.269-298. Cambridge University Press, Cambridge.

Fleming, T.H., Hooper, E.T., & Wilson, D.E. (1972) Three Central American bat communities: structure, reproductive cycles, and movement patterns. *Ecology,* **53**, 555-569.

Fleming, T.H., & Rauscher, R.J. (1978). On the evolution of litter-size in *Peromyscus leucopus. Evolution,* **32**, 45-55.

Freeland, W.J., & Janzen, D.H. (1974) Strategies in herbivory by mammals: the role of plant secondary compounds. *American Naturalist,* **108**, 269-289.

French, N.R., Maza, B.H., Hill, H.O., Aschwanden, A.P., & Kaaz, H.W. (1974) A population study of irradiated desert rodents. *Ecological Monographs,* **44**, 45-72.

French, N.R., Stoddart, D.M. & Bobek, B. (1975) Patterns of demography in small mammal populations. In: *Small mammals: their productivity and population dynamics.* Golley, F., Petrusewicz, K. & Ryszkowski, L. (Eds.) pp.73-102 Cambridge University Press, Cambridge.

Fuller, W.A. (1977) Demography of a subarctic population of *Clethrionomys gapperi*: numbers and survival. *Canadian Journal of Zoology,* **55**, 42-51.

Gagdil, M. (1971) Dispersal: population consequences and evolution. *Ecology,* **52**, 253-261.

Gaines, M.S., & Rose, R.K. (1976) Population dynamics of *Microtus ochrogaster* in eastern Kansas. *Ecology,* **57**, 1145-1161.

Gebczynski, M. (1975) Heat economy and the energy of growth in the bank vole during the first month of life. *Acta Theriologica,* **20**, 379-434.

Ghiselin, J. (1970) Edaphic control of habitat selection by kangaroo mice (*Microdipodops*) in three Nevadan populations. *Oecologia,* **4**, 248-261.
Gill, A.E. (1977) Food preferences of the California vole, *Microtus californicus. Journal of Mammalogy,* **58**, 229-233.
Goertz, J.W. (1971) An ecological study of *Microtus pinetorum* in Oklahoma. *American Midland Naturalist,* **86**, 1-12
Goodman, D. (1974) Natural selection and a cost ceiling on reproductive effort. *American Naturalist,* **108**, 247-268.
Gordon, M.S. (1977) *Animal physiology: principles and adaptations.* Third edition. MacMillen Publishing Company, New York.
Górecki, A., & Grygielska, M. (1975) Consumption and utilization of natural foods by the common hamster. *Acta Theriologica,* **20**, 237-246.
Grant, P.R. (1970) Experimental studies of competitive interaction in a two species system. II. The behaviour of *Microtus, Peromyscus* and *Clethrionomys* species. *Animal Behaviour,* **18**, 411-426.
Grant, P.R. (1971) The habitat preference of *Microtus pennsylvanicus,* and its relevance to the distribution of this species on islands. *Journal of Mammalogy,* **52**, 351-361.
Grant, P.R. (1975) Population performance of *Microtus pennsylvanicus* confined to woodland habitat, and a model of habitat occupancy. *Canadian Journal of Zoology,* **53**, 1447-1465.
Hamilton, W.D. (1964) The genetical evolution of social behaviour. I. *Journal of Theoretical Biology,* **7**, 1-16.
Hamilton, W.D., & May, R.M. (1977) Dispersal in stable habitats. *Nature, London,* **269**, 578-581.
Healey, M.C. (1967) Aggression and self-regulation of population size in deermice. *Ecology,* **48**, 377-392.
Heinrich, B. (1977) Why have some animals evolved to regulate a high body temperature? *American Naturalist,* **111**, 623-640.
Heithaus, E.R., Fleming, T.H., & Opler, P.A. (1975) Foraging patterns and resource utilization in seven species of bats in a seasonal tropical forest. *Ecology,* **56**, 841-854.
Hill, R.W. (1972) The amount of maternal care in *Peromyscus leucopus* and its thermal significance for the young. *Journal of Mammalogy,* **53**, 774-790.
Hill, R.W. (1976a) *Comparative physiology of animals: an environmental approach*. Harper and Row Publishers, New York.
Hill, R.W. (1976b) The ontogeny of homeothermy in neonatal *Peromyscus leucopus. Physiological Zoology,* **49**, 292-306.
Hirshfield, M.F., & Tinkle, D.W. (1975) Natural selection and the evolution of reproductive effort. *Proceedings of the United States National Academy of Sciences,* **72**, 2227-2231.
Hoffmeyer, I. (1973) Interaction and habitat selection in the mice *Apodemus flavicollis* and *A. sylvaticus. Oikos,* **24**, 108-116.
Howard, W.E. (1960) Innate and environmental dispersal of individual vertebrates. *American Midland Naturalist,* **63**, 152-161.
Howell, D.J. (1976) Weight loss and temperature regulation in clustered versus individual *Glossophaga soricina. Comparative Biochemistry and Physiology,* **53**, 197-199.
Humphrey, S.R. (1975) Nursery roosts and community diversity of nearctic bats. *Journal of Mammalogy,* **56**, 321-346.
Humphrey, S.R., & Cope, J.B. (1977) Survival rates of the endangered Indiana bat, *Myotis sodalis. Journal of Mammalogy,* **58**, 32-36.

Husar, S.L. (1976) Behavioral character displacement: evidence of food partitioning in insectivorous bats. *Journal of Mammalogy,* **57**, 331-338.

Iverson, S.L., & Turner, B.N. (1972) Winter coexistence of *Clethrionomys gapperi* and *Microtus pennsylvanicus* in a grassland habitat. *American Midland Naturalist,* **88**, 440-445.

Janzen, D.H. (1971) Seed predation by animals. *Annual Review of Ecology and Systematics,* **2**, 465-492.

Joule, J., & Cameron, G.N. (1975) Species removal studies. I. Dispersal strategies of sympatric *Sigmodon hispidus* and *Reithrodontomys fulvescens* populations. *Journal of Mammalogy,* **56**, 378-396.

Kenagy, G.J. (1972) Saltbush leaves: excision of hypersaline tissue by a kangaroo rat. *Science,* **178**, 1094-1096.

Kenagy, G.J. (1973a) Adaptations for leaf-eating in the Great Basin kangaroo rat, *Dipodomys microps. Oecologia,* **12**, 383-412.

Kenagy, G.J. (1973b) Daily and seasonal patterns of activity and energetics in a heteromyid rodent community. *Ecology,* **54**, 1201-1219.

King, J.A. (1973) The ecology of aggressive behavior. *Annual Review of Ecology and Systematics,* **4**, 117-138.

Kleiman, D.G. (1977) Monogamy in mammals. *Quarterly Review of Biology,* **52**, 39-69.

Krebs, C.J., & Meyers, J.H. (1974) Population cycles in small mammals. *Advances in Ecological Research,* **6**, 267-399.

Krebs, J.R. (1971) Territory and breeding density in the great tit *Parus major* L. *Ecology,* **52**, 2-22.

Levin, D.A. (1976) The chemical defenses of plants to pathogens and herbivores. *Annual Review of Ecology and Systematics,* **7**, 121-160.

Levins, R. (1968) Evolution in changing environments. *Monographs in Population Biology,* **2**, 1-120.

Lewontin, R.C. (1970) The units of selection. *Annual Review of Ecology and Systematics,* **1**, 1-18.

Lidicker, W.Z., Jr. (1973) Regulation of numbers in an island population of the California vole, a problem in community dynamics. *Ecological Monographs,* **43**, 271-302.

Lidicker, W.Z., Jr. (1975) The role of dispersal in the demography of small mammals. In: *Small mammals: their productivity and population dynamics,* Golley, F., Petrusewicz, K., & Ryskowski, L. (Eds.) pp.104-120. Cambridge University Press, Cambridge..

Lord, R.D., Jr. (1960) Litter size and latitude in North American mammals. *American Midland Naturalist,* **64**, 488-499.

MacArthur, R.H., & Pianka, E.R. (1966) On optimal use of a patchy environment. *American Naturalist,* **100**, 603-609.

MacArthur, R.H. & Wilson, E.O. (1967) The theory of island biogeography. *Monographs in Population Biology,* **1**, 1-203.

Maynard-Smith, J. (1976) Group selection. *Quarterly Review of Biology,* **51**, 277-283.

Maynard-Smith, J. (1977) Parental investment: a prospective analysis. *Animal Behaviour,* **25**, 1-9.

M'Closkey, R.T. (1975) Habitat dimensions of white-footed mice, *Peromyscus leucopus. American Midland Naturalist,* **93**, 158-167.

McManus, J.J. (1971) Early postnatal growth and the development of temperature regulation in the Mongolian gerbil, *Meriones unguiculatus. Journal of Mammalogy,* **52**, 782-792.

McNab, B.K. (1969) The economics of temperature regulation in neotropical bats. *Comparative Biochemistry and Physiology,* **31**, 227-268.
Meredith, D.H. (1976) Habitat selection by two parapatric species of chipmunks (*Eutamias*). *Canadian Journal of Zoology,* **54**, 536-543.
Mertz, D.B. (1970) Notes on methods used in life history studies. In: *Readings in ecology and ecological genetics.* Connell, J., Mertz, D., and Murdoch, W. (Eds.) pp.4-17. Harper and Row, New York.
Meserve, P.L. (1971) Population ecology of the prairie vole, *Microtus ochrogaster,* in the western mixed prairie of Nebraska. *American Midland Naturalist,* **86**, 417-433.
Meserve, P.L. (1974) Ecological relationships of two sympatric woodrats in a California coastal sage scrub community. *Journal of Mammalogy,* **55**, 442-447.
Meserve, P.L. (1976) Food relationships of a rodent fauna in a California coastal sage scrub community. *Journal of Mammalogy,* **57**, 300-319.
Meslow, E.C., & Keith, L.B. (1971) A correlation analysis of weather versus snowshoe hare population parameters. *Journal of Wildlife Management,* **35**, 1-15.
Michener, G.R., & Michener, D.R. (1977) Population structure and dispersal in Richardson's ground squirrels. *Ecology,* **58**, 359-368.
Millar, J.S. (1970a) The breeding season and reproductive cycle of the western red squirrel. *Canadian Journal of Zoology,* **48**, 471-473.
Millar, J.S. (1970b) Variations in fecundity of the red squirrel, *Tamiasciurus hudsonicus* (Erxleben). *Canadian Journal of Zoology,* **48**, 1055-1058.
Millar, J.S. (1972) Timing of breeding of pikas in southwestern Alberta. *Canadian Journal of Zoology,* **50**, 665-669.
Millar, J.S. (1973) Evolution of litter-size in the pika, *Ochotona princeps* (Richardson). *Evolution,* **27**, 134-143.
Millar, J.S. (1977) Adaptive features of mammalian reproduction. *Evolution,* **31**, 370-386.
Millar, J.S., & Zwickel, F.C. (1972a) Characteristics and ecological significance of hay piles of pikas. *Mammalia,* **36**, 657-667.
Millar, J.S., & Zwickel, F.C. (1972b) Determination of age, age structure, and mortality of the pika, *Ochotona princeps* (Richardson). *Canadian Journal of Zoology,* **50**, 229-232.
Morrison, P. (1960) Some interrelations between weight and hibernation function. *Bulletin of the Museum of Comparative Zoology, Harvard,* **124**, 75-91.
Morse, D.H. (1974) Niche breadth as a function of social dominance. *American Naturalist,* **108**, 818-830.
Negus, N.C., & Berger, P.J. (1977) Experimental triggering of reproduction in a natural population of *Microtus montanus. Science,* **196**, 1230-1231.
Nicols, J.D., Conley, W., Batt, B. & Tipton, A.R. (1976) Temporally dynamic reproductive strategies and the concept of *r*- and *K*-selection. *American Naturalist,* **110**, 995-1005.
O'Farrell, T.P., Olson, R.J., Gilbert, R.O., & Hedlund, J.D. (1975) A population of Great Basic pocket mice, *Perognathus parvus*, in the shrub-steppe of south-central Washington. *Ecological Monographs,* **45**, 1-28.
Olsen, R.W. (1973) Shelter-site selection in the white-throated woodrat, *Neotoma albigula. Journal of Mammalogy,* **54**, 594-610.
Orians, G.H. (1969) On the evolution of mating systems in birds and mammals. *American Naturalist,* **103**, 589-603.
Pianka, E.R. (1972) *r* and *K* selection or *b* and *d* selection? *American Naturalist,* **106**, 581-588.

Pianka, E.R. (1974) *Evolutionary ecology*. Harper and Row, New York.
Pianka, E.R., & Parker, W.S. (1975) Age-specific reproductive tactics. *American Naturalist,* **109**, 453-464.
Pidduck, E.R. & Falls, J.B. (1973) Reproduction and emergence of juveniles in *Tamias striatus* (Rodentia: Sciuridae) at two localities in Ontario, Canada. *Journal of Mammalogy,* **54**, 693-707.
Pyke, G.H., Pulliam, H.R. & Charnov, E.L. (1977) Optimal foraging: a selective review of theory and tests. *Quarterly Review of Biology,* **52**, 137-154.
Ralls, K. (1977) Sexual dimorphism in mammals: avian models and unanswered questions. *American Naturalist,* **111**, 917-938.
Randolph, J.C. (1973) The ecological energetics of a homeothermic predator, the short-tailed shrew. *Ecology,* **54**, 1166-1187.
Rapport, D.J., & Turner, J.E. (1977) Economic models in ecology. *Science,* **195**, 367-373.
Redfield, J.A., Krebs, C.J. & Taitt, M.J. (1977) Competition between *Peromyscus maniculatus* and *Microtus townsendii* in grasslands of coastal British Columbia. *Journal of Animal Ecology,* **46**, 607-616.
Reichman, O.J. (1975) Relation of desert rodent diets to available resources. *Journal of Mammalogy,* **56**, 731-751.
Reichman, O.J. (1977) Optimization of diets through food preferences by heteromyid rodents. *Ecology,* **58**, 454-457.
Ricklefs, R.E. (1977) On the evolution of reproductive strategies of birds: reproductive effort. *American Naturalist,* **111**, 453-478.
Rosenzweig, M.L. (1973) Habitat selection experiments with a pair of coexisting heteromyid rodent species. *Ecology,* **54**, 111-117.
Rosenzweig, M.L. (1974) On the evolution of habitat selection. *Proceedings of the First International Congress of Ecology,* 401-404.
Rothstein, B.E., & Tamarin, R.H. (1977) Feeding behavior of the insular beach vole, *Microtus breweri. Journal of Mammalogy,* **58**, 84-85.
Schaffer, W.M., & Tamarin, R.H. (1973) Changing reproductive rates and population cycles in lemmings and voles. *Evolution,* **27**, 111-124.
Schoener, T.W. (1971) Theory of feeding strategies. *Annual Review of Ecology and Systematics,* **2**, 369-404.
Sherman, P.W. (1977) Nepotism and the evolution of alarm calls. *Science,* **197**, 1246-1253.
Slade, N.A., & Balph, D.F. (1974) Population ecology of Uinta ground squirrels. *Ecology,* **55**, 989-1003.
Slobodkin, L.B. (1961) *Growth and regulation of animal populations.* Holt, Rinehart, and Winston, New York.
Smith, A.T. (1978) Comparative demography of pikas: effect of spatial and temporal age-specific mortality. *Ecology,* **59**, 133-137.
Smith, C.C. (1968) The adaptive nature of social organization in the genus of tree squirrels *Tamiasciurus. Ecological Monographs,* **38**, 31-63.
Smith, C.C. & Follmer, D. (1972) Food preferences of squirrels. *Ecology,* **53**, 82-91.
Smith, M.H., & McGinnis, J.T. (1968) Relationships of latitude, altitude, and body size to litter size and mean annual production of offspring in *Peromyscus. Researches in Population Ecology,* **10**, 115-126.
Soholt, L.F. (1976) Development of thermoregulation in Merriam's kangaroo rat, *Dipodomys merriami. Physiological Zoology,* **49**, 152-157.
Southwood, T.R.E. (1976) Bionomic parameters and population parameters. In: *Theoretical ecology, principles and applications.* May, R. (Ed.) pp.26-48. W.B. Saunders Company, Philadelphia.

Southwood, T.R.E. (1977) Habitat, the templet for ecological strategies? *Journal of Animal Ecology*, **46**, 337-365.

Southwood, T.R.E., May, R.M., Hassell, M.P., & Conway, G.R. (1974) Ecological strategies and population parameters. *American Naturalist,* **108**, 791-804.

Spencer, A.W., & Steinhoff, H.W. (1968) An explanation of geographic variation in litter size. *Journal of Mammalogy,* **49**, 281-286.

States, J.B. (1976) Local adaptations in chipmunk (*Eutamias amoenus*) populations and evolutionary potential at species' borders. *Ecological Monographs,* **46**, 221-256.

Stearns, S.C. (1976) Life-history tactics: a review of ideas. *Quarterly Review of Biology,* **51**, 3-47.

Stearns, S.C. (1977) The evolution of life history traits. *Annual Review of Ecology and Systematics,* **8**, 145-171.

Stebbins, G.L. (1977) *Processes of organic evolution*. 3rd ed. Prentice-Hall Inc., Englewood Cliffs, New Jersey.

Stoddart, D.M. (1970) Individual range, dispersion and dispersal in a population of water voles (*Arvicola terrestris* (L.)). *Journal of Animal Ecology,* **39**, 403-425.

Stoecker, R.E. (1972) Competitive relations between sympatric populations of voles (*Microtus montanus* and *M. pennsylvanicus*). *Journal of Animal Ecology,* **41**, 311-329.

Stout, I.J., & Sonenshine, D.E. (1974) Ecology of an opossum population in Virginia, 1963-69. *Acta Theriologica,* **19**, 235-245.

Studier, E.H., Procter, J.W., & Howell, D.J. (1970) Diurnal body weight loss in five species of *Myotis*. *Journal of Mammalogy,* **51**, 302-309.

Tinkle, D.W., & Ballinger, R.E. (1972) *Sceloporus undulatus*: a study of the intraspecific comparative demography of a lizard. *Ecology,* **53**, 570-584.

Tinkle, D.W., Wilbur, H.M., & Tilley, S.G. (1970) Evolutionary strategies in lizard reproduction. *Evolution,* **24**, 55-74.

Trivers, R.L. (1971) The evolution of reciprocal altruism. *Quarterly Review of Biology,* **46**, 35-57.

Trivers, R.L. (1972) Parental investment and sexual selection. In: *Sexual selection and the descent of man*. Campbell, B. (Ed.) pp. 136-179. Aldine Publishing Company, Chicago.

Trivers, R.L. (1974) Parent-offspring conflict. *American Zoologist,* **14**, 249-264.

Trune, D.R., & Slobodchikoff, C.N. (1976) Social effects of roosting on the metabolism of the pallid bat (*Antrozous pallidus*). *Journal of Mammalogy,* **57**, 656-663.

Van Valen, L. (1971) Group selection and the evolution of dispersal. *Evolution,* **25**, 591-598.

Van Valen, L. (1975) Group selection, sex, and fossils. *Evolution,* **29**, 87-94.

Wecker, S.C. (1963) The role of early experience in habitat selection of the prairie deermouse, *Peromyscus maniculatus bairdii*. *Ecological Monographs,* **33**, 307-325.

West-Eberhard, M.J. (1975) The evolution of social behavior by kin selection. *Quarterly Review of Biology,* **50**, 1-34.

Westoby, M. (1974) An analysis of diet selection by large generalist herbivores. *American Naturalist,* **108**, 290-304.

Whitaker, J.O., Jr., & Martin, R.L. (1977) Food habits of *Microtus chrotorrhinus* from New Hampshire, New York, Labrador, and Quebec. *Journal of Mammalogy,* **58**, 99-100.

Whitney, P. (1976) Population ecology of two sympatric species of subalpine microtine rodents. *Ecological Monographs,* **46**, 85-104.
Wiens, J.A. (1976) Population responses to patchy environments. *Annual Review of Ecology and Systematics,* **7**, 81-120.
Wilbur, H.M., Tinkle, D.W., & Collins, J.P. (1974) Environmental certainty, trophic level, and resource availability in life history evolution. *American Naturalist,* **108**, 805-817.
Williams, G.C. (1966) *Adaptation and natural selection*. Princeton University Press, Princeton.
Wilson, D.E., & Findley, J.S. (1970) Reproductive cycles of a neotropical insectivorous bat, *Myotis nigricans. Nature, London,* **225**, 1155.
Wilson, D.S. (1977) Structured demes and the evolution of group-advantageous traits. *American Naturalist,* **111**, 157-185.
Wilson, E.O. (1975) *Sociobiology. The new synthesis.* Belknap Press, Cambridge, Massachusetts.
Wynne-Edwards, V.C. (1962) *Animal dispersion in relation to social behaviour.* Oliver and Boyd, Edinburgh.
Wynne-Edwards, V.C. (1965) Self-regulating systems in populations of animals. *Science,* **147**, 1543-1548.

Population processes in small mammals 2

H.N. SOUTHERN

2.1 Introduction

In this and the following chapter I shall try to review what we now know about the distribution, abundance and turnover of small mammal populations and to indicate their importance in animal communities by virtue of their great numbers and their dominant position in food chains.

This is no light task because the economic influence of small mammals on agriculture and forestry and their ubiquity have made them popular subjects for researchers working both on applied and fundamental problems of population dynamics. The results of these researches are multitudinous and spread widely through ecological literature; therefore, I have found it necessary to be highly selective in presenting evidence within the scope of these two chapters.

Many previous workers have tended to use the term 'small mammals' to cover species between mouse and rat size. The Russians designate these as 'mouse-like rodents' and the French as 'micromammalia.' The decision by the editor of this volume to extend this designation to an upper limit of 5 kg in weight widens the amount of literature to be covered but enables us to bring in some valuable material which would otherwise be excluded. Thus we can consider another world-wide group of herbivores, the Lagomorpha, and we can bring in many smaller Carnivora and examine the impact of their predation upon the herbivores.

It is no coincidence that the dominant (in terms of numbers and species) groups of the two higher classes of vertebrates, birds and mammals, are small and short-lived. A range of weights of between 10 and 100 g embraces the most successful and the most highly evolved order in both classes, the Passeriformes among birds and the Rodentia among mammals. Of some 8600 species of living birds 5100 are passerines (Mayr and Amadon, 1951) and of 4237 species of mammals 1729 are rodents (Simpson, 1945; Morris,

1965). These figures are, of course, approximate because, even in these well-worked higher vertebrates, taxonomists are constantly breaking new ground; however, they emphasize the advantages of being small.

As a rough indication of abundance we may cite the numbers of mice and voles in the British Isles. Estimates vary between 5 and 500 per ha and, since there are few habitats where these small creatures do not occur, the lowest estimate for the whole country will be 125 million and the highest 12 500 million. Similarly, to take an example from birds, colonies of the dioch, *Quelea quelea*, in Africa may contain over 1 million individuals.

Small size means large appetites and it is not surprising that these animals often pose difficult problems for farmers and foresters. In the case of the above mentioned dioch, some 125 million nestlings were killed in one control effort in West Africa without any great alleviation of damage to crops.

Small size also means a short life and this, in turn, means a rapid turnover of individuals in a population. Thus in mice and voles the mean expectation of further life at weaning may be measured in months rather than in years (see details later) though most of these species, if cosseted in captivity, can live for several years.

A further consequence is that this rapid turnover implies an equally rapid response to changes in selective pressure. In general, biologists now recognize that wild animal populations are far more heterogeneous genetically and selection pressures much greater than had previously been supposed (*see*, e.g., Ford, 1971; Berry, 1977). Therefore, evolutionary change may be relatively fast. Berry (*loc. cit.*) has given evidence that the pattern of variation in British wood mice (*Apodemus sylvaticus*) may have taken shape during the last 1000 years rather than in the last 10 000 years as previously supposed. So, on a longer time scale, the spread of grasslands, especially those with species containing large amounts of silica, since the last retreat of glaciers may have fostered the evolution and proliferation of voles with their complex and angular cheek teeth (Kurten, 1968).

Most of these small mammals are herbivorous and their great abundance in almost all of the world's ecosystems creates a broad basic layer of primary consumers in the pyramid of numbers, which characterizes all animal communities. There are exceptions; insectivores – hedgehogs, moles and shrews – are primarily carnivorous and are widely-spread among the habitats of the world though their numbers do not rival those of the herbivores. Also, the specialized order of the Chiroptera, which must be included among small mammals, has adopted an aerial way of life and preys largely on flying insects; at least this is true of the widespread and dominant suborder of the Microchiroptera – the familiar bats of our woodlands and fields.

But to return to the rodents – nothing can convey more impressively the explosion of their adaptive radiation than a bald statement of their

Table 2.1 *Classification of myomorph rodents (after Morris, 1965).*

ORDER RODENTIA (1729 spp.)					
Suborder Myomorpha (1183 spp.)					
Superfamily	Muroidea (1116 spp.)		Gliroidea (31 spp.)	Dipodoidea (36 spp.)	
Family	Cricetidae	Muridae	Gliridae	Zapodidae	Dipodidae
	voles, lemmings, New World 'mice and rats', gerbils (602 spp. +21 mole rats)	Old World rats and mice (493 spp.)	Platacanthomyidae, Seleveniidae, dormice (31 spp.)	jumping mice (11 spp.)	jerboas (25 spp.)
	e.g. *Peromyscus* *Sigmodon* *Clethrionomys* *Lemmus* *Dicrostonyx* *Microtus* *Meriones*	e.g. *Apodemus* *Mus* *Rattus* *Praomys* *Acomys*	e.g. *Glis* *Muscardinus* *Graphiurus*	e.g. *Sicista* *Zapus*	e.g. *Dipus* *Jaculus*

classification (Table 2.1). The table deals only with the Myomorpha, the largest of the three suborders of the rodents, and is a rationalization of previous classifications (e.g., that of Ellerman and Morrison-Scott, 1951) which have blown up the list by excessive 'splitting'. Even so, it is striking testimony to the way in which these 'voles, rats and mice' have occupied the earth. In a book about ecology it may seem strange to stray into classification but I hold with Fager (1965) "that one cannot adequately study something until it can be fairly objectively identified and described".

The ecology of any group of animals (and of plants, for that matter) is concerned with two main questions – what factors promote or limit their distribution and what controls their numbers? It will be clear that these two problems interdigitate but it is convenient here to treat them separately. It will also be clear that the ecologist's focus of interest is in the stability or otherwise of distribution and numbers – in other words he studies processes and dynamic relationships. This can be done successfully only at the population level and this implies the use of demographic techniques, the measurement of density, of birth rates and death rates and of movements. These are difficult to make upon wild populations but already considerable progress has been made in what is still a young discipline. The difficulties are many and may be readily visualized; they cannot begin to be overcome without an intimate and sympathetic knowledge of the ways of the animals to be studied. Ecology must be based on sound natural history.

2.2 Distribution

Two aspects of distribution interest the ecologist. One is the extent and shape of a species' geographical range; the other is the pattern and contouring of densities on a smaller scale. For convenience we can label these Macrodistribution and Microdistribution. There is no strict dividing line between them but, again, it is convenient to treat them separately.

2.2.1 'Macrodistribution'

When one contemplates the geographical ranges of animal species, one is tempted to wonder why many of them are so small. Why do they not inherit the earth? 'Why,' in the words of Hutchinson (1959), "are there so many kinds of animals?" This is true on a large scale as well as on a smaller scale. The popularity of surveys and mapping of distribution made possible, in recent decades, by the growing up of networks of enthusiastic observers (e.g., Arnold (1978) for mammals in the British Isles) has revealed a marked circumscription of many species and notable gaps within the main range.

The answer is to be found in the relationships between species and between species and their environment – briefly in their evolution. But this answer is not easy to specify in individual instances. Broadly, we know that there are tracts of the world, biomes – to be more specific, where species are few and others where they are many and this is connected with the complexity of the environment. There are many more species in a tropical forest than there are on the tundra. There are more species in a tropical forest than there are in a temperate forest. Fleming (1973) lists 15–16 species of forest mammals in Alaska, 31–35 in eastern U.S.A. and 70 species in Panama.

This contrast presumably arises because of the greater opportunities for specialization in an environment which is complex structurally as well as botanically. Such fineness of specialization has been analysed convincingly for birds by, e.g. MacArthur (1958) and Lack (1944), and there is no reason to suppose the same principles do not apply to mammals. The upshot of this argument is that an animal species does not inherit the earth because conditions and/or other species prevent it from doing so.

So the geographical range of a species is dictated not so much by its powers of expansion as by the barriers to this expansion which we will examine in more detail below. First, it is instructive to observe what happens if barriers are removed, either by accident or by misguided introductions by man.

Elton in his *Ecology of Invasions by Animals and Plants* (1958) has documented the astonishing speed with which introduced animals and plants can spread over a new continent. His animal examples are drawn mainly from invertebrates but the spread of the muskrat (*Ondatra zibethicus*) over Eurasia is mapped in some detail. The first release of this Nearctic rodent was of five individuals in Czechoslovakia in 1905 and by 1927 it had reached Leipzig and Breslau northwards and Salzburg southwards – an average rate of spread of some 15 km/year. By 1956 further introductions had enabled it to colonize three great areas extending well over Eurasia. New Zealand is a country which is wide open to invasion by terrestrial mammals since, owing to its long isolation from other land masses, it has only two native species, both bats. Among the welter of animals which the homesick Anglo-Saxons introduced perhaps the two most notable inheritors of that part of the earth are the Australian opossum (*Trichosurus vulpecula*) and the European hedgehog (*Erinaceus europaeus*). The opossum, first imported in 1858 (Wodzicki, 1950), has now occupied almost all areas of forest and much open country as well and the damage it does to vegetation is heavy. The hedgehog was released towards the end of the nineteenth century and is now abundant in open country, especially those areas opened up for settlement and agriculture. Its density can be much greater than in its native lands as revealed by an index of road casualties (Brockie, 1960) and such crowding produces a substantial mortality from mange (Brockie, 1974).

The tale of these take-over bids could be multiplied almost indefinitely, such has been the folly of man. Rats, mice and rabbits indicate that world conquest is possible – with disastrous results. There are few reliable measurements of the speed at which such conquests have taken place. Averages do not mean much in this context except for comparison. Caughley (1977) cites some rough figures, varying from 64 km/year for the rabbit in Australia and 48 km/year for the feral horse in South America down to *c*. 1 km/year for various ungulates introduced into mountainous terrain in New Zealand. Also Haeck (1969) has shown that the newly reclaimed polders in Holland were colonized by moles (*Talpa europaea*) at a rate of 2–3 km/year. More accurate details are known for bird invaders. The house sparrow and the starling made spectacular progress westwards across the United States (Skinner, 1905), and interesting information is provided for New Zealand by Guthrie-Smith (1953) from observations during his lifetime. The sparrow, released at Auckland in 1867 had reached Hawkes Bay in fifteen years (33 km/year) travelling across the (then) inhospitable high ground of the North Island. By contrast, blackbirds and song thrushes, also liberated at Auckland, followed the coast and took 24 years to reach Hawkes Bay. Two final and recent examples which are well documented may be cited. The first is of the bank vole (*Clethrionomys glareolus*) which was first recorded in south-west Ireland (where there are no native voles) in 1964. Since then it has spread and occupied an area some 150 × 100 km (Fairley, 1971). It seems certain that this was either an inadvertent or mischievous introduction by man.

The second is of the American mink (*Mustela vison*) which has escaped from fur farms in many European countries during this century. Its spread in the British Isles still continues (Thompson in H.B.M.*); in Sweden, where it was introduced in the late 1920's, it had occupied the whole country by 1967 except for the mountainous areas (Gerrell, 1967).

These few examples are sufficient to show the explosive possibilities of animals introduced into new countries. Let us now examine the factors which, in the absence of man's interference, prevent such explosions.

The most obvious barrier, especially to land mammals, is the ocean. This applies both to large land masses, such as Australia and New Zealand, provided they have been isolated a long time, and to more remote island groups. This is emphasized not only by their vulnerability to introductions but also by the occupation of niches by somewhat improbable ecological equivalents. New Zealand has already been mentioned as remarkable in having no native land mammals except two bats. Before the arrival of man the small herbivore niche was partly filled by an assemblage of wetas – large, wingless Orthoptera, which can justifiably be referred to as

*H.B.M. = Corbet, G.B. and Southern, H.N. (Eds.) (1977). *The Handbook of British Mammals*. 2nd. edn. Blackwell Scientific Publs., Oxford.

'invertebrate mice or, even, rats'. A half grown giant weta (*Deinacrida heteracantha*) that I weighed scaled 30 g, so a full grown one would go to 50 g or over. Even their faeces, to superficial examination, look like those of rats and mice. Needless to say, since the coming of 'real' rats and mice, wetas have declined in numbers.

There is, by now, a vast literature, stemming from the pioneer work of Wallace (1902), recording the depauperate faunas of oceanic islands and demonstrating their importance as forcing houses of speciation in the case of archipelagoes (Darwin, 1886; Lack, 1947), or as areas where the processes of colonization and extinction and of competition and natural selection can be studied (MacArthur and Wilson, 1967; Lack, 1976) but there is no space to go into that fascinating subject here.

Perhaps more can be learned about the immediate problem by scrutiny of continental islands (in Wallace's terminology, i.e., those that lie fairly close to land masses and where the intervening seas have been only a partial filter to colonization). A typical example of differential filtering may be seen in the number of small mammal species in western Europe compared with those that have reached Great Britain and those that have penetrated beyond, to Ireland (Table 2.2). Note that introduced species are not included. The greatest difference, as concerns Great Britain, is seen in the rodents; only a third of European species has reached our shores, while less than a quarter of the species in Great Britain has penetrated to Ireland. The deficit is largely in voles which have never reached Ireland under their own steam.

Table 2.2 *Number of native species of small mammals in Western Europe, Great Britain and Ireland (based on van den Brink (1967), omitting eastern and Mediterranean species.*

	Europe	Great Britain	Ireland
Insectivora	16	7	2
Chiroptera	27	14	7
Lagomorpha	3	3	2
Rodentia	27	9	2
Carnivora	8	4	2
Total	81	37	15

With land mammals it is reasonable to assume that the seas are a strict barrier which can be overleaped only by man's aid. Yet there is a parallel condition with birds which may be surprising. Here also many species found in Great Britain are missing from Ireland (Lack, 1969). Since birds are more mobile than mammals, this leads us to the question of what barriers of a second order there are to colonization by a species that has overcome the restraint of a sea crossing and whether these may apply to mammals as well as to birds.

Clearly, all animals must have a dispersal phase whether it is great or small. In many species it has now been shown that distances moved fall into a bimodal scale with many individuals staying put and some other dispersing a considerable distance, a point which will be taken up later. What concerns us here is why these long distance movements are usually but not always abortive.

The operation of such restraints to expansion of range is best detected at the margin of a species' distribution. In the British Isles several species of small mammals reach their northern limit in England and they show curious contrasts. The yellow-necked mouse (*Apodemus flavicollis*) does not extend beyond a line from the Mersey to the Humber (H.B.M., 1977). On the continent it is widespread and abundant and shares out the habitats with the wood mouse (*Apodemus sylvaticus*), the latter preferring wood margins and scrub, the former continuous woodland. In England, however, the yellow-necked mouse is found in patches within the range of the wood mouse and, as far as our records go, seems to be contained within these patches.

On the other hand, the harvest mouse (*Micromys minutus*), whose present distribution covers England and extends to the lowlands of Scotland (H.B.M., 1977), has shown, during this century, notable fluctuations in abundance and perhaps also in range. There are records from Scotland as far north as Perthshire in the last century but none so far during this century. During the war extensive researches on the small mammal populations of corn ricks by the Bureau of Animal Population in Oxfordshire and Berkshire revealed only a few individuals (B.A.P., unpubl. records) but in the early 1950's they were found abundantly in this same area (Southwick, 1956). After that they became scarcer again, as testified by intensive trapping for small mammals near Oxford (B.A.P., unpubl. records). Yet again these trappings revealed increasing numbers in the same area during the 1960's and a survey conducted during the early 1970's for the Mammal Society by Stephen Harris demonstrated an extension of range westwards as far as Pembrokeshire and northwards up to Edinburgh and an increase of abundance generally, though this may include an element of sharpened awareness of signs by observers.

The dormouse (*Muscardinus avellanarius*) is another species whose northward boundary crosses England from, roughly, the Mersey to the Humber but, in contrast to the yellow-necked mouse, whose range appears to be stabilized, and the harvest mouse, which ebbs and flows, its range is contracting (Corbet, 1974). In the last century it was found almost up to the border but since 1960 there have been no records north of the Humber. This decrease may be linked with the disappearance of woodlands with a rich shrub layer, such as coppice with standards so beloved by nineteenth century landowners.

These three instances illustrate the dynamics of the boundary of an animal's range but only in the case of the yellow-necked mouse is it likely that it is restricted by competition with another species. This is, of course, one of the most obvious and widespread sources of range restriction to which we shall turn below.

For the moment we shall consider some further examples in which the spread of a species is prevented by either physical factors or habitat patterning. Introductions, being natural experiments, however deplorable, throw light on these matters. The coypu (*Myocastor coypus*), a large South American rodent, established in fur farms in Great Britain in the late 1920's, escaped and spread along water bodies until it had built up large populations by the later 1950's especially in the Broads of East Anglia. Its destructiveness to drainage embankments and agricultural crops instigated a trapping campaign to reduce its numbers, but the effects of cold weather were even more dramatic. During the winter of 1962–63 some 80–90 per cent of the population died (Gosling in H.B.M., 1977). Owing to this poor resistance to frost, the species does not spread naturally into eastern Europe.

When we come to examine the effect of habitat patterning on distribution, we are edged automatically into the question of competition between species, for the simple reason that different habitats support different specialists which can normally resist invasion. The relationships of the two hare species in the British Isles illustrate this neatly. The brown hare (*Lepus capensis*) occupies most of the lower ground up to about 300 m, while the blue hare (*L. timidus*) lives mainly above this altitude and is confined to mountainous regions except in Ireland where there are no native brown hares. In New Zealand, in the absence of this competitor, the brown hare can live above the timber line which is at about 1400 m in the instance quoted by Gibb and Flux, in Williams (1973). On a smaller scale the same principle is displayed on the Isle of Man (Fargher, 1977). The brown hare was introduced there long ago, the blue hare only recently. There are two areas of high ground, a large one in the centre and a small one in the south. In the former the blue hare occupies the heather moor above 180 m, the brown hare the agricultural land below this altitude. In the southern area the blue hare is, so far, absent and the brown hare alone occupies it up to its highest altitude of 275 m.

In contrast to remote islands, large land masses will have a much richer fauna in which most of the ecological niches will be filled by specialists. This will obviously make invasion by new species more difficult, if not impossible. It is instructive to compare the performance of mammal invaders with that of bird invaders in New Zealand, which, as already mentioned, has virtually no indigenous mammals. Although the native bird fauna is far from rich, the many European passerine species brought there are mainly confined to the areas where forest has been cleared for agriculture; they make little headway into the uncleared forest [Gibb and Flux, in Williams (1973)]. In contrast, introduced mammals, such as the opossum, rats and mice and various large ungulates, have penetrated deeply into the southern beech and podocarp forests and have created havoc over vast tracts of country. Similar resistance to invasion has been shown by native habitats and their occupants in Fiji (Pernetta and Watling, *in press*); here it is fortunate that only one introduced mammal, the mongoose (*Herpestes auropunctatus*), has really made an impact upon the native fauna.

A very notable sequence of events, following the introduction of predators, is recorded for the island of Terschelling, off the Dutch coast (van Wijngaarden and Mörzer-Bruijns, 1961). Early in this century 603 ha were planted with conifers, which were heavily damaged by water voles (*Arvicola terrestris*) and rabbits (*Oryctolagus cuniculus*). Stoats (*Mustela erminea*) and weasels (*M. nivalis*) were released in 1931 but the latter soon died out. The impact of the stoats was, however, spectacular; within 5 years the water voles had been exterminated and rabbits were reduced to very low numbers for the ensuing 12 years. The stoats, meantime, had risen to a big peak and then declined to a more modest level. By 1961 rabbits had recovered; subsequent events have not been reported but would be of great interest.

Frequently, competition between native and introduced species has been more evenly balanced. A well documented story is that of the spread of the American grey squirrel (*Sciurus carolinensis*) over the British Isles at the expense of the native red squirrel (*S. vulgaris*). Numerous introductions of the grey squirrel were made between 1876 and 1929 but its spread did not gain momentum until the 1920s and early 1930s, during which time it colonized most of central and southern England and two large areas in Scotland and Ireland. After this its momentum relaxed, though it has continued less energetically to the present day (Shorten, 1957; Lloyd, 1962; Tittensor in H.B.M., 1977). The red squirrel has mainly withdrawn to its present stronghold of conifer forest, where it still holds its own. It is a species subject to fluctuations in numbers and to epidemics and was never very numerous in deciduous woodland; it may be that, when or where it is in a poor way, the grey squirrel can edge forward. Certainly the grey is much more abundant and flourishing in deciduous woods than the red ever was. Densities of red squirrels in conifer habitats vary from 0.8/ha in Scotland (Tittensor in H.B.M., 1977) to 0.5/ha in East Anglia (Shorten, 1962), whereas in deciduous woodland near Oxford Mackinnon (1976) found an average over 3 years of 5.1/ha for the grey squirrel.

The story of the fat dormouse (*Glis glis*) in Britain provides an illuminating contrast. This species, which was a gastronomic titbit for the Romans, was released at Tring in Hertfordshire in 1902 and over the subsequent three-quarters of a century it has not penetrated more than 30 km in any direction from Tring. The reason for this is obscure but it may be significant that its range in Europe does not reach to the north coast of France, nor to the shores of the Baltic (van den Brink, 1967).

The outlying islands of the British Isles provide natural experiments on competition between species of voles and shrews because some have one species and some have two (many, of course, have none). On Jersey in the Channel Islands both the common shrew (*Sorex araneus*) and the lesser white-toothed shrew (*Crocidura suaveolens*) co-exist and they share out the habitats between them, *Crocidura* preferring the coastal areas (Godfrey, 1978). On Guernsey only the common shrew is present and occupies all the habitats.

Similarly among the Hebridean islands to the west of Scotland some have both bank voles (*Clethrionomys glareolus*) and short-tailed voles (*Microtus agrestis*), others have only one of these species. On the mainland the bank vole prefers scrub and woodland, the short-tailed vole grasslands and, indeed, its sharply angled cheek teeth are adapted to dealing with coarse, siliceous grasses. On islands, such as Mull, where both species are present, the situation is much the same as on the mainland; however, where the bank vole only is present (e.g., Raasay), it occupies the grasslands as well and has a high incidence of the 'complex' condition of the molars which has an extra inner ridge on the third upper molar (Corbet, 1964). Simply it is trying its best to be a short-tailed vole in the absence of this species.

All this evidence emphasizes the specialization that evolves when species are in competition and this has been abundantly demonstrated for birds (e.g.,Lack, 1971). For small mammals the evidence is not always so clear cut but a classic instance investigated by Miller (1964) is impressive and revealing. In Colorado four species of pocket gophers (Geomyidae) co-exist in the same broad region. All are burrowers and the most highly adapted to this mode of life is *Geomys bursarius* which is confined to deep sandy-loamy soils. *Cratogeomys castanops* is a large species preferring much the same conditions. On the other hand, *Thomomys bottae* is a smaller animal which can live in a wider range of soil types and *T. talpoides* is the smallest and will tolerate quite poor and shallow soils. Thus the two larger species, being aggressive, occupy the deep and workable soils while the two *Thomomys* spp. have to put up with the less ideal fossorial habitats. The species with a narrower niche are surrounded by those with less exacting requirements, though the latter would move to the better habitats if they were allowed. Experiments which lend some support to this analysis have since been conducted by Vaughan and Hanson (1964). This introduces us to the concept of specialists and generalists sharing out a major habitat.

Introductions apart, man's modification of native habitats has frequently been on a vast scale which has encouraged an equally vast extension of range by native animals. The house mouse (*Mus musculus*) probably originated in the steppe zone of Asia (Schwarz and Schwarz, 1943) and spread with the growing of grains (cornfields being a tolerable imitation of steppes). Another great agricultural transformation has been in the great areas of the world now covered by pasturelands into which rabbits and hares have extended and flourished. Again, man's buildings have attracted an array of commensals, among which small mammals are pre-eminent.

Compare with these great tides of change the stability of a woodland ecosystem. In 1954 the myxoma virus, in its rampage over the British Isles, arrived at the University of Oxford's Wytham estate (which contains large areas of deciduous woodland) and in a month had practically eliminated a considerable population of rabbits. I was fortunate to be able to follow the consequences of this large spanner hurled into the works of this animal community. An immediate result was observed in the following spring. Mice and voles became extremely scarce because predators of rabbit turned to them. For yards I could follow their runways dug up by foxes, and tawny owls (*Strix aluco*) were very unsuccessful in their breeding. An unexpected consequence was an unexampled richness of spring flowers, indicating that small rodents usually eat a large proportion of buds. By the next winter, however, and subsequently, practically all the members of the community were back to their normal levels with the sole exception of the stoat (*Mustela erminea*), which, being a predator specializing on rabbits, remained scarce for many years. Thus a change of this magnitude was adjusted to within a year. The only long term effect, apart from the stoat's decline, was the improved crop of tree seedlings.

Changing climate may also influence the distribution of animals. Movements of whole faunas on a vast scale were obviously involved during the succession of Ice Ages. Recently a warming of the climate in Europe from the last decades of the nineteenth century to the middle of the twentieth allowed many species to push their boundaries northwards. This is well documented for many birds and for some mammals. Kalela (1949) gives the history of the polecat (*Mustela putorius*) from 1879 when it occupied only the small south-eastern corner of Finland up to 1939 by which time it had spread north-westwards about 400 km and almost reached the shores of the Gulf of Bothnia. In Canada Macpherson (1964) records that the red fox (*Vulpes vulpes*) extended northwards only to the south of Baffin Island around 1918 but by 1962 it had reached Ellesmere Island.

Finally, we must mention the massive persecutions launched by man against certain groups of mammals, mainly predators and fur-bearing species, during the last century and into this. Seals and whales hardly count as small mammals but the diminution in the British Isles alone of such predators as the pine marten (*Martes martes*), the polecat and the wild cat (*Felis silvestris*) shows what can be done by persistent slaughter (Langley and Yalden, 1977).

Fortunately this trend has been, in a small way, reversed by a more enlightened view of the function of predators in an animal community. All three of the above-mentioned species have made a partial comeback during this century, aided by the reforestation of vast tracts of country in the north and west. Similar policies beyond our shores have paid off; Myrberget (1967) records that in Norway the population of beavers (*Castor fiber*) was reduced to about 100 by 1880 but since then has recovered to 5000 (1965).

Despite this spread of enlightened ideas, we have to face the fact that the multiplication of man and of his activities by itself threatens to reduce the ranges of mammals. This is not simply a question of the destruction or depauperization of habitats but of the disturbance to an animal's way of life and well-being from the mere presence of man in great numbers. Many species of bats are on the decline, both in Europe and North America (Racey and Stebbings, 1972) and, in particular, the greater horseshoe bat (*Rhinolophus ferrumequinum*) in Britain has decreased alarmingly in recent years. Disturbance of roosts has certainly contributed to this. Similarly, otters (*Lutra lutra*) have become rarer now because of (among other things) increasing human pressure on our waterways.

The study of the pattern of animals' ranges (what I have called *Macrodistribution*), whether they are expanding, stationary or contracting, reveals the end product of population processes, to which we must now give closer attention.

2.2.2 'Microdistribution'

Whereas macrodistribution is mapped by extensive survey methods, microdistribution is amenable to study by much more intensive methods, such as are used by individual workers in the field, e.g., live trapping on grids or lines with or without marking and recapture of individual animals. A population here is a term of convenience – in Krebs' (1972) phrase, 'a group of organisms of the same species occupying a particular space at a particular time'. Such groups constitute samples of the whole species' range and may be considered, with adequate safeguards, as representing the behaviour and population ecology of the species as a whole. It is the most practical unit for examining the course of evolution and adaptation and can yield information on the genetic composition of a population and on the equilibrium, or lack of it, which determines numbers. This means that we can now neglect the heading – 'inability to get there' – which bulked so large as a determinant of range, and concentrate on adaptations to habitats and their interplay with intra- and interspecific competition and with the effects of predation.

An example, studied in some depth in Wytham Woods, near Oxford (Southern and Lowe, 1968) may be an instructive introduction to this

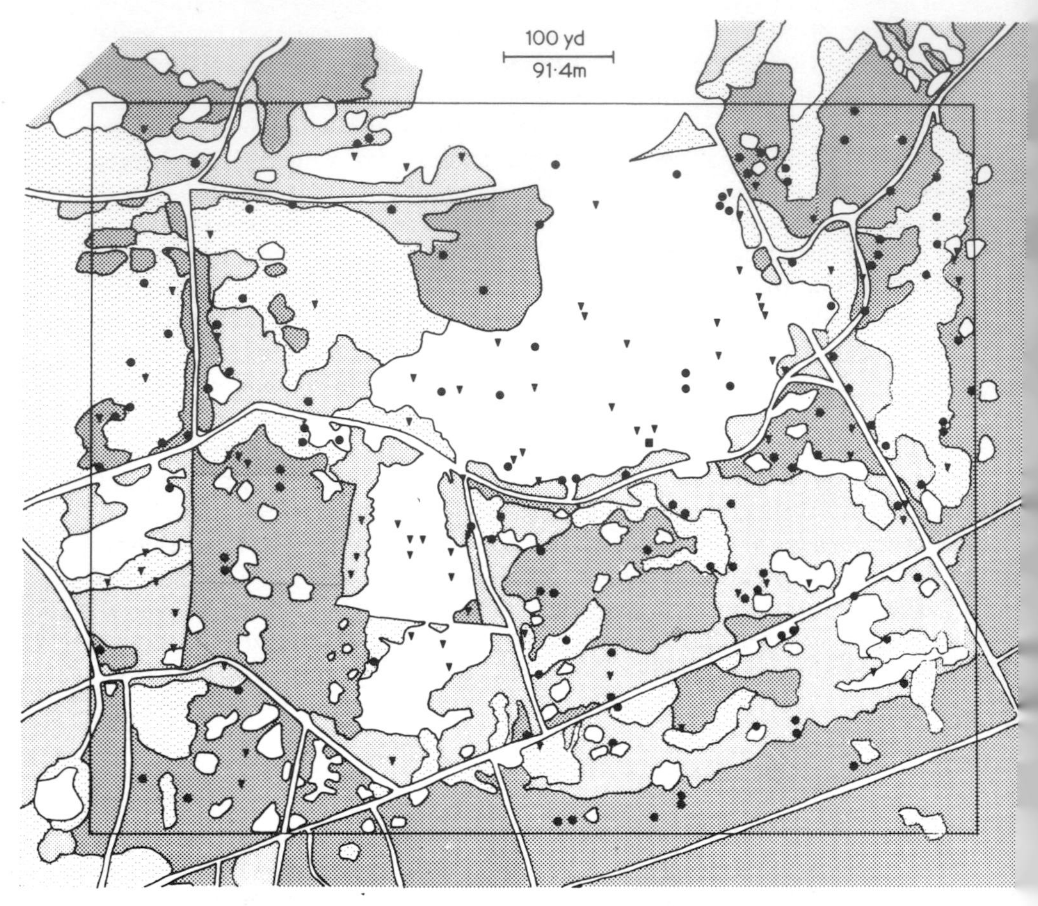

Fig. 2.1 Map of Southern and Lowe's study area in Wytham Woods near Oxford, showing four categories of density of ground cover. The white areas (type 1) are under heavy canopy and the ground is almost bare; the lightest shading (type 2) indicates sparse ground vegetation, the intermediate shading (type 3) fairly thick ground vegetation and the darkest shading (type 4) shows areas of very dense bracken and bramble cover. The trapping grid is the rectangle reaching just short of the margins of the map. Symbols indicate where the prey, whose marking rings were recovered in owl pellets, were originally captured, marked and released. Black triangles represent wood mice (*Apodemus sylvaticus*), black circles bank voles (*Clethrionomys glareolus*) and the few black squares short-tailed voles (*Microtus agrestis*).

complex subject, which has a vast literature. The area trapped from October 1954 to June 1956 covered 49 ha of oak-ash-sycamore woodland on a north-facing slope where soils varied from thin, over limestone at the top, to rich over Oxford clay at the bottom. The grid was divided into 0.4 ha squares and an estimate made of the predominant cover type in each square. The general trend was from thin shrub and herb cover under heavy

Table 2.3 *Mice and voles trapped in Wytham Woods in relation to cover type (see Fig. 2.1). Adapted from Southern and Lowe (1968).*

	Grid squares dominated by cover-type				
	1	2	3	4	Total
	Open cover			Dense cover	
No. squares in grid	27	25	30	38	120
A. Wood mouse					
No. trapped	156	143	158	169	626
No. trapped per square	5.8	5.7	5.3	4.5	
B. Bank vole					
No. trapped	159	282	394	531	1366
No. trapped per square	5.9	11.3	13.1	14.0	

tree canopy on the limestone to dense cover of bracken and bramble with more isolated trees and many shrubs on the clay. There was much interspersion of cover types, which are characterized and their patterns mapped in Fig. 2.1.

The two predominant species of small rodents studied were the wood mouse (*Apodemus sylvaticus*) and the bank vole (*Clethrionomys glareolus*). It was known in a general way that the vole was fonder of cover, being diurnal as well as nocturnal, than the mouse, which is nocturnal, but the trapping results revealed this tendency in a more precise way (Fig. 2.1 and Table 2.3).

The numbers of wood mice caught per 0.4 ha square were approximately the same in all four cover types, the small difference between type 4 and the rest being non-significant statistically. In contrast, bank voles were on a level with mice only on the more open ground (type 1), and progressively out-numbered them as the cover grew denser until in the really thick vegetation they were three times as abundant as the mice. The symbols on Fig. 2.1 show where those mice and voles were marked, whose rings subsequently turned up in owl pellets. They show the same pattern visually as is evidenced in Table 2.3.

This sharing out of habitats by closely related species is by now an ecological axiom derived from the principle that species with the same ecological demands will compete *à outrance* – one or the other must go to the wall. A vast amount of research has, therefore, been devoted to the detection of ecological divergencies between species which appear to have similar requirements but manage to co-exist.

It is difficult to pick out the best examples from all the research that has been done on this intriguing problem and much significant work must, perforce, be passed over in this brief review. The case of the common shrew (*Sorex araneus*) and the pygmy shrew (*S. minutus*) has attracted some very hard and intelligent labours. Croin Michielson (1966) had already revealed that the larger species spent much of the winter below ground, being an expert burrower, while the tiny pygmy shrew remained on the surface. Pernetta (1976a) pinpointed a significant difference in diet of the two species; common shrews on the Wytham estate, where he studied them, eat many earthworms, which pygmy shrews ignore, while of harvestmen, beetles and spiders common shrews take larger species than do pygmy shrews. We may note here, in passing, that the very obvious way of avoiding competition between two species by their taking different foods has instigated a large body of research on methods of identifying prey remains in stomach contents and faeces. Predators, such as owls, which regurgitate the undigested parts of their prey in the form of pellets allow a fairly straightforward evaluation of the composition of their diet, by volume or by weight (e.g., Uttendörfer, 1939; Southern, 1954); other studies, such as that by Pernetta (1976a) cited above, require more elaborate techniques. The common occurrence of this divergence in the size of food items taken by closely related, sympatric species presumably underlies the co-existence of a chain of species of increasing size, e.g., in Europe, the woodpeckers which range from the lesser spotted (*Dryobates minor*) (14.4 cm) to the black (*Dryocopus martius*) (45 cm) and the owls, which range from the pygmy (*Glaucidium passerinum*) (16.3 cm) to the eagle (*Bubo bubo*) (67.5 cm). This sequence can be demonstrated more neatly in birds but can be seen also in related groups of mammals, e.g. the Mustelidae, where the smallest is the weasel (body length 16 to 23 cm) and the largest the wolverine (*Gulo gulo*) (body length 70 to 82.5 cm).

Some curious facts have recently come to light about the selection of foods by predators. It is well known that shrews are distasteful to many predators though the barn owl (*Tyto alba*) will eat them in large quantities. But choosiness by predators between their small rodent prey had not previously been suspected until King (1971) found that weasels in deciduous woodland near Oxford preyed mostly upon bank voles and neglected wood mice which were also numerous. Furthermore, Moors (1975) showed that his farmland weasels in Aberdeenshire took mainly short-tailed voles, though there were plenty of bank voles in this habitat. Finally, it has been convincingly demonstrated by Macdonald (1977), in a series of elegantly planned experiments with semi-tame foxes, that short-tailed voles head the preference list, followed by bank voles, and wood mice come last of all.

So it is not simply a matter of the differential availability of food resources that encourages predatory species to share out habitats, but the predators have, themselves, a say in the matter.

Grant (1972) has deployed a carefully designed experimental approach to test how competition between different species of small rodents leads them normally to occupy different habitats. He used 0.4 ha enclosures, in which half was woodland and half grassland, and manipulated populations of woodland species, *Clethrionomys gapperi* and *Peromyscus maniculatus*, and of a grassland species, *Microtus pennsylvanicus*, to show that a woodland species would occupy the grassland in the absence of *Microtus* and would vacate it when *Microtus* was reintroduced. These results echo the conditions on Hebridean islands already mentioned but add the necessary experimental rigour.

One of Grant's results, namely that a woodland species may not move out into grassland if its numbers are stationary, leads us to consider a very significant aspect of population processes, namely seasonal movement between habitats. Much recent research has been concentrated on this, particularly on the characteristics of that part of a population that moves. We shall consider first the demographic elements involved in such movements and then the genetical elements about which knowledge is only just beginning to accumulate in the case of small mammals.

It is now established (and the subject is treated in greater detail in the next chapter) that social systems among mammals are important in influencing the pattern of their distributions and densities. At the one extreme are species like the wild cat which are territorial in the true sense, i.e., individuals or pairs defend an area against all comers; at the other extreme are species which are highly colonial like seals and many bats. The majority of mammals fall between these two extremes with a wide variety of ways of living together, based on family groups, on kinship or on dominance hierarchies or a combination of these. Hence the area occupied by a group is more truly a 'home range' rather than a territory, since it may be shared by a number of individuals or even with other groups.

A solitary way of life is less common in mammals than it is in birds. Even the fox, which has long been held up as an example of a solitary hunter, has now been shown by Macdonald (1978) to form in some habitats social groups which are dominated by one male, and may contain several vixens, of which only the highest in rank may breed. I hope I may be forgiven for making the fox an honorary 'small mammal' because the organization of its populations, as so admirably elucidated by Macdonald, is probably typical of many small mammals.

The relevant point to note about this arrangement is that 'surplus' individuals may either stay put and omit breeding, except for the long-term chance of rising in the hierarchy, or move out and seek new homes, which may be very hard to find. The first choice will restrict gene flow, the second will promote it. There is a delicate balance between these alternatives and either holds out a slim chance of contributing to future generations. But, as far as the species or population is concerned, it is advantageous that both these processes should occur; a closed group ensures that what breeding does occur is successful and the individuals that are expelled or emigrate exert a constant pressure for the discovery of unoccupied or new habitats. Of course, most of these explorers or roamers will either land up in unsuitable habitats or will keep on going until they drop. The harsher the pressure of natural selection, the better for the species.

Mackinnon (1976) studied a population of grey squirrels (*Sciurus carolinensis*) on 15 ha of deciduous woodland near Oxford and found that it was occupied by 6 social groups consisting of both sexes with a strictly linear dominance order in which males tended to dominate females. These groups remained fairly stable over the three years of her study and fed mainly on tree seeds from late summer through to spring but during June and July before the mast ripened they ate large quantities of cambium tissue and lost weight. During this same summer period she also trapped squirrels in a pole-stage plantation of beech (*Fagus sylvatica*) a kilometre or so distant, and of 15 whose age was determined 12 were juveniles or subadults. Clearly these were the explorers moving out to second-class habitats and having a difficult time. Of trees over 30 cm in diameter 93 per cent were badly stripped of bark owing to the squirrels' search for cambium tissue. This damage, for which grey squirrels are notorious in Great Britain, is presumably due to the lack of other resources in this habitat. These records support the view of Taylor (1970) that bark stripping is done by vagrant young squirrels not only for food but as an aggressive activity in an unfamiliar environment.

Alibhai (1976) made the same kind of comparison on the same estate in his work on bank voles. Since 1947 the ebb and flow of wood mouse and bank vole populations on this estate have been followed (see later) in the deciduous woodlands and the pattern of changes in numbers over the seasons usually shows a steep increase in autumn followed by a sharp decrease in spring with continuing scarcity until the next autumn's upsurge. Alibhai studied the bank voles in the plantations and found a diametrically opposite swing with high numbers in spring and summer and low numbers in winter. Of the new individuals which came into the plantations in spring and summer 80–100% were immigrants and not recruits from local breeding. The conclusion seems inescapable that considerable movement takes place between habitats in the form of an overspill from the mature deciduous woodland to the plantations. Such movement has not been confirmed by marking individuals but the possibility that the plantation voles are wanderers colonizing second-class habitat is suggested by one or two facts. They build up temporarily populations much denser than those in the mature woodland (29–58 per ha compared with 5–13 per ha); they breed at a greater rate (667 young produced per ha/year compared with two estimates of 270 and 357); and their diet contains largely green vegetation in the absence of tree seeds. These characteristics recall the difference between *r* and *K* selection treated in the last chapter.

Investigations of other small mammals show that the situation is not always so simple as this. Kozakiewicz (1976), studying in Poland the same species of vole as did Alibhai, found that, when he trapped out an area of 5.75 ha in a mixed conifer and deciduous forest, the animals that filtered into this area consisted mainly of sexually mature voles, though there were fewer pregnant females among the immigrants and their weights were lighter than those in the surrounding settled areas. This may have been simply because the experimental area was part of a continuous, homogeneous habitat. Yet Stoddart (1970) has proved that female water voles (*Arvicola terrestris*) may abandon their home ranges and settle in another quite distant one to produce a second litter.

The interspersion of habitats is clearly a powerful influence on the dispersal of surplus individuals. Kikkawa (1964), studying the popualtions of wood mice in an area of mature deciduous woodland, surrounded by arable fields near Oxford, recorded a notable exodus of mice into the cornfields in summer. Whether or not they returned in the winter is not clear. Leigh-Brown (1977) detected similar movements by wood mice into cornfields from woodland in summer and was able to demonstrate their return in late summer. Furthermore, using the recent technique of starch gel electrophoresis to identify genetic variability in blood proteins, he was able to show that mice returning from the fields had a higher incidence of the *c* genotype at the phosphoglucomutase locus than those that remained in the woodland. Nevertheless, this high incidence rapidly decreased to the woodland level during the winter, suggesting a potent action by natural selection. Thus, evidence is accumulating about the patterning of intraspecific variation and about the changes that may take place in sections of a population that are mobile.

Other researches on these biochemical polymorphisms in small mammals, e.g., Semeonoff and Robertson (1968) on *Microtus agrestis*, Tamarin and Krebs (1969), Gaines and Krebs (1971) on *M. ochrogaster* and *M. pennsylvanicus* and Rasmussen (1970) on *Peromyscus* have shown that the proportions of genotypes can vary with time and place. There has, however, been as yet no clear linking of these variations with behavioural traits subject to 'pendulum' natural selection such as Chitty (1960) has postulated. The most convincing evidence about this problem to date has been that produced by Berry (1977) on an island population of house mice (*Mus musculus*). This illuminating story will be followed out in more detail in the next chapter in connection with the regulation of populations. It is sufficient here to note that the mice on this small island are divided into residents, which occupy the peripheral cliff slopes round the year, and vagrants which overflow from these bases during the summer and establish themselves temporarily on the central part of the island. The migrants and the residents differ in details of skeletal structure but also in bio-chemical genotypes and those with a higher basal metabolic rate are the ones that survive the rigours of winter in the cliff refuges. The 'two-tiered' structure of many small mammal populations is amply confirmed on this island and, in addition, the seasonal selective advantage of one genotype is demonstrated.

It is difficult to know how far this ebb and flow from a central reservoir both of animals and of genes are dependent on the heterogeneity of habitats. The work of Armitage and Downhower (1974) on yellow-bellied marmots (*Marmota flaviventris*) in Colorado suggests that diversity of habitats may not always be the determining factor in such contouring. Here the central elements of the population comprise colonies with one or more males and their harems. In the interstices are so-called satellite colonies which may consist of up to 10 subordinate animals. The main difference between the main and the satellite colonies is that the production of young in the satellites is 1.1 per female per year, whereas the parallel figure for the focal colonies is 2.0 per year. This condition may be a halfway house between populations that are totally resident and those in which resident and vagrant sections are spatially separated.

Enough evidence has now been given to show the variation in microdistribution that may be imposed upon populations of small mammals by the interspersion of habitats and by the type of social system. Before we go on to examine the finer details of variations in abundance within and between species we may well draw this discussion of microdistribution to a close by a brief review of the different patterns of dispersion that are to be found among small mammal population – in other words we must hark back to 'knowing our animals'.

A striking example of how quite closely related animals may diverge radically in their dispersion patterns and social structure is displayed by the hares on the one hand, the rabbit on the other. Both brown hares and blue hares in the British Isles live above ground (blue hares do make short burrows but do not use them as refuges), are dispersed as solitary individuals or pairs, though with much overlap of home ranges, and move around a large area which may be considered as a home range, e.g, 28 ha recorded by Hewson (1976) for the blue hare. In hilly country both species may have widely separated 'home' areas and feeding areas, involving a journey of as much as 2 km.

In contrast, rabbits, as everybody knows, live in colonies with extensive systems of burrows (hence the generic name *Oryctolagus*, literally the 'digging hare'). Their colonies are organized in a clear-cut social hierarchical system with bucks dominating all animals in their group and being aggressive to the bucks of neighbouring groups. Among the does there is one which is dominant in each group but subordinates may live fairly peaceably with them. They may, however, be edged away from the main warren for breeding (Myers and Poole, 1959). Rabbits range over much smaller areas than hares and their normal foraging does not extend more than 150–200 m from the warren, though, where populations are sparse or when there is some particularly attractive source of food, they may move further away (up to 400 m recorded, *see* Lloyd (1977) in H.B.M.).

An interesting point brought out by several studies is that rabbits do not flourish if they are thinly distributed. Their best habitats are characterized by burrowable soil and surrounding grassland which can be grazed to a find sward. The delay in the recovery of rabbit numbers after myxomatosis was partly due to the abundant growth and flowering of grasses which their absence promoted. Wodzicki and Roberts (1960) compared the body weights of samples of rabbits in New Zealand from populations varying in density from 0.5 to 92 per ha and found that those from the densest areas were in the best condition judged by weight and reserves of fat. Gibb *et al.* (1969) demonstrated that, where rabbits in New Zealand had been reduced in numbers by poisoning, they tended to be kept scarce by predators – feral cats and ferrets with some help from stoats [*see* also Gibb and Flux, in Williams (1973)]. This recalls the discovery by Elton (1953) that farmyard cats were unable to deal with an established infestation of brown rats (*Rattus norvegicus*) but, if the rats had been cleared out by other methods, then the cats could prevent reinfestation.

Since patterns of dispersion and home range may vary so much between closely related species and even within a species at different times of year, it is vital to obtain information on the details of such patterning for any species whose populations are being studied. The measurement of the size of home ranges is far from straightforward because it involves not only the delineation of a boundary but also some knowledge of the intensity of use within that boundary. The methods that have been proposed (and some of them applied) have been reviewed recently by Flowerdew (1976) and I give simply an outline here, which concentrates on relating the various techniques to the kind and precision of the information they supply.

Probably the most widely used means for making a gross estimation of home range is to set out live traps in a grid, mark the catch so that individual animals can be identified and record, over a period, the different trapping points at which individuals turn up. Providing the trap spacing is appropriately fine, and the trapping carried on for a sufficient period, this information will emerge automatically from studies of density and turnover of a population. The results are admittedly rough and various ways of joining up the peripheral trap points at which an animal has been captured have been suggested (*see* Flowerdew, *op. cit.*) but they are adequate for making comparisons between sexes and ages, between habitats and times of year.

By and large live trapping will underestimate a home range for one obvious reason – that an animal which has been trapped cannot move any further until it has been released. Beyond this the effect of providing bait *ad libitum* is difficult to gauge.

Again the topographical pattern of the frequency of recaptures will show the areas of the home range most assiduously used – whether there is a core, e.g., a nest site from which excursions radiate or several cores separated by ground that is rapidly traversed. But the answers to these questions by this method are extremely rough – rather like a photograph reproduced through a very coarse screen.

Several methods have been devised to achieve a finer resolution. For mice and voles especially the use of tunnels or shelters floored with smoked paper to take footprints allows the animals studied compelte freedom of movement. If this is combined with prior live trapping and toe clipping individual animals can be identified from their footprints. If, however, the population is very dense, individual footprints become obscured.

Research on home ranges of wood mice by Randolph (1973, 1977) used an ingenious technique. Trapped mice were removed to the laboratory and each individual fed overnight with differently marked bait, i.e., wool, feathers, hair, dyed with different colours; in the morning each was returned to its point of capture. At the same time a finely spaced grid of baited shelters was set out on and around the trapping grid. Faeces collected subsequently from these shelters were examined and the markers identified. Analysis of this material from January to August (1977 paper) showed that males increased their ranges, with the onset of breeding, from 1.06 ha to 2.18 ha, while those of the females remained stable to 0.7 to 0.8 ha. There were wide overlaps in both sexes but, during breeding, there was a tendency for a male and a female range to be associated. This has confirmed previous knowledge but with much greater precision.

The tracing of a home range by attaching a radio-active marker and following an animal's movements, pioneered by Godfrey on *Microtus agrestis* (1954), gives even greater precision but on a much smaller scale. Progress in the use of this method has been reviewed by Bailey, Linn and Walker (1973). Again, the injection of radio-active substances makes possible the detection and mapping of radio-active faeces. In this way Stoddart (1970) was able to reveal that, in the case of the water vole (*Arvicola terrestris*), tracer-revealed home ranges were larger by some 12.5 per cent than trap-revealed ranges.

An even more elaborate (and expensive) technique, now very popular, is to attach a radio transmitter to an animal and so follow its movements at a distance. No such transmitter has yet been produced so small that it would not discommode a mouse but useful results have been obtained with larger animals, e.g., grey squirrels, raccoons, and exceptionally comprehensive and detailed information for the fox (again cited as an honorary small mammal) has been obtained by Macdonald (1978). The chief snag in using radio transmitters is that one can follow an animal's movements with considerable precision, barring technical hitches which not infrequently occur with such delicate apparatus, but one cannot tell what it is doing. Macdonald overcame this difficulty by locating an animal by a radio fix and then following it on foot and observing it with an infra-red viewer. An intermediate aid is to affix Betalights (*see* Twigg, 1975) to collar or ear tag, whose glow can be picked up at a distance of 100–200 m.

So there is an ever growing body of knowledge about the size and use of the home ranges of small mammals but most of this refers to species that are abundant. This is not surprising because such species are the ones which impinge on man's economy and call forth the funds for research.

Far less is known about species that are thinly distributed or rare. Obviously these present great difficulties to a would-be researcher, yet rarity in itself is a problem that raises many biological queries. I am not concerned here with species that are going downhill from persecution or destruction or habitat or thinning out at the edge of their range, though these are just as troublesome to study. Rather, I wish to draw attention to the biological oddity of rareness as exhibited by many species. How do members encounter each other for breeding? Are their home ranges larger than those of commoner animals? Are there great gaps between home ranges and so on? A hint of what may happen is given by the work of King (1975) on a species that is not rare but is thinly distributed – the weasel. She found that the home ranges of females in an area of deciduous woodland were of the order of 1–4 ha and had large gaps between them. In this instance the males covered much greater areas (7 to 15 ha), so there was no problem in the sexes encountering each other.

But in really rare species the situation is more acute. Many examples will spring to mind but one which I have come across in my own experience has been a constant puzzle to me. The water shrew (*Neomys fodiens*) is patchily distributed even in its favoured waterside habitats but it is well known to trappers of small mammals that it can be found in very sparse numbers and successfully breeding miles away from water. Over four years trapping on a grid of 94 ha in Wytham Woods, near Oxford, involving some 12 000 trap nights, only five water shrews were caught. Similarly on a smaller grid trapped at 6-monthly intervals between 1952 and 1970 and involving some 5000 trap nights, the total of water shrews was again five. Nevertheless, even such an extenuated population was capable of breeding, since one of a few trial nest boxes sunk in the ground by Dr J. Godfrey produced a litter of water shrews. Admittedly, individuals have been recorded moving considerable distances [over 150 m (Shillito, 1963)], but the kind of organization and use of home ranges on this vastly extended scale is difficult to envisage and will be even more difficult to elucidate.

2.3 Abundance

Having travelled down the scale from geographical to home ranges, we must now turn to scrutinize measurements of the abundance of small mammal populations, for these are the bricks we need to build a representation of the shape and size of any population. There is not space to go into great detail and it is unnecessary, since excellent and up-to-date acocunts of the subject have recently been published, e.g., Krebs (1972), Flowerdew (1976), Caughley (1977). Rather, I propose to concentrate on the way in which different life styles influence population processes and their stability or instability, thus leading on to the subject matter of the following chapter.

Abundance may, for convenience, be treated under the headings of density, the number of animals present per unit area, and turnover, the rates at which animals enter and leave the population.

2.3.1 Density

This is a basic measurement which underlies any further calculations of movements and rates of increase and decrease. Its expression must be formulated in the light of knowledge about the biology of the animal (again the need for a sound natural history background) and the heterogeneity of its habitat. Is it dispersed in a clumped (e.g., where the family group is the unit) or in a random fashion? Are there parts of the study area where the species is thin on the ground or absent?

Absolute counts are the ideal way of measuring density (and, as one ecologist put it to me, 'to hell with statistics') but are rarely feasible with small mammals. In the days before the combine harvester, when corn ricks were a common feature of the landscape, complete populations of house mice could be obtained when the ricks were threshed (e.g., Southern and Laurie, 1946). On open moorland total numbers of mountain hares can be counted, especially when they are still partly in their white winter pelage (Hewson, 1965). Some species of bats which roost communally can be counted, e.g. the common long-eared bat (*Plecotus auritus*) (Stebbings, 1966), though there are problems here, e.g., Dumitresco and Orghidan (1963) estimated that there were between 80 000 and 100 000 pipistrelles (*Pipistrellus pipistrellus*) in a limestone cave in Roumania.

Obviously, in a case like this, the observers would have recourse to sampling, i.e., counting the numbers of bats on randomly chosen areas and multiplying the mean of these by the total area occupied. In these kinds of situations one is well and truly launched on statistics and it is vital to consult one of the standard text-books before tackling the job. This type of sampling, however, is more applicable to populations of invertebrates and, among vertebrates, to birds rather than to mammals.

With small mammals, it is more usual to apply a proportionate method of estimating numbers. This is based on the existence or creation of two classes in a population. At its simplest this calculation can be based on the proportion of the two sexes; if the sex ratio is first determined, then a stated number of one sex removed and a second determination of sex ratio made, then the original number can be estimated from the shift in the sex ratio (Kelker's method, *see* Caughley, 1977). This can be useful where selective culling of the sexes of game animals is employed.

More frequently an estimate of this kind is based on capturing a sample of the population, marking and releasing them and observing the ratio of marked to unmarked animals in a subsequent sample. From this, one can calculate the number of animals at the first sampling that escaped capture and hence the total captured and uncaptured which composed the population at that time. There are many snags in arriving at these estimates and a huge literature has accumulated largely dealing with the statistical problems involved (*see* the text-books already cited) but also with the biological pitfalls which are probably more serious – are the marks permanent and do marked and unmarked animals react to subsequent sampling in the same way and so on? Again it comes back to the maxim of 'know your animal'. This method of capture-mark-recapture, or, as it has been designated, the Peterson or Lincoln index, is a particularly powerful tool in population dynamics because, from a continued chain of trappings, estimates can be made not only of the numbers present at each trapping but also of the numbers of animals entering and leaving the population between each trapping (*see* later).

The greater the proportion of animals marked in the population studied, the more precise will be the estimation of total numbers, and so another method originally proposed by Hayne (1949), which may be designated the removal method, is now frequently used. With small mammals, especially mice and voles, such intensive work can be applied only to relatively small areas (a few ha), unless unlimited manpower is available. On a grid of traps the removal consists either of dead trapping (e.g., with snap traps) over a period until catches fall nearly to zero or of live trapping, marking and releasing until nearly all the catch is marked. Marking and releasing are here equivalent to removal with the advantage that no vacuum is created to draw in outsiders. The increase in the proportion of marked animals from day to day usually follows a fairly steady trend, so that plotting each day's catch of unmarked animals against the cumulative total of those marked will give a straight line which will cut the abscissa at the estimated total number of animals present. An example (Southern, 1973) will make the procedure clear (Fig. 2.2).

With trapping at this intensity and with animals marked for individual identification, e.g., by numbered rings or tags or by toe-clipping [*see* Twigg

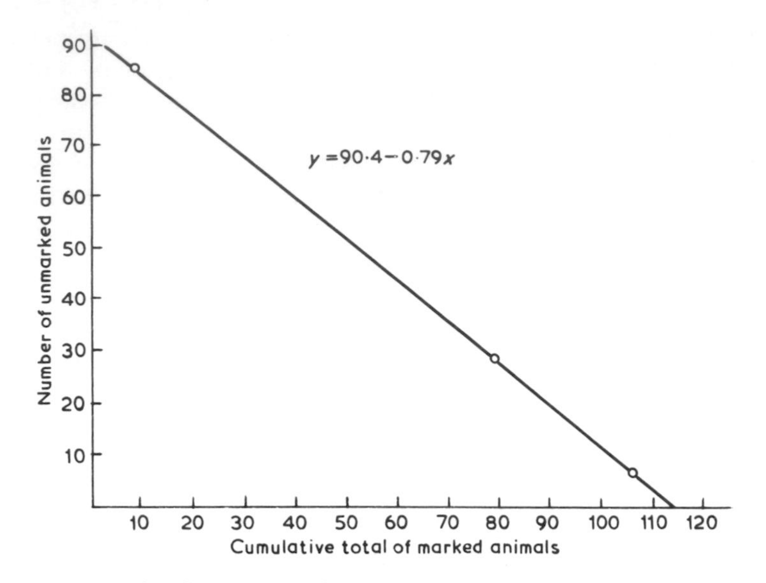

Fig. 2.2 Regression line calculated by Hayne's (1949) method from data from live trapping of the short-tailed vole (*Microtus agrestis*).

(1975) for a review of marking methods] a simple enumeration, or what Polish research workers have dubbed a 'calendar of captures' is often sufficiently near the truth. This comprises the total of animals caught during a trapping session plus any that must have been present as evidenced by their capture in subsequent trapping sessions.

For rough comparisons between times and places some kind of index to density is often adequate and may save much time and labour. Standardization of effort is necessary, e.g. lines of traps set for a standard period and at a standard density will produce catch figures which can be compared between habitats, seasons and years (e.g. Southern, 1965).

Traces of an animal's presence often form a valuable index to their numbers. Counts of faeces have been widely used since the pioneer work on cycles of abundance of *Microtus agrestis* in the hilly borders between England and Scotland (Elton *et al.*, 1935). By simply dropping a quadrat frame at stated intervals and determining the presence or absence of vole traces, extensive areas can be sampled. A similar technique enabled Gibb *et al.* (1969) to follow the changes in density of rabbits over 1580 ha at six-monthly intervals in New Zealand.

A good naturalist will readily determine what is the best sign or trace, left by the animal he is studying, to be subjected to quantitative analysis. Such an index may be a matter of straight-forward observation, e.g. the counting of nests in the case of squirrels or harvest mice, or it may involve some complication such as the preparation of areas to take footprints, the use of

Table 2.4 *Records of densities of selected species (for British Isles unless otherwise specified).*

Species	Numbers/ha	Habitat	Authority
Opossum (*Didelphis marsupialis*)	0.05 – 1.0	Fields and woodland. Virginia	Stout and Sonenshine (1974)
Possum (*Trichosurus vulpecula*)	0.75	Open country with tree groups. Canberra	Dunnet (1964)
Possum (*Trichosurus vulpecula*)	6.4 – 10.6	Forest nr. Wellington New Zealand	Crawley (1973)
Mole (*Talpa europaea*)	8 – 16	Grassland	Larkin (1948)
Hedgehog (*Erinaceus europaeus*)	4 – 8	Pastureland. New Zealand	Campbell (1973)
Common shrew (*Sorex araneus*)	12 – 18	Dunes with scrub. Holland	Michielson (1966)
Common shrew (*Sorex araneus*)	5.3 – 8	Ungrazed grassland.	Pernetta (1976b)
Common shrew (*Sorex araneus*)	0 – 12	Open fields. Sweden	Hansson (1968)
Pygmy shrew (*Sorex minutus*)	4 – 10	Dunes with scrub. Holland	Michielson (1966)
Pygmy shrew (*Sorex minutus*)	4 – 10	Ungrazed grassland.	Pernetta (1976b)
Rabbit (*Oryctolagus cuniculus*)	8 – 26	Pastureland.	Phillips (1955)
Rabbit ((*Oryctolagus cuniculus*)	0.3	Pasture and scrub. New Zealand	Gibb *et al.* (1969)
Rabbit (*Oryctolagus cuniculus*)	150	Run down pasture.	Southern (1940)
Cottontail (*Sylvilagus floridanus*)	9	Woodlot in farmland. Wisconsin.	Trent and Rongstad (1974)
Varying hare (*Lepus americanus*)	0.73	Range margin in forest. Colorado	Dolbeer and Clark (1975)
Varying hare (*Lepus americanus*)	0.34 – 5	Conifer. Alberta	Brand *et al.* (1976)
Varying hare (*Lepus americanus*)	0.9 – 6.9	Conifer. Alberta	Keith (1974)
Hare (*Lepus capensis*)	2.9 – 6.6	Farmland. Poland	Petrusewicz (1970)
Mountain hare (*Lepus timidus*)	<0.01 – 0.05	Arctic alpine zone.	Watson and Hewson (1973)

Table 2.4 *(continued)*

Species	Numbers/ha	Habitat	Authority
Mountain hare (*Lepus timidus*)	0.01 – 0.1	Grouse moors.	Watson and Hewson (1973)
Mountain hare (*Lepus timidus*)	0.5	Grouse moors.	Flux (1970)
Grey squirrel (*Sciurus carolinensis*)	5.1	Decid. woodland.	Mackinnon (1976)
Grey squirrel (*Sciurus carolinensis*)	11.9	Decid. and conifer.	Shorten and Courtier (1955)
Grey squirrel (*Sciurus carolinensis*)	1.9	Decid. Ohio	Nixon, McClain and Donohoe (1975)
Grey squirrel (*Sciurus carolinensis*)	0.6 – 3.4	Decid. woodland. N. Carolina	Barkalow, Hamilton and Soots (1970)
Liomys salvini	4 – 8	Dry tropical forest. Costa Rica	Fleming (1974)
Heteromys desmarestianus	7 – 18	Wet tropical forest. Costa Rica	Fleming (1974)
Wood mouse (*Apodemus sylvaticus*)	3 – 18	Decid. woodland.	Montgomery (1976)
Wood mouse (*Apodemus sylvaticus*)	16 – 24	Deciduous woodland. Ireland	Fairley and Comerton (1972)
Wood mouse (*Apodemus sylvaticus*)	10 – 140	Deciduous woodland.	Kikkawa (1964)
Wood mouse (*Apodemus sylvaticus*)	0 – 6	Open fields. Sweden	Hansson (1968)
Yellow-necked mouse (*A. flavicollis*)	9 – 40	Deciduous woodland	Montgomery (1976)
Yellow-necked mouse (*A. flavicollis*)	3.9 – 10.9	Beech forest. Poland	Bobek (1969)
Bank vole (*Clethrionomys glareolus*)	10 – 60	Deciduous woodland	Kikkawa (1964)
Bank vole (*Clethrionomys glareolus*)	4.5 – 15.6	Beech forest. Poland	Bobek (1969)

Table 2.4 *(continued)*

Species	Numbers/ha	Habitat	Authority
Bank vole (*Clethrionomys gapperi*)	2.7	Conifer swamp. Connecticut	Miller and Getz (1977)
Brown lemming (*Lemmus trimucronatus*)	12 – 38 (high) 5 (low)	Tundra. Alaska	Banks, Brooks and Schnell (1975)
Short-tailed vole (*Microtus agrestis*)	2 – 61	Open fields. Sweden	Hansson (1968)
Short-tailed vole (*Microtus agrestis*)	100 – 327	Fields and orchards. Finland	Myllymäki (1970)
Californian vole (*M. californicus*)	62 – 395	Grassland. California	Batzli and Pitelka (1970)
Californian vole (*M. californicus*)	1 – 240	Scrub. California	Lidicker (1973)
Montane vole (*M. montanus*)	10 – 127	Two areas of alpine meadows. California	Fitzgerald (1977)
Stoat (*Mustela erminea*)	3.5	Two areas of alpine meadows. California	Fitzgerald (1977)
Weasel ♂ (*M. nivalis*)	0.2 – 1	Young plantation.	Lockie (1966)
Weasel ♂ (*M. nivalis*)	0.07 – 0.14	Deciduous woodland	King (1975)
Weasel ♂ (*M. nivalis*)	0.04 – 0.1	Farmland	Moors (1974)

Note: Records are for British Isles unless stated otherwise

automatic recording apparatus to count and time passages through a fence line, e.g. Bayfield and Hewson (1975) on mountain hares, or the automatic photography of small rodents passing along a runway (Pearson, 1959).

Finally, vast reservoirs of information about changes in density can be tapped from record books of game bags, e.g. Hewson (1970) on mountain hares, or from returns, made by the trappers of fur-bearing animals. A classical instance of the last is the analysis by Elton and Nicholson (1942) of the remarkably regular fluctuations of the lynx (*Lynx canadensis*), echoing the changes in density of its prey *Lepus canadensis* in the coniferous forests of North America. These figures come mainly from the records of the Hudson's Bay Company and extend back to 1820 (*see* ch. 3).

Since so much information has now accumulated about the densities at which many species of small mammals live, I have presented only a small selection in Table 2.4 but one which, I hope, covers a wide variety of mammal groups and habitats. The wide brackets of numbers/ha given for many species represent mainly seasonal peaks and troughs but with the voles and the varying hare they indicate the swing of their population cycles (*see* ch. 3).

I would first draw attention to records I have selected for two species both in their native haunts and in countries to which they have been introduced. The grey squirrel densities for America are from an exploited population (Ohio) and an unexploited one (N. Carolina); both are considerably smaller than the measurements in England. The same holds for the possum (*Trichosurus vulpecula*) though some influence must be attributed here to difference of habitat.

Overall, we can detect that larger species achieve smaller densities, especially when they live in open habitats, while predators are naturally much less abundant than their prey. The figures for the rabbit show an interesting contrast: the lowest is from a population that had been reduced from a higher level by poisoning and was being held low by predators; the highest is from English derelict farmland during the inter-war years. The numbers of the insectivores suggest that they are relatively stable but, when we come to the small rodents there is great variation. On the whole the mice and the bank voles fluctuate less than the species of *Microtus* and it is interesting to note that the two species of heteromyid rodents studied in Costa Rica live at densities much the same as do European mice, though, of course, rain forest will have more species. *Microtus*, wherever it occurs, is the one that hits plague proportions during its cycles and can be a serious economic threat.

2.3.2 'Income and expenditure'

The estimation of densities at one point in time is the first step to the appreciation of the population dynamics of a species. As with a bank account, the balance at any one time is a product of the amount paid in and the amount drawn out. Again, it is convenient to assess these two components separately, though the operation of both simultaneously is the essence of the system.

Growth of a population

As Darwin pointed out, the capacity for increase of all animal populations is potentially exponential and we have already encountered examples of animals introduced into a foreign environment, where this potential is realized for a time. But we know that, in nature, this is exceptional; otherwise the world would be submerged under a flood of animals.

Yet, before we go on to consider the factors that limit this increase, we must seek some way of expressing and comparing capacities for increase between species and circumstances. The standard measurement, now universally agreed, is the rate at which a population increases when food and space are unrestricted. This is normally determined in the laboratory, where conditions can be standardized. It allows for the physiological wear and tear of a population, i.e., it does not treat individuals as being immortal, but it does not subject them to shortage of food or predation. It can, therefore, compare species with species from the same base line and it can also compare the performance of one species under different regimes of temperature, humidity, day length and so on. These last have more influence with invertebrates than with vertebrates. Perhaps more important with vertebrates, it can compare the rates of increase of sub-sections of a population with different genetic constitutions.

This basic rate of increase is measured by the parameter r, which is the exponent in the differential equation for population growth in an unlimited environment under specified physical conditions – $dN/dt = rN$ (Odum, 1971). It can take various values according to the structure of the population, but if the age distribution is stationary and stable, then r_{max}, or the *intrinsic rate of natural increase*, will give a figure for comparing species. Thus r for laboratory populations of *Microtus agrestis* has been determined as 4.5 per year (Leslie and Ranson, 1940). This comparison is, perhaps, more easily comprehended if it is expressed in terms of the time it would take for the population to double its numbers (7.9 weeks for the vole).

In wild populations r_{max} is, of course, stringently curbed at the point where it comes up against environmental resistance (K) and population growth levels off. Thus, r may take almost any value but tends towards zero. Its calculation is described in many standard text books on ecology and the matter has been dealt with in the previous chapter.

This, however, will refer to a population which is undergoing both output and input. It is still desirable to have values put on these parameters separately. The potential growth of a population in the wild can be assessed from breeding rates and embryo rates. Ornithologists are fortunate in that they can start with the egg and the size of clutch; for small mammals a sample of the population must usually be sacrificed to discover the mean and variation of the number of embryos and, indeed, whether some individuals are not breeding at all. In many studies, where this sacrifice cannot be made, one has to be content with recording litter rates or, even more crudely, as with mice and voles, the number of young that enter the trappable part of the population. This matter is simplified if a species produces only one young at a birth, as with most bats and some marsupials and ungulates.

With the best available knowledge of the number of young produced and with information about the age structure of the population, we can then set up an age-specific fertility table. This is usually expressed as the number of female offspring produced per female of each reproductive age group per unit of time (usually one year with a seasonally breeding animal). This is the best estimate one can obtain of the 'income' of a population.

Mortality

Estimates of the expenditure of a population usually have to lump mortality and emigration because the latter demands much more energy to trace than straightforward disappearance of marked individuals. It is some consolation that, when a sample area within a widespread population is being studied, movement into and away from that area will be roughly equivalent.

The rate at which individuals leave a population is best expressed in life tables and survivorship curves. The life table is taken over from human demography and needs much labour to construct with such unco-operative subjects as wild animals. At its simplest it starts with a cohort of animals born into a population at approximately the same time, e.g. during one breeding season or during one surge within a breeding season, and traces how many of these are still present at appropriate intervals until the last one has disappeared. The columns under which these data are recorded are labelled x, the time interval between each record, l_x the number of the cohort surviving at each interval (either the actual number or, for convenience the number calculated from the starting point of 1000), d_x, the number dying in the interval, q_x the proportion of those entering the time interval that died during it and e_x the average number of time intervals that an individual may expect to survive from the beginning of each interval. This last is

Table 2.5 *Life table for the grey squirrel based on 1023 individuals of both sexes marked as nestlings, brought to a starting cohort of 1000 (after Barkalow* et al., *1970).*

Age in years (x)	Survivors (l_x)	Deaths (d_x)	Death rate (q_x)	Expectation of further life (e_x) in years
0 – 1	1000	753	75%	0.99
1 – 2	247	135	55	1.82
2 – 3	112	30	27	2.41
3 – 4	82	26	32	2.10
4 – 5	56	15	26	1.84
5 – 6	41	25	59	1.32
6 – 7	16	1	2	1.48
7 – 8	15	15	100	–

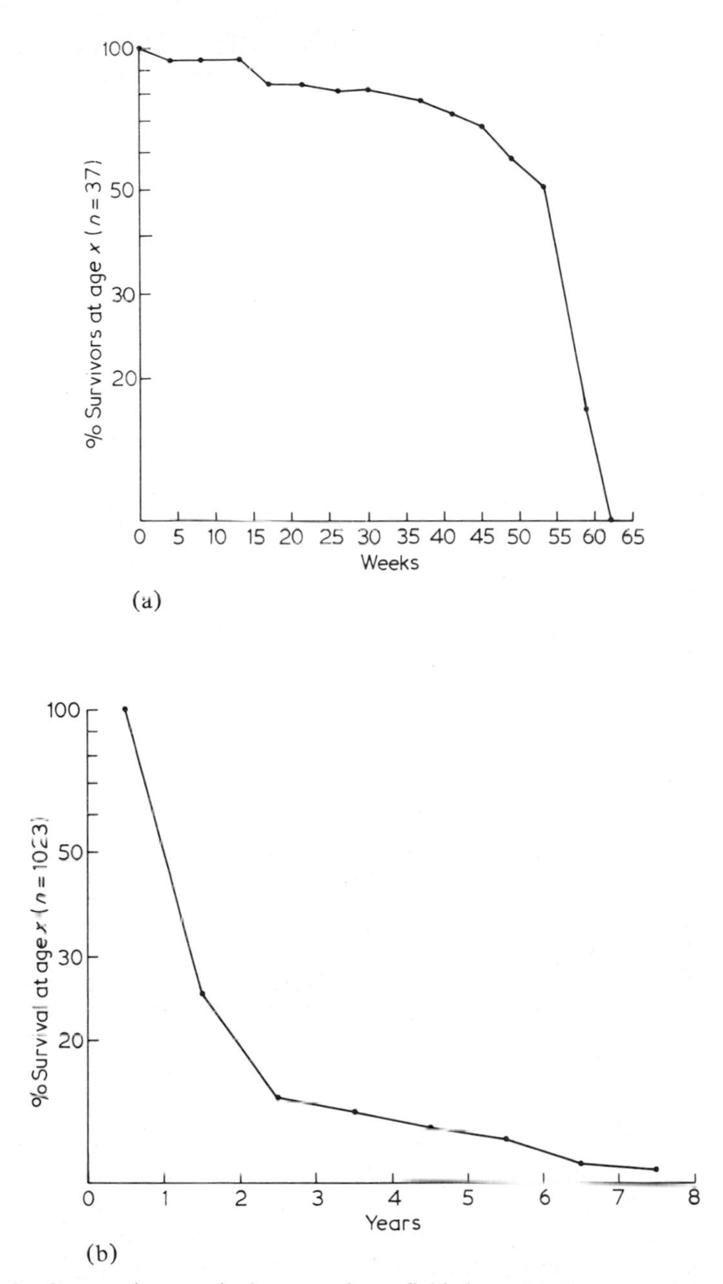

Fig. 2.3 Contrasting survival curves from field data. (a) Common shrew (*Sorex araneus*) in Holland (Michielson, 1966). This represents the fate of a single cohort of shrews, 37 in number, but omits mortality in the nest. (b) Grey squirrel (*Sciurus carolinensis*) in North Carolina (Barkalow *et al.*, 1970). This is a composite survival curve of 1023 squirrels marked as nestlings and brought to a common starting point.

known as the mean expectation of further life and is arrived at by adding up all the years, or whatever, in the l_x column below the start of any interval and dividing the total by the survivors at that starting point. There are some niceties in the calculation of e_x which are lucidly explained in Krebs (1972). This layout will be more easily understood by reference to the sample in Table 2.5.

Clearly the l_x column of the life table gives us the basic data from which a survivorship curve can be drawn and the q_x and e_x columns pinpoint the variations in the mortality rate during the history of the cohort concerned. It has long been recognized that the varying impact of mortality results in three main types of survivorship curve (Pearl, 1928) (*see* Fig. 2.3). Rarely a species may survive well until old age and fade out suddenly (Pearl's type 1 curve). Something approaching this has been recorded for the common shrew (Michielson, 1966) (Fig. 2.3A), but this starts with weaned and trappable young. Most likely an initial steep drop in the curve should ideally be included to cover embryo and nestling mortality. By far the most frequent type of survival curve for small mammals is one which drops steeply through the juvenile and subadult stages and then levels out. A typical example of this is shown by the grey squirrel (Fig. 2.3B, Barkalow *et al.*, 1970). This curve begins with marked nestlings and, therefore, is that much more complete than the curve for the shrew. Note that Michelson's curve is for an actual cohort of shrews entering the population at approximately the same time. That of Barkalow *et al.*, includes nestling squirrels from several years all scaled back to a common starting point. Pearl's third type of curve hardly concerns us here, since it refers mainly to invertebrates and some lower vertebrates, e.g., fish, that produce enormous quantities of young, of which very few survive to maturity. The result is an exaggeratedly concave survival curve.

The following of marked individuals throughout their lifetimes may, of course, be an arduous business, if the species concerned is long-lived. There is a story, probably apocryphal, of the student who set out to make these measurements on the unique rhynchocephalian reptile of New Zealand, *Sphenodon,* and gave up when he found that it took eighteen years to reach sexual maturity.

Tracing a cohort is labelled a horizontal life table ('cohort' or 'dynamic' life table of some authors). A short cut takes the form of a vertical life table ('static' and 'time-specific' are other designations), in which an instantaneous sample is taken of a population and the age of each individual in the sample is determined. This is not always feasible but methods of determining the age of mammals are improving [see Morris (1972) for a review]. The frequency of the age classes in such a cross section of the population constitutes, in effect, an l_x series from which a life table can be constructed. Obviously, a life table of either kind can give a true picture of population proceses only if the population is stationary and the age distribution stable, but we have to accept many roughnesses in dealing with populations of wild animals.

Population turnover

Turnover is an awkward term but it is useful shorthand for the speed at which individuals enter and leave a population and composite pictures of birth and death rates can compare the turnover of populations of different species at different times and places. To illustrate this, as a basic example, I take again figures for the grey squirrel from Barkalow *et al.*, since similar figures are scarce for small mammals (Table 2.6). 'Small mammalogists'

Table 2.6 *Turnover rate of a grey squirrel population (females only considered; after Barkalow* et al., *1970).*

Age in years (x)	Survivors (l_x), n = 530	Productivity (m_x)	$l_x \cdot m_x$
0 – 1	1000	0.05	50
1 – 2	253	1.28	324
2 – 3	116	2.28	264
3 – 4	89	2.28	203
4 – 5	58	2.28	132
5 – 6	39	2.28	89
6 – 7	25	2.28	57
7 – 8	22	2.28	50

who are prepared to stray outside these boundaries will find more complete pictures for the red deer (*Cervus elaphus*) in Lowe (1969) and for the thar (*Hemitragus jemlahicus*) in Caughley (1977).

The number of females recorded is again scaled to a starting point of 1000 and the proportionate decline in their numbers followed down the survivors column (l_x). For each age group the number of female offspring per female is given in the column labelled m_x. The product of columns 2 and 3 gives the total of female young produced by each age group. The salient points to notice are that the fertility of female squirrels increases up to their third year, when they make their maximum contribution to the 'income' of the population, and that the subsequent decline in this contribution is due entirely to the diminution of their numbers as they grow older (*see* l_x column).

Thus for 1000 female squirrels that have died over the period concerned, 1169 (the sum of the $l_x.m_x$ column) have entered the population. r_0 is therefore positive (1.169) and it appears that the numbers have been slightly

Table 2.7 *Mean expectation of further life* (e_x) *for selected species.*

Species	Locality and circumstances	e_x at weaning	max	Authority
Common shrew (*Sorex araneus*)	Holland – from trappable age, cohort 1	–	10 ms	Michielson (1966)
	Holland – from trappable age, cohort 2	–	*c.* 6.6 ms	Michielson (1966)
Pygmy shrew (*S. minutus*)	Holland – from trappable age, cohort 2	–	*c.* 5.2 ms	Michielson (1966)
Vampire bat (*Desmodus rotundus*)	S. America	2.1 yrs	7.1 yrs	Lord, Muradali and Lazaro (1976)
Long-eared bat (*Plecotus* spp.)	England – as adults	–	3.5 yrs	Stebbings (1966)
Whiskered bat (*Myotis mystacinus*)	Holland – as adults	–	3.5 yrs	Bezen, Sluiter and van Heerdt (1960)
Daubenton's bat (*M. daubentoni*)	Holland – as adults	–	4.5 yrs	Bezen, Sluiter and van Heerdt (1960)
Pika (*Ochotona princeps*)	U.S.A.	0.6 yrs	4.5 yrs	Smith (1974)
Rabbit (*Oryctolagus cuniculus*)	Wales	–	3.3 ms	Phillips (1955)
Hare (*Lepus capensis*)	Poland	–	1.1 yrs	Petrusewicz (1970)
Blue hare (*L. timidus*)	Scotland	–	1.5 yrs	Flux (1970)
Cottontail (*Sylvilagus floridanus*)	Wisconsin	–	0.75 yrs	Trent and Rongstad (1974)
Grey squirrel (*Sciurus carolinensis*)	England	–	1.1 – 1.3 yrs	Mackinnon (1976)

Table 2.7 *(continued)*

Species	Locality and circumstances	e_x at weaning	max	Authority
Grey squirrel (*Sciurus carolinensis*)	N. Carolina – unexploited popn. from nestling stage	0.99 yrs	2.41 yrs	Barkalow *et al.*, (1970)
Grey squirrel (*Sciurus carolinensis*)	Ohio – exploited popn.	–	0.75 yrs	Nixon, McClean and Donohoe (1975)
Yellow-necked mouse (*Apod. flavicollis*)	Poland, high density	2.9 ms	–	Bobek (1969)
Yellow-necked mouse (*Apod. flavicollis*)	Poland, low density	3.6 ms	3.8 ms	Bobek (1969)
Wood mouse (*A. sylvaticus*)	England – summer from trappable age	–	*c.* 2 ms	Flowerdew (1974)
Yellow-bellied marmot (*Marmota flaviventris*)	Colorado, females only	1.7 yrs	2.8 yrs	Armitage and Downhower (1974)
Bank vole (*Clethrionomys glareolus*)	Poland – high density	2.2 ms	3.1 ms	Bobek (1969)
Bank vole (*Clethrionomys glareolus*)	Poland – low density	3.2 ms	4 ms	Bobek (1969)
Water vole (*Arvicola terrestris*)	Scotland, yg. to 30 June	6.5 ms	–	Stoddart (1971)
Water vole (*Arvicola terrestris*)	Scotland, after 30 June	1.5 ms	–	Stoddart (1971)

Note: ms = months

increasing. This is to be taken warily because (a) the m_x figures were not drawn from a truly random sample of the female population (*see* Barkalow *et al.*, for details), and (b) the apparent surplus may have dispersed to less suitable habitats, i.e., movement in and movement out may not have been equivalent. This is the kind of biological distortion that can bedevil statistical calculations; the picture revealed may be near to the truth and this is the best we can achieve.

Looked at in another way these figures indicate that a cohort of grey squirrels will be replaced in some 6 years (*see* Barkalow *et al.*, for details of this calculation) and this may be regarded as the rate of turnover of the population. This would be an average rate for an animal of this size and with its expectation of further life as given in Table 2.5.

Given that most small mammals exhibit the same order of caring for their young, we would expect that the average length of life of a species would be shorter, the smaller the animal. This turns out to be approximately true, since there is a progression from the shrews, mice and voles up to squirrels, rabbits and hares (*see* Table 2.7). Of course, there are variations but they do not greatly diverge from the main trend, except in the bats. They are, indeed, systematically very far apart from the main groups we have been considering under the heading of small mammals, yet it comes as a surprise that their average length of life is 4–5 years and there are by now many records of individuals living for more than 20 years.

As an index to rate of turnover and longevity (however, that may be interpreted), I have thought it best to use the figure for mean expectation of further life (e_x) and to compile a selected list of records. Other authors prefer to use, for comparison, the percentage mortality per unit of time and there is probably not much to choose between them.

Naturally, these e_x figures are rough and difficult to compare in detail. Except for the records for the squirrel by Barkalow *et al.*, where the starting point is the nestling, weaning is taken broadly as the time of appearance of young animals in traps. Where a full life table is given, showing the usual rise and fall of life expectation with age, the maximum figure is clear; where only a blanket proportionate mortality rate per month or year is given, I have assigned the corresponding life expectation to the 'maximum' column. Where e_x is short, as with mice and voles, it is difficult to know how much time of year or the status of the population influences the result. Stoddart's figures for the water vole emphasize this. These circumstances are so protean in *Microtus* spp. that I was forced to omit them from the table.

When all this is said, there are still certain broad trends to be detected in the table. The life expectancy of the common shrew, whose population turns over in a year, is surprisingly long because of the form of its survival curve. If, as Michielson suggests, its burrowing propensities enable it to spend most of the winter underground, thus avoiding competition with the pygmy shrew, it is understandable that the latter species has an expectation of further life nearer to that of the small rodents.

The fact revealed that some rodents, e.g., yellow-necked mice and bank voles, survive longer at low than at high densities is not surprising but it is valuable to have this documented. Similarly the shorter life expectation of the grey squirrel in England than in North Carolina, where its density in general is lower (*see* Table 2.7), follows the same trend. On the other hand, the effect of exploitation on grey squirrels is notably to reduce their life expectation. In Ohio 55 per cent of the annual mortality is due to hunting, which keeps the population turning over fast. One would guess that the production rate and survival of nestlings were high and the popualtion was, in a sense, being 'kept happy'. One may recall that in England during a campaign against this squirrel, in one year 257 644 were shot without noticeable effect on their numbers (Forestry Commission, 1953).

By and large Table 2.7 confirms the increase of life expectation with the size of the species, with shrews and small rodents at the bottom of the scale and hares at the top (always excepting the bats). The small figure for rabbits came from a very dense population before the days of myxomatosis and, furthermore, from an area that was being intensely exploited.

With these thoughts in mind about the dynamics of small mammal populations, we can turn in the next chapter to examine the reasons that some are stable and others very unstable.

The stability and instability of small mammal populations

3

H.N. SOUTHERN

3.1 Introduction

Stability and instability are not, of course, characteristics peculiar to small mammal populations. The central problems of why and how do animal populations remain at an equilibrium density or depart, sometimes violently, from it have been researched and debated in all corners of the animal kingdom. It is, however, possible that, after insects, small mammals may be the group with the highest economic importance to agriculture and forestry.

In general, it seems that ecologists fall into two camps as to whether they are more impressed with the stability or the instability of animal populations. Darwin, long ago, pointed out that all animals were endowed with massive potential powers of increase, yet these were hardly ever fulfilled [at the moment it looks as if man were the only animal that is approaching its intrinsic rate of natural increase (*see* Chapter 2)]. He drew the conclusion that all animals suffer an enormous mortality rate which allows them to evolve by natural selection.

Ecologists are more interested in instant rather than evolutionary problems of population control though, as we shall see, the two aspects grade into one another. If I had to point to one among the legion of ecologists who have argued that the stationary nature of most populations is the basic fact that demands investigation and explanation, I should choose David Lack. In his many papers and, pre-eminently in two of his books (1954, 1966), he has championed the theory that the numbers of animals are regulated by density-dependent mortality factors, i.e., those whose impact tends to cancel any departure from an equilibrium level upwards or downwards as the case may be. His evidence was drawn almost exclusively from birds but similar arguments have been advanced in other groups of animals.

On the other hand many ecologists are impressed by the fluctuations in numbers that are observed to occur in some populations and attribute the fact that they do not increase indefinitely to the arbitrary effects of extrinsic factors in the environment, such as weather. Long ago Elton (1927) insisted "that no animal population remains the same for any great length of time, and that the numbers of most species are subject to violent fluctuations." A whole school of thought on these lines was later built up by ecologists (notably Andrewartha and Birch, 1954) and considerable controversy was generated.

The whole question is complicated by the fact the numbers of animals change with changes in habitat. Thus, an animal's numbers may be steady in the central part of its range and fluctuate on the periphery where it is occupying less and less suitable habitat (Richards and Southwood, 1968). Since so many parts of the world have environments that fluctuate violently from season to season and year to year, either in the nature of things, e.g., Australia where Andrewartha and Birch worked, or because of the vast transformations that the activities of man have produced, e.g. replacing forest cover by arable and pasturelands, it is not surprising that the animals occupying them may do the same. On the same line of thought, the explosions of species introduced not only to new habitats but to new continents, outlined in the preceding chapter, are to be expected.

Nevertheless, there are instances where exaggerated changes in numbers occur in areas that have not been modified by man. Most spectacular are the locusts but some forest insects, e.g. the pine looper moth and the spruce budworm, can, at intervals, devastate vast areas. Among small mammals, the 3–4 year cycles of lemmings and voles and the 10 year cycle of the showshoe rabbit, sometimes of remarkable amplitude, appear to be quite natural (more of these anon). At any rate, this judgment must apply to the tundra and boreal forest zones where these cycles classically take place. There is some doubt about the vole cycles in temperate grasslands (the products of man) as Tapper (1976) has pointed out.

All this suggests the possibility that violent fluctuation in numbers of periods greater than a year may be the exception rather than the rule but we must now turn to examine the evidence put forward by these differing schools of thought. The matter is of central importance in population dynamics because only with rigorously tested hypotheses can we make predictions about future population trends.

3.2 The regulation of small mammal populations

I follow here the terminology proposed by Solomon (1964) and others. A 'control factor' is anything that influences the numbers of an animal population – not one that reduces its numbers to an economically acceptable level. A 'regulating factor' is one that acts in a density-dependent way; at its crudest, if numbers rise, proportionately more are killed; if numbers drop, proportionately fewer are killed. Thus an equilibrium is maintained.

I shall glance only briefly at the theoretical background to these views about control. This is dealt with ably and clearly by Odum (1971), Krebs (1972), Varley *et al.* (1973), Solomon (1969) and a host of other ecologists. This will enable me to give more space to the examples of case histories which follow.

3.2.1 Theoretical background

The proponents of the density-dependent theory maintain that the norm is for an animal species to maintain its numbers at a general average value with fluctuations above and below this value being counteracted by varying mortality rates. By and large, these regulating mortality factors come under the headings of starvation, predators and parasites, disease and competition for space. This equilibrium value will be different in different environments; mice and voles, on the whole, remain more abundant in deciduous than in conifer woodlands. The value will also shift to a new level if the habitat is changed.

It is, of course, equally possible for adjustments of this kind to be made by variation in the birth rate as well as in the death rate. Among birds, where the matter has been copiously researched, the number of eggs produced seems to remain fairly standard for species and habitat. In fact, Lack and others have maintained that the size of a clutch is determined by the number of young that a pair of birds can rear to a free-flying state in good condition. Those that lay a larger than average clutch may, in fact, produce fewer young than those laying a more modest clutch and natural selection will act against them. It has always seemed odd to me that so many ornithologists disregard the members of a population whose clutch size is zero and which are prevented, in one way or another, from breeding at all. This is hard to measure but, in many territorial species, the speed with which any casualty in the territory owners is replaced indicates a reservoir of non-breeding birds. Furthermore, there are many predatory species of birds that are well known to omit breeding if their prey is scarce. My own research on the tawny owl (*Strix aluco*) in woodland (Southern, 1970) has shown that the proportion of pairs that actually launch themselves on breeding activity may vary from 0 to 80 per cent in different years according to the number of mice and voles present.

Among small mammals, there is more scope for varying the birth rate than among birds. Not only may the number of ova ovulated and the number of embryos implanted vary but a section of the population may achieve sexual maturity in the year of their birth or may postpone this until next year. Indeed, in the common European vole (*Microtus arvalis*), which frequently reaches plague numbers, females may become sexually mature before they are weaned – a somewhat astonishing example of precocity.

With all these considerations, it does seem, by and large, that the death rate is a more immediately flexible adjuster of numbers among vertebrates than the birth rate. Insects are a somewhat different matter but need not concern us here.

So far I have, as a starting point, merely considered the idea of density-dependence in a crude and simple way, in which its action is immediate (*direct density-dependence*). In some systems, e.g. predator and prey, the reaction may be delayed because the predator needs time to build up its numbers (*delayed density-dependence*). Many other influences must be at work, including those, e.g. climate, storms, frosts, floods, etc., whose action is not adjusted, or only indirectly, to the density of the population affected. Such *density-independent* factors are frequently the ones which disturb an equilibrium and, therefore, create 'work' for density-dependent factors to do. Often a mortality factor may operate in relation to density but in the opposite direction to that implied by density-dependence. Thus, if the numbers of a prey population increase, a predator may step up its impact proportionately, i.e. act in a classical density-dependent way, but only up to a certain point. If the prey can increase beyond this point, it may escape from the predator's control and go on soaring upwards, so that with every step of increase, the predator will be killing proportionately less, instead of proportionately more, of the prey. The condition is now *inversely density-dependent* and the prey population will continue to increase until it is curbed by another mortality factor, probably more stringent than the predator, e.g. starvation. Some forest insects, whose larval and pupal stages are preyed upon by small mammals, illustrate this switch; Holling (1959) has elegantly analysed this process in the case of the deer mouse (*Peromyscus*) hunting for the cocoons of the pine sawfly over a range of densities.

These three fundamental processes can be preliminarily detected from a series of censuses over a number of years or generations (Fig. 3.1) (after Solomon, 1964). If the logarithm of the numbers is plotted against time, three hypothetical curves or lines are generated: A depicts a constant density- independent geometric increase; B shows the form of change in a density-dependent system; and C that under conditions of inverse density-dependence. The references cited at the beginning of this section will show how analysis can proceed beyond this point.

We must now turn to a brief examination of an opposing hypothesis which attributes the control of an animal's distribution to climatic factors. Andrewartha and Birch (1954) maintained that, since there are only limited periods when the rate of increase of any animal is positive, numbers must

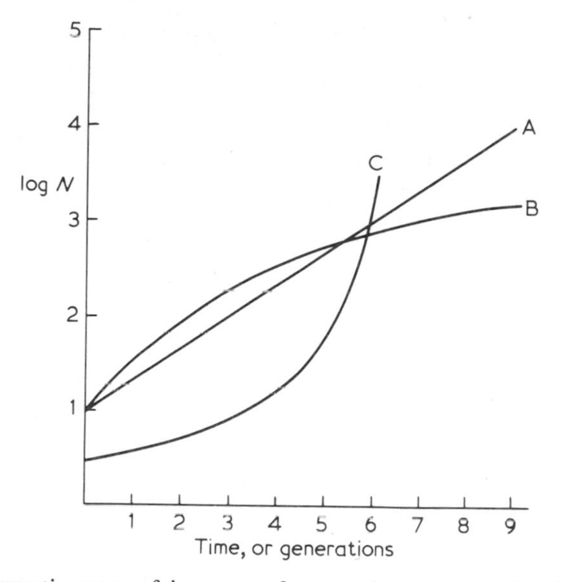

Fig. 3.1 Schematic rates of increase of a population under density-independent conditions (A), density-dependent (B) and inversely density-dependent conditions (C). (After Solomon, 1964.)

fluctuate up and down in consonance with these periods. Thus it is, in their view, the environment which controls the numbers of animals and of this the most important is climate, though their analysis embraces other factors as well – food availability, other animals and a place to live. They totally reject what they call the dogma of density-dependence.

This is no place to go into the storm of controversy evoked by these hypotheses. The pros and cons are clearly set out in the appendix to Lack's (1966) book; furthermore, hindsight reveals this to have been rather a storm in a teacup. Andrewartha and Birch were studying, mainly, invertebrates in the very variable and often semi-arid habitats of Australia, whereas the proponents of density-dependence drew their examples from the more stable conditions of the north temperate zone. Once it had been demonstrated [and this was done very convincingly by Whittaker (1971) from his research on spittlebugs of the genus *Neophilaenus*] that, on the edge of an animal's range, numbers did fluctuate violently, often in consonance with weather, whereas in the centre of its range numbers were relatively stable and were regulated by factors such as parasitism and predation, the controversy was shown up as a somewhat sterile one. Both conditions exist and in some habitats the one predominates and in some the other. Perhaps one of the most significant facts that Andrewartha and Birch drew attention to is the patchiness of habitats and the effect this has on the population dynamics of the animals that share them out.

All this seems pretty logical; because they are rarely observed to eat out their food supply, herbivores must be regulated by predators and parasites ('the world', in Slobodkin's phrase, 'is green'), whereas the predators and parasites are regulated by the abundance of their prey and hosts. But the fact that the world is green worried ecologists in another way which was to produce a fruitful line of thought, namely that animals, especially vertebrates, can, through intra-specific competition, regulate their own numbers at a level which does not normally endanger their food supply.

This concept of self-regulation has gathered to itself more and more evidence during the last two or three decades and may, by now, be regarded as 'respectable'. Chitty (1960) was one of the first to study intensively the 3-4 year cycle of numbers in species of *Microtus*; he tested the changes he observed against all the possible extrinsic factors, such as we have considered above, *viz*. climate, food shortage, predation and parasitism, and could not find that any of them gave a satisfactory answer. So he turned his attention to what may be called 'intrinsic' factors, i.e. variation in the animal itself, and came up with the hypothesis that, as densities rose, natural selection favoured aggressive genotypes (or phenotypes) of voles with the result that less aggressive voles were dispersed and suffered heavy mortality and reduced reproductive success. But, as the density of the population thinned out, selective pressures were reversed in favour of animals that were tolerant of each other's presence. Since very strong selective pressures are now known to operate in nature (Ford, 1971) and since populations have been shown to contain great genetic variability (Berry, 1977), Chitty's hypothesis obviously merits consideration as applicable to a wider field than microtine cycles. It has stimulated a vast amount of research and some controversy, but, to date, no very clear answer has emerged about vole cycles.

The question of self-regulation was approached from a different angle by Wynne-Edwards (1962). He relied mainly on birds for his evidence but his general conclusions are of wide application. Briefly stated, his thesis was that the numbers of animals are ultimately controlled by the availability of food, but that various intra-specific mechanisms prevent the food supply from being catastrophically exhausted. The possession of a defended territory is known widely among birds and, rather less widely, among mammals; this achieves a spacing out so that starvation among territory holders is rare except during emergencies, e.g. long spells of exceptionally hard weather. This condition was clearly illustrated by the tawny owls that I studied; variation in the numbers of mice and voles from year to year was strongly correlated with breeding success but had no effect on the number of territory holders. In one year, when rodent prey was so scarce that not a single pair of territory holding adults even attempted to breed, there was no decline in the numbers of the latter. Their motto evidently was 'I'm all right, Jack and can live to breed another year'.

Perhaps more significant with mammals is the kind of social system or hierarchy that many of them establish, whereby subordinate individuals are either forced out into second-class habitats with increased chances of perishing or remain in the social group but do not breed. Females may even become 'aunts' and assist in rearing the young of those that do breed.

The most convincing demonstration of how such a two-tiered population structure operates is for a bird and I shall make no apology for quoting it. Carrick (1963) studied the Australian 'magpie' (*Gymnorhina tibicen*) in open country with groups of trees near Canberra and found the population sharply divided into two categories. The aggressively territorial groups, usually consisting of one dominant male and one or two females were the only ones that were able to breed. Non-breeding flocks occupied the poorer grasslands and scrub and the surplus from the breeding groups was forced out to join these flocks. The ratio of breeders to non-breeders was of the order of 1:1.5. The most significant thing about this arrangement was that whereas, when a breeding female died, she was replaced by one from the flock, when a male died, the whole breeding group broke up and was replaced by a new one, the breeding females retiring to a non-breeding condition in the flock.

There is nothing quite so clear cut in mammals. Table 3.1 gives selected examples of what is known about social structure and its variation. The following tendencies may be tentatively discerned in these somewhat heterogeneous data.

(1) Some mammals are territorial, i.e. individuals defend exclusive areas, e.g. shrews, weasels. Most have home ranges which show all degrees of exclusiveness, whether defended by individuals (usually the males) or by groups as a whole, e.g. possum, ground squirrel.

(2) Males usually have larger home ranges than females, e.g. hares, squirrels, voles.

(3) Social structure may vary in different habitats, e.g. possum.

(4) The general tendency of all these systems is for lower ranking animals to be forced out into areas where they are at much greater risk.

(5) This may result in a two-tiered population reminiscent of Carrick's magpies except that the overflow, far from refraining to breed, may do so exuberantly but with high mortality. This suggests an analogy with the *r* and *K* selection conditions treated in Chapter 1, e.g. house mouse.

(6) It also indicates that surplus animals may have the alternative possibility of 'staying put and lying low', i.e. not breeding but perhaps assisting the breeders to rear their young, as Macdonald (1978) has so elegantly demonstrated with his foxes. The employment by mammals of a wide range of submissive postures, movements, and vocalizations shows that selection

Table 3.1 *Selected examples showing social structure and size of territories, or home ranges, in small mammals.*

Species	Social structure	Habitat and home range	Authority
Brush-tailed possum (*Trichosurus vulpecula*)	Male defends territory with 1 – 2+ females.	Australia. Tree groups in grassland. *c.* 8 ha.	Dunnet (1964)
Brush-tailed possum (*Trichosurus vulpecula*)	Home ranges in both sexes with much overlap.	Podocarp-broadleaved forest, New Zealand. Male 0.8 ha, female 0.5 ha.	Crawley (1973)
Common shrew (*Sorex araneus*)	In winter both sexes defend exclusive territories which break down in summer.	Dunes in Holland. Winter territories 0.04 to 0.06 ha.	Croin-Michielson (1966)
Pygmy shrew (*S. minutus*)	As above	0.05 to 0.19 ha.	Croin-Michielson (1966)
Rabbit (*Oryctolagus cuniculus*)	Mutually exclusive groups with dominance hierarchy in males and females. Subordinates less exclusive	Australia, enclosure experiments.	Myers and Poole (1959)
Mountain hare (*Lepus timidus*)	Both sexes have similar home ranges with wide overlap.	Scottish moorland. Males *c.* 16 ha, females *c.* 10 ha.	Flux (1970)
Grey squirrel (*Sciurus carolinensis*)	Mutually exclusive social groups, dominance order within groups, males dominating females. Ranges overlapping within groups.	England, deciduous woodland. Males *c.* 4 ha, females *c.* 2 ha.	Mackinnon (1976)
Richardson's ground squirrel (*Spermophilus richardsonii*)	Groups dominated by female. Overlapping home ranges.	Short-grass prairie, S. Saskatchewan.	Michener (1973)

(continued)

Species	Social structure	Habitat and home range	Authority
Columbian ground squirrel (*S. columbianus*)	Mutually exclusive groups with dominant male and several females.	Alpine region, Rocky Mountains.	Steiner, quoted in Michener (1973)
Short-tailed vole (*Microtus agrestis*)	Males strongly territorial, females with smaller, overlapping home ranges.	Enclosure experiments, Finland	Myllymäki (1970)
Water vole (*Arvicola terrestris*)	Linear home ranges along streams. Male ranges overlap widely and are constant. Females overlap less and may shift.	Scottish heathland. Males 84 to 183 m, females 56 to 126 m.	Stoddart (1970)
Wood rat (*Neotoma fuscipes*)	Females have small home ranges with some overlap. Males have larger home ranges and wander widely in breeding season.	N. California, riparian woodland. Males 0.2 to 0.4 ha, females 0.1 to 0.2 ha.	Cranford (1977)
Wood mouse (*Apodemus sylvaticus*)	As for wood rat but greater overlap in both sexes. Male and female home ranges tend to be associated in breeding season.	England, mixed woodland. Males 1.1 to 2.2 ha, females 0.7 to 0.8 ha.	Randolph (1977)
House mouse (*Mus musculus*)	Mutually exclusive social groups in refuges. Expelled surplus forms summer breeding populations in temporary habitats.	Canadian prairies.	Anderson (1970)
Weasel (*Mustela nivalis*)	Both sexes defend exclusive territories; those of females smaller than those of males and not contiguous.	England, deciduous woodland. Males 7 to 15 ha, females 1 to 4 ha.	King (1975)

may, up to a point, favour 'staying put' and encourage cohesion within a social group.

So far I have tried to state the case for self-regulation only in terms of behavioural variation and variability. What underlies this variability is a question that is only now being broached by the combined efforts of ecologists and geneticists. In 1968 Pimentel put forward the hypothesis that a predator and its prey or a herbivore and its food supply, when first introduced to each other, would fluctuate violently but, as time passed, genotypes would be selected for accommodation between the partners so that the fluctuations would be eventually ironed out and they would settle down to an equilibrium. This has been designated the genetic feedback mechanism and it has mainly been investigated in invertebrate populations. Nevertheless, it is probable that an experiment on a vast scale on natural populations has shown the working of this mechanism. In the early 1950's the virus of myxomatosis was introduced into the rabbit populations of Australia, Western Europe and the British Isles. The disease is more or less benign in its native hosts in South America, *Sylvilagus* spp., but in the strange environment of the rabbit, it produced an immediate, raging pandemic, which killed more than 99 per cent of the population. With time, two processes of selection have modified this extreme situation; less lethal strains of the virus have spread and so have rabbit genotypes resistant to the disease. At the moment the rabbit is staging a comeback, though nowhere has it reached its former superabundance. The end of the process is not yet in sight.

It is clear, therefore, that evolution has some hand in population regulation. As Krebs (1972) has well expressed it, 'self-regulatory systems have added an additional degree of freedom to the system, the individual with variable properties'. We can no longer think of animals as behaving like molecules, an error into which some ecologists, but hardly any naturalists, have fallen.

The elucidation of these individual properties is still rather in its infancy. Genetic heterogeneity has been demonstrated in many species, especially since the technique of starch gel electrophoresis has been deployed to detect biochemical polymorphisms in various proteins. In many small mammals, e.g. wood mice (Leigh-Brown, 1977), deer-mice, *Peromyscus* spp. (Rasmussen, 1970), the voles *Microtus agrestis* (Semeonoff and Robertson, 1968) and *M. ochrogaster* and *M. pennsylvanicus* (Tamarin and Krebs, 1969; Gaines and Krebs, 1971), such polymorphisms have now been verified, sometimes correlated with seasonal or cyclical phases of population abundance. No clear relationship, however, has yet been demonstrated with behavioural or physiological differences except possibly in the house mouse (Berry, 1977), a case which is described in more detail later.

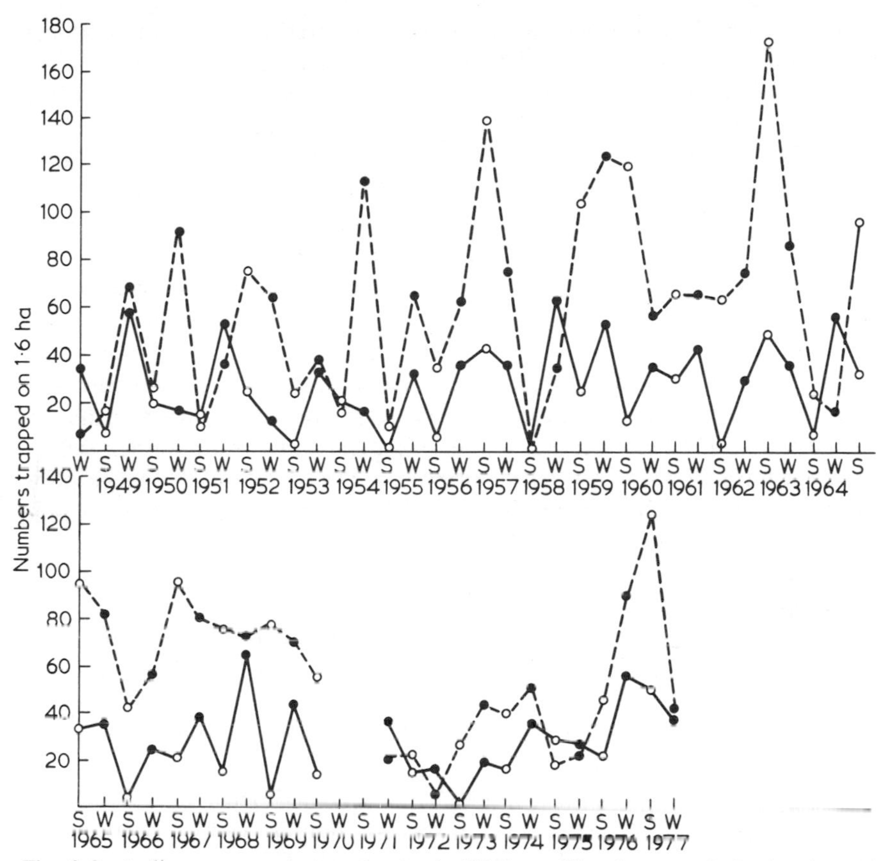

Fig. 3.2 Indices to population density in Wytham Woods, near Oxford, of wood mice (solid line) and bank vole (dashed line) from 1949 to 1977. Records are at 6-monthly intervals; open circles = summer (May-June) and solid circles = winter (Nov.-Dec.) (Southern, unpublished data.)

3.3 Case studies of change in small mammal populations

The following examples are chosen from a multitude of competitors either because they cover a long series of generations (so many studies now are limited to the three years necessary to qualify for a doctorate!) or because they are full enough to allow an illuminating analysis. Usually both these reasons have contributed to their selection.

3.3.1 Change in numbers of mice and voles in Wytham Woods over 30 years

These data were originally collected to discover the effect of rodent density on the numbers and breeding success of their main predator, the tawny owl.

Two areas, each of 0.8 ha, in the middle of extensive oak-ash-sycamore woodland on the University of Oxford's Wytham estate were trapped at intervals of six months, once in spring when numbers are usually low and once in the late autumn when numbers are at their seasonal peak. On each occasion 150 Longworth live traps were placed on the two trapping grids and left for 48 hours with the doors pinned up and whole oats in the traps ('prebaiting'). They were then set in the morning and the catch examined in the evening of the same day and again on the following morning, after which the traps were taken up. Voles caught the first evening were marked so that they could be discounted, if they turned up again the following morning. The figures given in Fig. 3.2 are simply the numbers of wood mice (solid line) and bank voles (broken line) caught at each trapping session and bear no implications about density. The research on tawny owls was discontinued after 1959 but I though it was worth carrying on this index, which involved little labour and it still continues at the present day. The employment of so high a density of traps on these relatively small areas was to ensure that no animal was deprived of the chance of being caught; thus the catch could be regarded as a true index to the numbers present on each occasion.

The results (Fig. 3.2) seem to me abundantly to justify the long period of effort. If we consider first the period from 1948 to 1970 (the gap between summer 1970 and winter 1971 was unfortunate), the sequence seems to follow a reasonably consistent pattern which can be interpreted roughly as follows.

(1) Both species fluctuate around a mean level, which is higher for bank voles than for wood mice (62 per trap session compared with 28). There is a suggestion that voles reach high peaks at intervals of 2–4 years but this can hardly be interpreted as cyclic behaviour. Rather the peaks reflect the seasonal high numbers each autumn. In any case from 1967 onwards the pattern, such as it is, fades out.

(2) Superimposed on this is an occasional reversal of the spring low-autumn high pattern and in each instance the unexpected spring high came after an abundant crop of tree seeds in the previous autumn, which extended the breeding season into the winter (Smyth, 1966).

(3) From the consistent way in which 'what goes up comes down' we may conjecture that the system is a classical one with irregular fluctuations around different means for each species and this implies regulation. Furthermore, we may recall from the work of Alibhai (1976) cited in the previous chapter that there was evidence for movement from the deciduous woodland into plantations, so the system whose operation is indicated in Figure 3.2 may be concerned with the resident section of a two-tier population.

So much for the long period of records from 1948 to 1970. If the story had ceased here, we could well have been convinced that the system would go ticking over like this for ever. There followed, however, a series of years in which both species were notably depressed until 1976 when the original system seemed to have re-asserted itself. During this time the average level around which the catches fluctuated fell to 22 per trapping session for wood mice and 31 for bank voles – not a great drop for the mice but a halving of the figure for the voles. Theory suggests that the habitat had altered somehow during this time but unfortunately changes in the habitat were not monitored, even had we known what aspects to monitor. We have just to accept the fact that after 22 years of fairly consistent results, there are still surprises in store, if we persist long enough.

3.3.2 Regulation of wood mice in Wytham woods 1948–66

A more intensive analysis of the changes in numbers of wood mice at Wytham was made by Watts (1969). This was based on 3-year periods of trapping at monthly intervals by himself and by two other students. These results were placed against the background of the data presented in 3.3.1 above over the period 1948–66.

Watts discerned some intriguing seasonal changes in numbers which can be roughly summarized as follows (the schematic diagram in Fig. 3.3 will help in comprehending these).

(1) Numbers of mice tended to level off at about the same figure each winter but the events during the rest of the year were strikingly different.

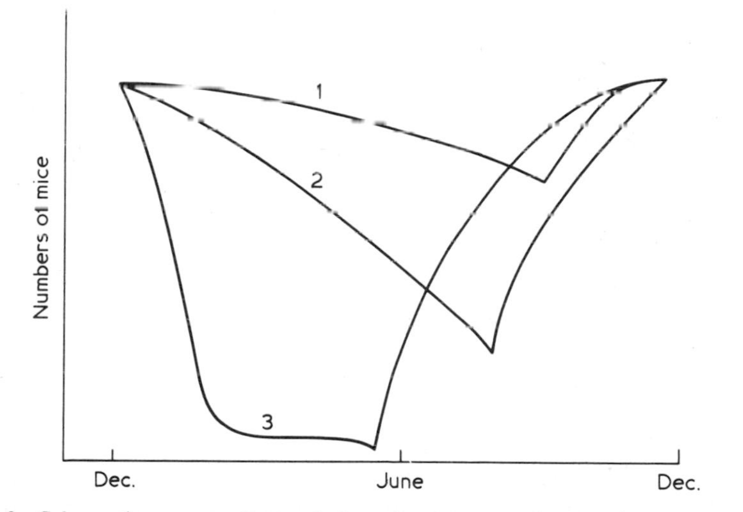

Fig. 3.3 Schematic course of population changes over the year in wood mice in Wytham Woods, near Oxford (1) when winter food is abundant; (2) when winter food is moderate; and (3) when winter food is extremely scarce. (After Watts, 1969.)

(2) In all years there was a drop in numbers during the spring. If the crop of acorns had been good, numbers declined slowly (curve 1 in Fig. 3.3) but, if the crop was poor, the spring decrease was much more emphatic (curve 2). This was the most frequent course of events. On one occasion only the spring decline was catastrophic, as illustrated by curve 3.

(3) Although, in most years, breeding started in April, there was a long pause before numbers began to climb up again. Except in the curve 3 year this recovery was postponed until the autumn.

(4) The timing of this autumn upsurge was strongly correlated with the level of the summer population (and consequently with the severity of the spring decline). The greater the numbers in summer, the later did the increase leading to the winter peak manifest itself. A comparison of curves 1 and 2 illustrates this relationship and the pattern is further confirmed in the aberrant year of curve 3 – the only year when the mice began their increase as early as June.

Watts' interpretation of these events is as follows:

(1) Survival of mice from winter to spring is linked with the availability of food, especially acorns. This is the starting point which dictates the compensatory processes which finally return the numbers of mice to a consistent winter peak. Attempts to verify this experimentally by providing food have not produced very clear cut results, though both Watts and, later, Flowerdew (1973) were able to advance the start of breeding by 2–3 weeks.

(2) The failure of the earlier part of the breeding season to promote increase can be linked with the very poor survival of the first few litters, as demonstrated by Watts' figures. He attributed this tentatively to increased aggressiveness of the overwintered males at the onset of breeding and Flowerdew (1974) has since shown, in this same general area, that, if he removed these older males, the survival of the weanlings was improved. Whether, under normal circumstances, they die or move out is not clear but it is evident that behavioural changes are here again making for a two-tiered population of residents and transients (a self-regulatory phase).

(3) Finally, there is the key question of the timing of the sudden upsurge of the population in autumn, as if some repressive factor, perhaps the waning aggressiveness of the older males or their elimination from the population, had been removed at a blow. But the timing of this event is such that it occurs later in the autumn, the higher the summer population. Thus numbers of mice level off at about the same winter peak, whatever has been the course of events during the summer. This adjustment is made clear in Watt's graph, reproduced here as Fig. 3.4, in which the data used are only from populations trapped at intervals frequent enough (monthly) to trace accurately what happened. What we see here is an example of a perfect density-dependent relationship, though what factor triggers this change from repression to an exuberant bound upwards in numbers remains a

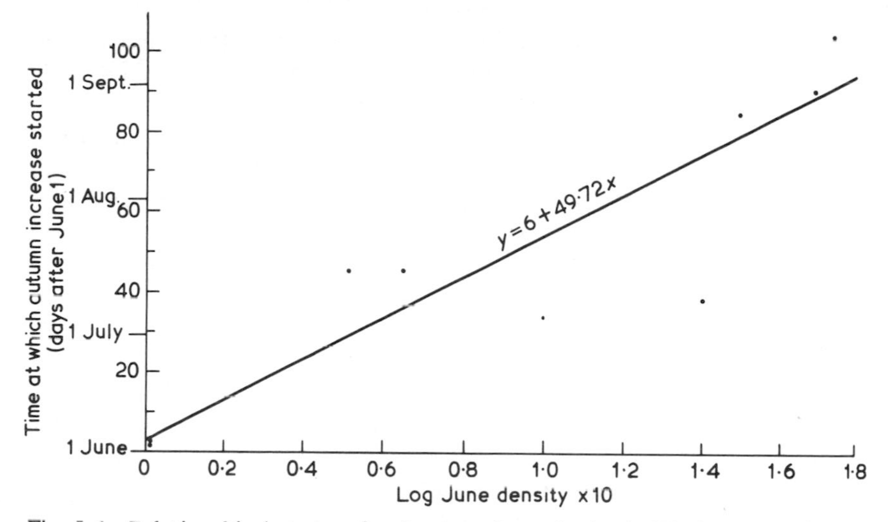

Fig. 3.4 Relationship between the density of wood mice in Wytham Woods, near Oxford, in June and the time at which the autumn increase started. (After Watts, 1969).

matter for speculation. It is possible, maybe probable, that older males are eliminated more rapidly, the smaller the summer population but this awaits verification.

The virtue of this research on the wood mouse is that it shows a complex of regulatory factors at work – the food supply dictating spring survival, intra-specific aggression maintaining low numbers in the summer and the timing of the autumn increase returning the population to a consistent winter peak.

3.3.3 The story of the house mouse (*Mus musculus*)

This hardy and ubiquitous commensal of man has attracted a long series of fruitful researches, partly because of the damage it can do to man's economy and partly because it is so numerous and has been introduced to so many and so varied habitats. During the last war, crops of grain were still stored in ricks (this was before the spread of the combine harvester) and these ricks often remained standing in the fields for long periods because of the pressure of work on threshing tackles and their crews. These ricks provided happy homes for house mice and they bred and increased very rapidly within the protection of such units of habitat. In the investigations that fell to my lot at that time, two significant factors were revealed. Firstly, these rick populations levelled off through some form of intra-specific strife while there was still plenty of food present and, secondly, that a peripheral population occupied the surrounding fields and hedgerows, bred up in summer and declined in winter (Southern and Laurie, 1946). These results foreshadowed later discoveries about self regulation in mice and their division into residents and transients.

After the war, work by Crowcroft and his colleagues in the Ministry of Agriculture, Fisheries and Food on confined populations of house mice, reported in a very diverting manner in Crowcroft (1966), established that there was a very strict dominance order among the mice and that the family group was a basic unit in their social organization. When the numbers in a group, or in a constellation of groups, grew beyond a certain point, a surplus of non-breeding animals formed what the behaviourists would call a sink with no territorial rights. The condition recalls the non-breeding flocks of Carrick's Australian magpies. Presumably, in an unenclosed population, this surplus would have emigrated and endeavoured to maintain itself in second-class habitats.

The story was developed to a further stage by several research workers but pre-eminently by Anderson (1970). He studied the organization of house mouse populations on the grain-growing areas of the Canadian prairies and found that they were divided into (a) stable, small groups living in the good cover of grain stores and (b) fluctuating large assemblages in the open country. The latter were composed of the overflow from the former, being mainly the early born young. Later born young remained in the 'establishment' and bred in the following year. The small, resident groups of mice were, under these conditions, very isolated, since there was no return of the expelled surplus, and, as a consequence, were liable to evolve genetic differences from inbreeding and genetic drift. The environment for most house mouse populations is obviously man-made but Anderson suggests that this two-tiered structure, consisting of residents and colonists and reminiscent of the contrast between r and K selection, is widespread among small rodents.

Finally, the penetrating researches by Berry (1977) and his colleagues on the house mice of Skokholm island, lying off the coast of Pembrokeshire, must be cited. The picture recalls the condition analysed by Anderson in that a population of resident groups of mice occupies the marginal cliffs of the island permanently, while, in the summer, the surplus from these spills over the central plateau of the island and is exterminated each autumn and winter. Berry has examined the incidence of variants both in skeletal characters and in blood proteins. These vary by season and by habitat, suggesting that different selective pressures are operating on the cliff residents and on the overspill on to the centre of the island at different seasons. It is clear that selection is stringent in a population that may increase eight to ten fold in the summer with a corresponding shrinkage in the winter. The results of the genetical analyses are, so far, somewhat bewildering though they do show seasonal swings which indicate differential selection between residents and colonists, even if no clear physiological connection has yet been demonstrated between genotypes and their summer and winter survival values. Berry suggests that this kind of 'endocyclic' selection (as he has labelled it) enables the house mice on Skokholm to achieve higher populations both in summer and winter than would be possible if they were genetically homogeneous. The operation of intrinsic factors is put beyond dispute in this work.

3.3.4 Cyclic species

Many species of animals fluctuate irregularly in numbers, often with such amplitude that the peaks may constitute plagues. Hewson (1976), for instance, has shown that on moorland in north-east Scotland, the mountain hare (*Lepus timidus*) may alternate between extreme scarcity and extreme abundance but without any regularity in these ups and downs; peak years recorded since 1930 were 1931, 1941, 1953, 1958 and 1971. In this fairly common situation, therefore, there is no regularity either in time or amplitude and fluctuation seems the appropriate term.

By contrast, a few species fluctuate with irregular amplitude but with the most remarkable constancy of the period between peaks, and these changes are dubbed 'cycles'. Cycles have puzzled ecologists for a great many years and have evoked a large body of research, which, however, has so far provided no complete answer to what controls this cyclic behaviour.

Cycles are by no means confined to mammals. Perhaps the most astonishing instance is that of the 17-year cicada which appears in immense numbers with exactly that periodicity. Cycles are recorded also for other insects, e.g. cockchafers, game birds, fish and for many birds and beasts of prey, though these last mainly reflect changes in numbers of their prey.

Among mammals the snowshoe rabbit or varying hare (*Lepus americanus*) cycles with a periodicity of about 10 years in the boreal forest of the New World and various voles and lemmings cycle with a periodicity of 3–4 years in the tundra zone all around the world and in some grasslands further south. These now demand attention.

The ten-year cycle

Only the one key species, *Lepus americanus*, is involved in this and the constancy of the period is, very fortunately, attested over more than a century from the fur returns of its predator, the lynx (*Lynx canadensis*), in the archives of the Hudson's Bay Company. If anybody doubts the reality of this cycle, a glance at Figure 3.5, from Elton and Nicholson (1942), will be immediately convincing. Figure 3.6 gives parallel returns for both predator and prey (1850–1900) from an area that is exceptionally well documented – stations on the north of Hudson's Bay. One of the striking

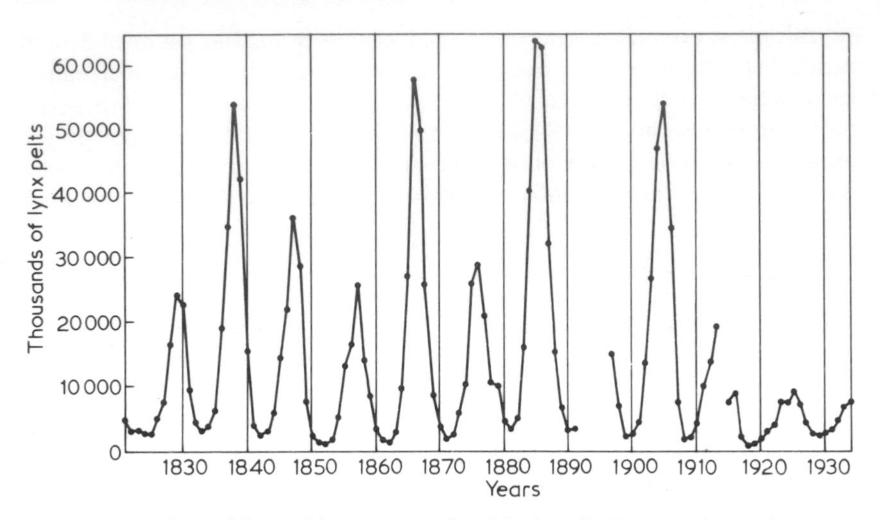

Fig. 3.5 Number of furs of lynx returned to Mackenzie River stations of Hudson's Bay Company from 1820 to 1945. (From Elton & Nicholson, 1942.)

features of these cycles is the rapidity with which the populations climb to and decline from the peak (the latter is often referred to as the 'crash') compared with the extended period of low numbers before they accelerate again to the next peak. Another feature to be noted is the synchrony of these peaks and troughs over very large areas of Canada. Finally, there are interesting anomalies: no other lagomorph in North America exhibits these regular cycles and none at all in the Old World [though there is some rather tenuous evidence that the mountain hare may cycle in Russia (Naumov, 1972)]; the regular cycles appear to be confined to the conifer forest zone in America and there are areas on the edge of the snowshoe rabbit's range where it does not cycle (Keith, 1974); on the other hand there are places where the lynx is absent, but snowshoe rabbits, nevertheless, cycle (Keith, 1963).

Over the years various explanations have been advanced of this cycle of numbers. A first glance suggested that it might be an example of the theoretical predator-prey oscillation outlined by Volterra and Lotka but the slow reaction of the predator (their impact during the phase of swift increase of the prey is inversely density-dependent and so they cannot be responsible for the crash) puts this possibility out of court. Climatic cycles, including that of sunspot activity, have been cited but eventually found not to correspond. Interaction with food supply was firmly advocated by Lack (1954) but mainly as a logical necessity without any field evidence. A change in viability – physiological stress, lowered reproductive performance, increased susceptibility to disease – due to crowding was put forward by those who favoured the operation of intrinsic factors and, indeed, Green *et al.* (1939) with their detection of hypoglycaemic shock were among the few

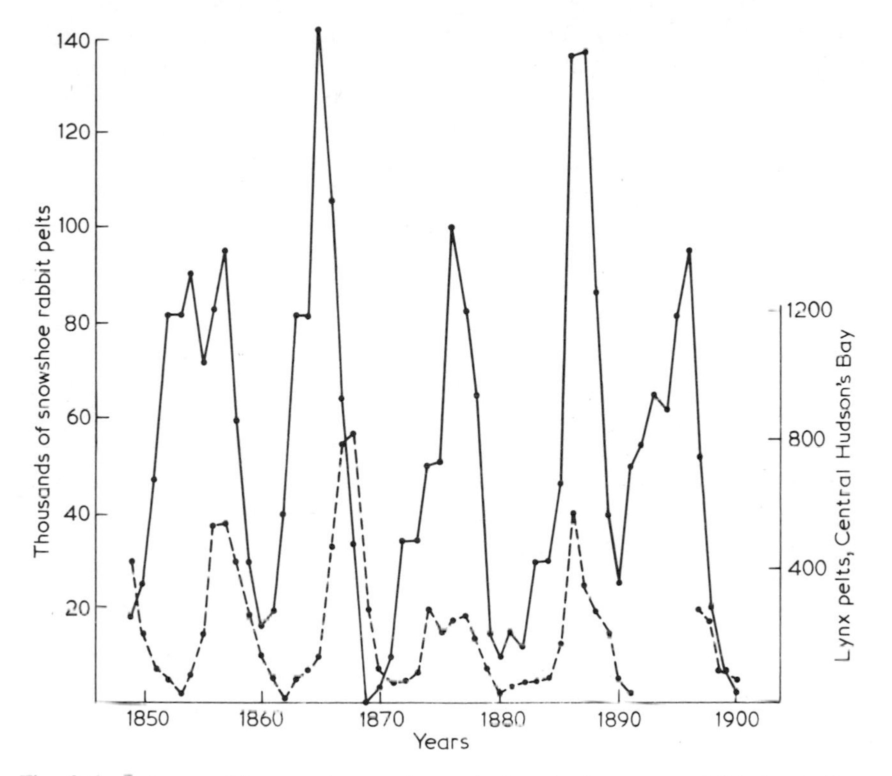

Fig. 3.6 Returns of lynx and snowshoe rabbit pelts from central Hudson's Bay, 1850–1900 (From Keith, 1963.)

who contributed real field evidence to the problem. But the difficulty remained of explaining the extreme tardiness of the recovery after numbers had been reduced to a low level by the crash.

Recent work, however, by Keith *et al.*, summarized in Keith (1974) and Keith and Windberg (1978) has thrown new light on the problem from field data. Their very detailed results were obtained from Alberta, right in the middle of the cyclic events, and combine censuses, not only of the snowshoe rabbits, but also of a group of predators concerned, lynx, coyote, red fox, horned owl and goshawk, and they show that the lag in recovery after the crash is probably due to an increasing proportion of predation, as the numbers of rabbits decline. Thus, when the rabbits are at a peak density of 7/ha predation accounts for no more than 5 per cent of their mortality, whereas, when their numbers have declined to 0.9/ha, this proportion increases to 41 per cent. This increased impact is both functional, e.g. great horned owls (*Bubo virginianus*) will step up the proportion of showshoe rabbits in their diet from 34 per cent to 81 per cent, and numerical in that the territorial pairs of owls on a study area rose from 5 to 16 and the proportion nesting from 20 per cent to 100 per cent. Similar information has now come to light about the cycle of some voles and lemmings (see later). An interesting parallel on a smaller scale is seen in the seasonally varying impact of tawny owls on wood mice and bank voles in Wytham Woods, near Oxford. During late summer and autumn, when the rodents are reaching their yearly peak of numbers, they compose only about 30–40 per cent of the owls' diet, whereas, in spring, when rodent numbers are lowest just before the breeding season, this contribution rises to about 70 per cent (Southern, 1954).

What, then, is the cause of the crash, which the predators follow up so efficiently? Keith makes the illuminating point that herbivores differ as to whether, like most rodents, they have a social organization which promotes self-regulation or whether, like lagomorphs and most ungulates, they are incapable of regulating their own numbers. In the latter condition, they will frequently come up against the exhaustion of their food supply, unless their numbers are stabilized by predators. The absence or removal of predators has frequently confirmed this and shown that the world is not always so green as Slobodkin maintained.

Keith and his co-workers, therefore, turned their attention to the food of the snowshoe rabbit (Windberg and Keith, 1976). Over the vast area of boreal forest, where the 10-year cycle is most consistent, the main food is woody shrubs and, after these have been heavily browsed, some years elapse before shoots within the reach of the rabbits can regenerate. The result is greatly lowered reproduction and very heavy winter mortality of juveniles. It seems, therefore, that Lack's prediction was partly correct.

It is possible that this condition may yield an explanation of the anomalies mentioned above. It would be interesting to know, for instance, what is the diet of the snowshoe rabbit on the edge of its range, where it does not cycle. Similarly, where the lynx is absent, yet the rabbits cycle, do other predators exert a sufficient pressure to delay the recovery of the rabbit?

The 3–4 year cycle

The cycle in numbers of microtine rodents, lemmings and voles, is a different matter and perhaps a more inscrutable one than that of the snowshoe rabbit. Its classic terrain is the tundra zone of both Old and New Worlds, though various species of voles, predominantly of the genus *Microtus*, exhibit cycles in grasslands further to the south. A salient difference between this and the 10-year cycle is that voles have the kind of social organization which can produce a two-tiered population such as I have already discussed earlier in this chapter. Thus it is unusual for them to eat out their food supply, though this does happen sometimes, as demonstrated for lemmings in Alaska by Pitelka and others. But, on the whole, the vole's world remains green, since it has a capacity for self-regulation.

Since Elton (1942) made his extensive documentation of the facts as then known about cycles, a massive amount of laborious research has been carried out, with and without experimentation, in an attempt to elucidate their nature. By and large, two schools of thought have led the attack, each concerned to establish its own point of view. It is again the dichotomy between those who maintain that extrinsic factors are responsible for the changes observed and those who favour intrinsic factors. It seems to me to be the same old argument that we have already encountered in considering general theories about the control of animal numbers and may similarly, in the end, prove to have been a storm in a teacup.

There are two main proponents of the intrinsic school of thought. Christian (1950 and other papers) maintains that as numbers of voles increase, the consequent crowding produces stress and malfunctioning of the adreno-pituitary system. Reproductive performance is lowered and mortality increased and so numbers fall again to a 'safe' level. Such a sequence of events is demonstrable in the laboratory but has not yet been pinpointed in the field.

On the other hand, D.H. Chitty, ably followed up by C.J. Krebs, after a long sequence of research, both in the field and the laboratory, during which he critically examined extrinsic factors, which might generate the vole cycle, concluded that the prime factor must be an intrinsic one and should be sought in changes of behaviour during the phases of the cycle rather than physiological disturbance. This could have a genetic basis involving a rhythmical swing of selection pressures on the allelomorphs at a locus, or what we might term pendulum polymorphism. Indeed, some biochemical variants have now been shown to change in proportion during the phases of the cycle but these have not yet been linked with changes in behaviour. The situation is reminiscent of Berry's results on house mice, cited previously. The whole long, laborious story is admirably summed up by Krebs and Myers (1974). The rationale of the behaviour hypothesis is that increasing numbers and contacts favour aggressive genotypes, whose preponderance causes widespread dispersal, increased susceptibility to factors of mortality and reduced reproduction. The consequent decline in numbers then swings back selection to favouring unaggressive genotypes and so the wheel comes full circle.

From the mass of experiments conducted or instigated by Chitty and Krebs we may highlight the following results which seem to be especially significant. (1) The phase of increase in the 3–4-year cycle is usually swift in contrast to the declining phase which can be swift but is often prolonged over one or even two years; (2) High density promotes large scale dispersal; in an enclosure experiment preventing dispersal, the voles ate out their food supply and starved; (3) High density shortens the breeding season and postpones sexual maturity in juveniles; (4) Mortality is low in the phase of increase and heavy in the decline; (5) Adults are usually larger by 20 – 50 per cent during an increase, which coincides with greater aggression.

All these results are compatible with the behavioural hypothesis advanced by Chitty and Krebs. Let us now turn to examine the results of those who favour the hypothesis that extrinsic factors, especially food supply and predation, at least play a part in the control of the 3 – 4 year cycle.

Some of the best authenticated evidence that vegetation may be seriously damaged by the peak numbers of this cycle comes from a long-term series of observations and experiments carried out at Point Barrow, Alaska, by F.A. Pitelka and his co-workers on the brown lemming (*Lemmus trimucronatus*). This and other species of lemming are remarkable in that they breed not only in the summer but also frequently beneath the snow in the depth of the Arctic winter – obviously an added threat to the sparse and fragile plant cover of the tundra. Batzli (1975) has vividly summarized the destruction that may occur at peak numbers of lemmings. Brown lemmings in these coastal regions eat about equal parts of mosses, sedges and grasses and their assimilation of these is poor. In a peak year from January to June they may eat 50 per cent of the vegetation and, if they are driven to uproot the plants to feed on the rhizomes, this figure may be increased to 90 per cent. Still further damage may result from burrowing activities. In addition to this direct damage there may be delay in the return of nutrients to the plant tissues. It is possible that the flat terrain may discourage dispersal in contrast to the classical behaviour of the European lemming (*Lemmus lemmus*) which, in the mountains of Scandinavia, disperses on a massive scale (the famous 'migrations') into habitats below the tree line.

A brief summary of the demographic changes in the brown lemming population is given by Pitelka (1973). From 1945 to 1965 numbers went up and down at 3 – 4 year intervals (125 – 250 lemmings/ha at peaks; 0.25/ha at troughs) with peaks in 1949, 1953, 1956, 1960 and 1965. From the trapping index figures given, it would appear that the decline, usually during the summer, was as abrupt as the preceding rise to the peak. The shape of the curves recalls those for the snowshoe rabbit, though, of course on a shorter time scale. Curiously, after 1965, these cycles faded out and numbers remained low until 1971 when the collared lemming (*Dicrostonyx groenlandicus*), previously an insignificant member of the fauna, suddenly achieved a peak. It is possible that a radical change in the vegetation had taken place because this species prefers dicotyledonous plants (Batzli, 1975).

So in the tender cover of tundra plants there is clear evidence that both food and cover are depleted and may cause the sharp decline of the lemmings from peak numbers. When we turn to examine the cycle in species of *Microtus* in grasslands at lower latitudes, we see a rather different picture. Measurements have been made of the amount removed by voles from the standing crop of grasses and this is usually quite small. Batzli (1975) quotes a range from 1 to 35 per cent and Krebs and Myers (1974) give instances from the literature ranging from 1 to 20 per cent and the higher figure is unusual. Here the importance of the vegetation may be for cover just as much as for the supply of food, since Birney, Grant and Baird (1976) report that, at widely distant sites in the western United States, populations of *Microtus ochrogaster* and *M. pennsylvanicus* cycle where the grasslands are dense (1.8/ha to *c.* 70/ha) but not where the sward is kept low by burning or grazing (general level of 2–5/ha).

Thus, in these temperate grasslands, the operation of the social system linked with changes of behaviour is likely to be the most potent influence in reducing peak numbers. Here we may note what seems to me to be a highly significant fact established by Chitty and Krebs on both sides of the Atlantic, namely that the decline from peak numbers of the voles may be sudden (a 'crash') but just as frequently it may take a wavering course lasting one or two years. This would hardly be expected if the cause of the decline were shortage of food. In addition, changes in behaviour would be likely to be reflected in changes in the amount of dispersal and Krebs and Myers (1974) amply document that this is so. In one study of *M. pennsylvanicus* 50–70 per cent dispersed at high densities but only 12–15 per cent at low densities. Apart from this, reproductive success was generally lower and mortality rates higher during the decline. This echoes the results that have been quoted earlier about seasonal changes in numbers and behaviour in, e.g. wood mice and house mice.

What all this amounts to is that the decline from the peak of numbers in the 3-4 year cycle may, in some habitats, be geared to the exhaustion of the food supply but, more frequently, may be a result of changes in behaviour. What, therefore, is the rôle of predators in this cycle of events?

Some years ago Pitelka *et al.* (1955) showed that avian predators, particularly the pomarine skua (*Stercorarius pomarinus*) and the snowy owl (*Nyctea scandiaca*), being highly mobile, could concentrate rapidly in areas where lemmings were abundant. Because of the necessary delay in their reproduction, however, their main impact came when lemmings were already on the decline and, again because of their mobility, they tended to move away when lemming numbers were low. The situation with resident mammalian predators, however, is different. Not only are they active under the snow during the winter but, when their prey is scarce, they must remain and redouble their hunting efforts. MacLean *et al.* (1974) found by examining winter nests of brown lemmings after the snow melt had uncovered them, that weasels (*Mustela nivalis*) had preyed upon 35 per cent of them; the same rate of predation occurred during the following winter, although lemmings were by then scarce.

This suggests that the maintenance of this rate of predation on a shrinking population of prey may be a significant factor in delaying the recovery of the prey.

If we now turn to vole populations in temperate latitudes, we find incontrovertible evidence that mammals actually increase their rate of predation as their prey diminishes in the work of Pearson (1966, 1971) and this has been further confirmed by other researches. The predators concerned were predominantly feral cats but some raccoons (*Procyon lotor*), grey foxes (*Urocyon cinereoargenteus*), striped and spotted skunks (*Mephitis mephitis* and *Spilogale putorius*) were also present. The vole was *Microtus californicus* and it reached peak numbers in 1961, 1963 and 1965. In each year, after the voles had declined, pressure of predation increased sharply: in 1961–62 Pearson estimated that as much as 88 per cent of the standing crop of voles was removed by predation; in 1963–64 and 1965–66 this figure was lower (25 per cent and 33 per cent), probably because of the elimination of some of the feral cats. In contrast, when vole numbers were increasing and at a peak, only about 5 per cent of them succumbed to predators. It seems that once a cat has 'got its eye in' for voles, it can hunt even a sparse population successfully.

Similarly, Fitzgerald (1977), by examining remains of prey in winter nests of *Microtus montanus* in the following spring estimated that 'weasels' (*Mustela erminea* and *M. frenata*) killed 20–25 prey per 100 nests in 1965–66 when voles were declining, 55/100 nests in 1966–67 when voles were at their lowest, 6/100 nests in 1967–68, when the voles were increasing and 20/100 nests in 1968–69, after the prey had crashed.

As a tailpiece to this account of predation I must mention briefly the results of Lidicker (1973) with a population of voles that had been introduced to a small island off the coast of California, which lacked predators. He found that on the usual seasonal swing of numbers there was superimposed a tendency for numbers to be higher in each alternate year. In other words, there was a two year cycle, though the amplitude was far smaller than in mainland populations.

The general conclusions must be (a) that predators have no braking effect on an expanding population of prey, and (b) that their main impact is to delay the recovery of the prey by keeping them at a lower level than they would otherwise reach.

I have dwelt on this subject of cycles at some length because it seems to me that there is now sufficient evidence to show the various ways in which they are maintained and to detect in them the same processes that influence the ups and downs of animals that do not cycle. Cycling is thus a special case of a universal system and is predominantly found in relatively simple environments and communities. I suggest that, as with non-cycling species, both extrinsic and intrinsic factors interdigitate and that the time is now past when a facile dictum like that of Pearson ('no predators, no cycles') had any meaning.

3.4 Conclusions

In fine, the evidence I have presented in this and the preceding chapter indicates that, whereas ultimate factors, such as food and other basic environmental resources, set a limit to the increase of small mammal populations, there are many proximate factors which curb increase before the limits of environmental resources are reached and not the least potent of these is the variability of the animal concerned. Concerted efforts between ecologists and geneticists open up a fruitful field of research, provided always that their attention is concentrated on what goes on in the field.

3.5 References to Chapters 2 and 3

Alibhai, S.K. (1976) *A study of factors governing the numbers and reproductive success of* Clethrionomys glareolus *(Schreber).* Unpublished D.Phil. thesis, University of Oxford.

Anderson, P.K. (1970) Ecological structure and gene flow in small mammals. *Symposia of the Zoological Society of London,* **26**, 299-325.

Andrewartha, H.G. & Birch, L.C. (1954) *The Distribution and Abundance of Animals.* Chicago: University Press.

Armitage, K.B. & Downhower, J.F. (1974) Demography of yellow-bellied marmot populations. *Ecology,* **55**, 1233-1245.

Arnold, H.R. (1978) *Provisional Atlas of the Mammals of the British Isles.* Natural Environment Research Council.

Bailey, G.N.A., Linn, I.J. & Walker, P.J. (1973) Radioactive marking of small mammals. *Mammal Review,* **3**, 11-23.

Banks, E.M., Brooks, R.J. & Schnell, J. (1975) A radiotracking study of home range and activity of the brown lemming (*Lemmus trimucronatus*). *Journal of Mammalogy,* **56**, 888–901.

Barkalow, F.S., Hamilton, R.B. & Soots, R.F. (1970) The vital statistics of an unexploited gray squirrel population. *Journal of Wildlife Management,* **34**, 489-500.

Batzli, G.O. (1975) The role of small mammals in arctic ecosystems. In *Small Mammals: their Productivity and Population Dynamics,* ed. Golley, F.B., Petrusewicz, K. & Ryszkowski, L. pp. 243-268. London: Cambridge University Press.

Batzli, G.O. & Pitelka, F.A. (1970) Influence of meadow mouse populations on California grassland. *Ecology,* **51**, 1027-1039.

Bayfield, N.G. & Hewson, R. (1975) Automatic monitoring of trail use by mountain hares. *Journal of Wildlife Management,* **39**, 214-217.

Berry, R.J. (1977) *Inheritance and Natural History.* London: Collins.

Bezem, J.J., Sluiter, J.W. & Heerdt, P.F. van (1960) Population statistics of five species of the bat genus *Myotis* and one of the genus *Rhinolophus*, hibernating in the caves of S. Limburg. *Archives néerlandaises de Zoologie,* **13**, 511-539.

Birney, E.C., Grant, W.E. & Baird, D.D. (1976) Importance of vegetative cover to cycles of *Microtus* populations. *Ecology,* **57**, 1043-1051.

Bobek, B. (1969) Survival, turnover and production of small rodents in a beech forest. *Acta theriologica,* **14**, 191-210.

Brand, C.J., Keith, L.B. & Fischer, C.A. (1976) Lynx responses to changing snowshoe hare densities in Central Alberta. *Journal of Wildlife Management,* **40**, 416-428.

Brink, F.H. van den (1967) *A Field Guide to the Mammals of Britain and Europe.* London: Collins.

Brockie, R.E. (1960) Road mortality of the hedgehog (*Erinaceus europaeus* L.) in New Zealand. *Proceedings of the Zoological Society of London,* **134**, 505-508.

Brockie, R.E. (1974) The hedgehog mange mite, *Caparinia tripilis,* in New Zealand. *New Zealand Veterinary Journal,* **22**, 243-247.

Brown, A.J. Leigh (1977) Genetic changes in a population of fieldmice (*Apodemus sylvaticus*) during one winter. *Journal of Zoology, London,* **182**, 281-289.

Campbell, P.A. (1973) The feeding behaviour of the hedgehog (*Erinaceus europaeus* L.) in pasture land in New Zealand. *Proceedings of the New Zealand Ecological Society,* **20**, 35-40.

Carrick, R. (1963) Ecological significance of territory in the Australian magpie (*Gymnorhina tibicen*). *Proceedings XIII International Ornithological Congress,* 740-753.

Caughley, G. (1977) *Analysis of Vertebrate Populations.* London: John Wiley.

Chitty, D. (1960) Population processes in the vole and their relevance to general theory. *Canadian Journal of Zoology,* **38**, 99-113.

Christian, J.J. (1950) The adreno-pituitary system and population cycles in mammals. *Journal of Mammalogy,* **31**, 247-259.

Corbet, G.B. (1964) Regional variation in the Bank vole *Clethrionomys glareolus* in the British Isles. *Proceedings of the Zoological Society of London,* **143**, 191-219.

Corbet, G.B. (1974) The distribution of mammals in historic times. In *The Changing Flora and Fauna of Britain*, ed. Hawksworth, D.L., pp. 179-202. London: Academic Press.

Corbet, G.B. & Southern, H.N. (1977) *The Handbook of British Mammals.* 2nd edn. Oxford: Blackwell Scientific Publications.

Cranford, J.A. (1977) Home range and habitat utilization by *Neotoma fuscipes* as determined by radiotelemetry. *Journal of Mammalogy,* **58**, 165-172.

Crawley, M.C. (1973) A live-trapping study of Australian brush-tailed possums, *Trichosurus vulpecula* (Kerr), in the Orongorongo Valley, Wellington, New Zealand. *Australian Journal of Zoology,* **21**, 75-90.

Crowcroft, W.P. (1966) *Mice All Over.* London: Foulis.

Darwin, C. (1886) *The Origin of Species.* 6th edn. London: John Murray

Dollbeer, R.A. & Clark, W.R. (1975) Population ecology of snowshoe hares in the central Rocky Mountains. *Journal of Wildlife Management,* **39**, 535-549.

Dumitresco, M. & Orghidan, T. (1963) Contribution à la connaissance de la biologie de *Pipistrellus pipistrellus* Schreber. *Annales de Spéléologie,* **18**, 511-517.

Dunnet, G.M. (1964) A field study of local populations of the brush-tailed possum *Trichosurus vulpecula* in Eastern Australia. *Proceedings of the Zoological Society of London,* **142**, 665-695.

Ellerman, J.R. & Morrison-Scott, T.C.R. (1951) *Checklist of Palaearctic and Indian Mammals.* London: British Museum (Natural History).

Elton, C. (1927) *Animal Ecology.* London: Sidgwick & Jackson.

Elton, C. (1942) *Voles, Mice and Lemmings.* Oxford: Clarendon Press.

Elton, C.S. (1953) The use of cats in farm rat control. *British Journal of Animal Behaviour,* **1**, 151-155.

Elton, C.S. (1958) *The Ecology of Invasions by Animals and Plants.* London: Methuen.

Elton, C., Davis, D.H.S. & Findlay, G.M. (1935) An epidemic among voles (*Microtus agrestis*) on the Scottish Border in the spring of 1934. *Journal of Animal Ecology,* **4**, 277-288.
Elton, C. & Nicholson, M. (1942) The ten-year cycle of the lynx in Canada. *Journal of Animal Ecology,* **11**, 215-244.
Fager, E.W. (1965) Communities of organisms. In *The Sea*, ed. Hill, M.N., pp. 415-37. New York: John Wiley.
Fairley, J.S. (1971) The present distribution of the bank vole *Clethrionomys glareolus* Schreber in Ireland. *Proceedings of the Royal Irish Academy,* **71B**, 183-189.
Fairley, J.S. & Comerton, J.E. (1972) An early-breeding population of fieldmice *Apodemus sylvaticus* (L.) in Limekiln Wood, Athenry, Co. Galway. *Proceedings of the Royal Irish Academy,* **72B**, 149-163.
Fargher, S.E. (1977) The distribution of the Brown hare (*Lepus capensis*) and the Mountain hare (*Lepus timidus*) in the Isle of Man. *Journal of Zoology, London,* **182**, 164-167.
Fitzgerald, B.M. (1977) Weasel predation on a cyclic population of the montane vole (*Microtus montanus*) in California. *Journal of Animal Ecology,* **46**, 367-397.
Fleming, T.H. (1973) Numbers of mammal species in north and central American forest communities. *Ecology,* **54**, 555-563.
Fleming, T.H. (1974) The population ecology of two species of Costa Rica heteromyid rodents. *Ecology,* **55**, 493-510.
Flowerdew, J.R. (1973) The effect of natural and artificial changes in food supply on breeding in woodland mice and voles. *Journal of Reproduction and Fertility, Supplement,* **19**, 257-267.
Flowerdew, J.R. (1974) Field and laboratory experiments on the social behaviour and population dynamics of the wood mouse (*Apodemus sylvaticus*). *Journal of Animal Ecology,* **43**, 499-511.
Flowerdew, J.R. (1976) Ecological methods. *Mammal Review,* **6**, 123-160.
Flux, J.E.C. (1970) Life history of the Mountain hare (*Lepus timidus scoticus*) in north-east Scotland. *Journal of Zoology, London,* **161**, 75-123.
Ford, E.B. (1971) *Ecological Genetics*. 3rd edn. London: Chapman and Hall.
Forestry Commission (1953) The campaign against the grey squirrel. Leaflet.
Gaines, M.S. & Krebs, C.J. (1971) Genetic changes in fluctuating vole populations. *Evolution,* **25**, 702-723.
Gerell, R. (1967) Food selection in relation to habitat in mink (*Mustela vison* Schreber) in Sweden. *Oikos,* **18**, 233-246.
Gibb, J.A. & Flux, J.E.C. (1973) Mammals. In *The Natural History of New Zealand*, ed. Williams, G.R., pp. 334-371. Wellington: A.H. & A.W. Reed.
Gibb, J.A., Ward, G.D. & Ward, C.P. (1969) An experiment in the control of a sparse population of wild rabbits (*Oryctolagus c. cuniculus* L.) in New Zealand. *New Zealand Journal of Science,* **12**, 509-534.
Godfrey, G.K. (1954) Tracing field voles (*Microtus agrestis*) with a Geiger-Müller counter. *Ecology,* **35**, 5-10.
Godfrey, G.K. (1978) The ecological distribution of shrews (*Crocidura suaveolens* and *Sorex araneus fretalis*) in Jersey. *Journal of Zoology, London,* **185**, 266-270.
Grant, P.R. (1972) Interspecific competition among rodents. *Annual Review of Ecology and Systematics,* **3**, 79-106.
Green, R.G., Larson, C.L. & Bell, J.F. (1939) Shock disease as the cause of the periodic decimation of the snowshoe hare. *American Journal of Hygiene,* **20B**, 83-102.

Guthrie-Smith, H. (1953) *Tutira, the Story of a New Zealand Sheep Station*. 3rd edition. Edinburgh: William Blackwood.

Haeck, J. (1969) Colonization of the mole (*Talpa europaea* L.) in the Ijsselmeer-polders. *Netherlands Journal of Zoology,* **19**, 145-248.

Hansson, L. (1968) Population densities of small mammals in open field habitats in south Sweden. 1964-1967. *Oikos,* **18**, 53-60.

Hayne, D.W. (1949) Two methods for estimating populations from trapping records. *Journal of Mammalogy,* **30**, 399-411.

Hewson, R. (1965) Population changes in the mountain hare, *Lepus timidus* L. *Journal of Animal Ecology,* **34**, 587-600.

Hewson, R. (1976) A population study of mountain hares (*Lepus timidus*) in north-east Scotland from 1956-1969. *Journal of Animal Ecology,* **45**, 395-414.

Holling, C.S. (1959) The components of predation as revealed by a study of small-mammal predation of the European pine sawfly. *Canadian Entomologist,* **91**, 293-320.

Hutchinson, G.E. (1959) Homage to Santa Rosalina, or why are there so many kinds of animals? *American Naturalist,* **93**, 145-159.

Kalela, O. (1949) Changes in geographical ranges in the avifauna of northern and central Europe in relation to recent changes in climate. *Bird Banding,* **20**, 77-103.

Keith, L.B. (1963) *Wildlife's Ten-year Cycle*. Madison: University of Wisconsin Press.

Keith, L.B. (1974) Some features of population dynamics in mammals. *Transactions of the XI Congress of the International Union of Game Biologists* (Stockholm), pp. 17-58.

Keith, L.B. & Windberg, L.A. (1978) A demographic analysis of the snowshoe hare cycle. *Wildlife monographs,* no. 58.

Kikkawa, J. (1964) Movement, activity and distribution of the small rodents *Clethrionomys glareolus* and *Apodemus sylvaticus* in woodland. *Journal of Animal Ecology,* **33**, 259-299.

King, C.M. (1971) *Studies on the ecology of the weasel,* Mustela nivalis *L.* Unpublished D.Phil. thesis, University of Oxford.

King, C.M. (1975) The home range of the weasel (*Mustela nivalis*) in an English woodland. *Journal of Animal Ecology,* **44**, 639-669.

Kozakiewicz, M. (1976) Migratory tendencies in population of bank voles and description of migrants. *Acta theriologica,* **21**, 321-338.

Krebs, C.J. (1972) *Ecology*. New York: Harper & Row.

Krebs, C.J. & Myers, J.H. (1974) Population cycles in small mammals. *Advances in Ecological Research,* **8**, 267-399.

Kurtén, B. (1968) *Pleistocene Mammals of Europe*. London: Weidenfeld & Nicholson.

Lack, D. (1944) Ecological aspects of species-formation in passerine birds. *Ibis,* **86**, 260-286.

Lack, D. (1947) *Darwin's Finches*. Cambridge: University Press.

Lack, D. (1954) *The Natural Regulation of Animal Numbers*. Oxford: Clarendon Press.

Lack, D. (1966) *Population Studies of Birds*. Oxford: Clarendon Press.

Lack, D. (1969) The numbers of bird species on islands. *Bird Study,* **16**,193-209.

Lack, D. (1971) *Ecological Isolation in Birds*. Oxford: Blackwell Scientific Publications.

Lack, D. (1976) *Island Biology*. Oxford: Blackwell Scientific Publications.

Langley, P.J.W. & Yalden, D.W. (1977) The decline of the rarer carnivores in Great Britain during the nineteenth century. *Mammal Review,* **7**, 95-116.

Langley, P.J.W. & Yalden, D.W. (1977) The decline of the rarer carnivores in Great Britain during the nineteenth century. *Mammal Review,* 7, 95-116.

Larkin, P.A. (1948) *Ecology of mole* (Talpa europaea *L.*) *populations.* Unpublished D.Phil. thesis, University of Oxford.

Leslie, P.H. & Ranson, R.M. (1940) The mortality, fertility and rate of natural increase of the vole (*Microtus agrestis*), as observed in the laboratory. *Journal of Animal Ecology,* **9**, 27-52.

Lidicker, W.Z. (1973) Regulation of numbers in an island population of the California vole, a problem in community dynamics. *Ecological Monographs,* **43**, 271-302.

Lloyd, H.G. (1962) The distribution of squirrels in England and Wales, 1959. *Journal of Animal Ecology,* **31**, 157-165.

Lockie, J.D. (1966) Territory in small carnivores. *Symposia of the Zoological Society of London,* **18**, 143-165.

Lord, R.D., Muradali, F. & Lazaro, L. (1976) Age composition of vampire bats (*Desmodus rotundus*) in northern Argentina and southern Brazil. *Journal of Mammalogy,* **57**, 573-575.

Lowe, V.P.W. (1969) Population dynamics of the red deer (*Cervus elaphus* L.) on Rhum. *Journal of Animal Ecology,* **38**, 425-457.

MacArthur, R. (1958) Population ecology of some warblers of northeastern coniferous forests. *Ecology,* **39**, 599 619.

MacArthur, R. & Wilson, E.O. (1967) *The Theory of Island Biogeography.* Princeton: University Press.

Macdonald, D.W. (1977) On food preference in the Red fox. *Mammal Review,* **7**, 7-23.

Macdonald, D.W. (1978) *The behavioural ecology of the red fox,* Vulpes vulpes: *a study of social organization and resource exploitation.* Unpublished D.Phil. thesis, University of Oxford.

Mackinnon, K.S. (1976) *Home range, feeding ecology and social behaviour of the grey squirrel* (Sciurus carolinensis *Gmelin*). Unpublished D.Phil. thesis, University of Oxford.

MacLean, S.F., Fitzgerald, B.M. & Pitelka, F.A. (1974) Population cycles in Arctic lemmings: winter reproduction and predation by weasels. *Arctic & Alpine Research,* **6**, 1-12.

Macpherson, A.H. (1964) A northward range extension of the red fox in the eastern Canadian Arctic. *Journal of Mammalogy,* **45**, 138-140.

Mayr, E. & Amadon, D. (1951) A classification of recent birds. *American Museum Novitates* no. 1496.

Michener, G.R. (1973) Intraspecific aggression and social organization in ground squirrels. *Journal of Mammalogy,* **54**, 1001-1003.

Michielson, N. Croin (1966) Intraspecific and interspecific competition in the shrews *Sorex araneus* L. and *S. minutus* L. *Archives néerlandaises de Zoologie,* **17**, 73-174.

Miller, R.S. (1964) Ecology and distribution of pocket gophers (Geomyidae) in Colorado. *Ecology,* **45**, 256-272.

Miller, D.H. & Getz, L.L. (1977) Comparisons of population dynamics of *Peromyscus* and *Clethrionomys* in New England. *Journal of Mammalogy,* **58**, 1-16.

Montgomery, W.I. (1976) On the relationship between Yellownecked mouse (*Apodemus flavicollis*) and Woodmouse (*A. sylvaticus*) in a Cotswold valley. *Journal of Zoology, London,* **179**, 229-233.

Moors, P.J. (1974) *The annual energy budget of a weasel* (Mustela nivalis *L.*) population in farmland. Unpublished Ph.D. thesis, University of Aberdeen.

Moors, P.J. (1975) The food of weasels (*Mustela nivalis*) on farmland in north-east Scotland. *Journal of Zoology, London,* **177**, 455-461.
Morris, D. (1965) *The Mammals.* London: Hodder & Stoughton.
Morris, P. (1972) A review of mammalian age determination methods. *Mammal Review,* **2**, 69-104.
Myers, K. & Poole, W.E. (1959) A study of the biology of the wild rabbit, *Oryctolagus cuniculus* (L.), in confined populations. I. The effects of density on home range and the formation of breeding groups. *C.S.I.R.O. Wildlife Research,* **4**, 14-26.
Myllymäki, A (1970) Population ecology and its application to the control of the field vole, *Microtus agrestis. European and Mediterranean Plant Protection Organization. Publications, Series A,* **58**, 27-84.
Myrberget, S. (1967) Den norske bestand au Bever. *Castor fiber. Meddelelser fra Statens Viltundersøkelser, Serie* 2, nr 26. 1-40.
Naumov, N.P. (1972) *The Ecology of Animals.* Urbana: University of Illinois Press.
Nixon, C.M., McClain, M.W. & Donohoe, R.W. (1975) Effects of hunting and mast crops on a squirrel population. *Journal of Wildlife Management,* **39**, 1-25.
Odum, E.P. (1971) *Fundamentals of Ecology.* 3rd edition. Philadelphia: W.B. Saunders.
Pearl, R. (1928) *The Rate of Living.* New York: Knopf.
Pearson, O.P. (1959) A traffic survey of *Microtus-Rheithrodontomys* runways. *Journal of Mammalogy,* **40**, 169-180.
Pearson, O.P. (1966) The prey of carnivores during one cycle of mouse abundance. *Journal of Animal Ecology,* **35**, 217-233.
Pearson, O.P. (1971) Additional measurements of the impact of carnivores on California voles (*Microtus californicus*). *Journal of Mammalogy,* **52**, 41-49.
Pernetta, J.C. (1976a) Diets of the shrews *Sorex araneus* L. and *Sorex minutus* L. in Wytham grassland. *Journal of Animal Ecology,* **45**, 899-912.
Pernetta, J.C. (1976b) Bioenergetics of British shrews in grassland. *Acta theriologica,* **21**, 481-497.
Pernetta, J.C. & Watling, D. (1978) The introduced and native terrestrial vertebrates of Fiji. *Pacific Science* (in press).
Petrusewicz, K. (1970) Dynamics and production of the hare population in Poland. *Acta theriologica,* **15**, 413-445.
Phillips, W.M. (1955) The effect of commercial trapping on rabbit populations. *Annals of Applied Biology,* **43**, 247-257.
Pimentel, D. (1968) Population regulation and genetic feedback. *Science,* **159**, 1432-1437.
Pitelka, F.A. (1973) Cyclic pattern in lemming populations near Barrow, Alaska. In *Alaskan Arctic Tundra*, ed. Britton, M.E. AINA Technical Paper no. 25, pp. 199-215.
Pitelka, F.A., Tomich, P.Q. & Treichel, G.W. (1955) Ecological relations of jaegers and owls and lemming predators near Barrow, Alaska. *Ecological Monographs,* **25**, 85-117.
Racey, P.A. & Stebbings, R.E. (1972) Bats in Britain – a status report. *Oryx,* **11**, 319-327.
Randolph, S.E. (1973) A tracking technique for comparing individual home ranges of small mammals. *Journal of Zoology, London,* **170**, 509-520.
Randolph, S. (1977) Changing spatial relationships in a population of *Apodemus sylvaticus* with the onset of breeding. *Journal of Animal Ecology,* **46**, 653-676.

Rasmussen, D.I. (1970) Biochemical polymorphisms and genetic structure in populations of *Peromyscus. Symposia of the Zoological Society of London,* **26**, 335-349.

Richards, O.W. & Southwood, T.R.E. (1968) The abundance of insects: introduction. In *Insect Abundance*, ed. Southwood, T.R.W., pp. 1-7. Oxford: Blackwell Scientific Publications.

Schwarz, E. & Schwarz, H.K. (1943) The wild and commensal stocks of the house mouse, *Mus musculus* Linnaeus. *Journal of Mammalogy,* **24**, 59-72.

Semeonoff, R. & Robertson, F.W. (1968) A biochemical and ecological study of plasma esterase polymorphism in natural populations of the field vole, *Microtus agrestis* L. *Biochemical Genetics,* **1**, 205-227.

Shillito, J.F. (1963) Field observations on the growth, reproduction and activity of a woodland population of the common shrew *Sorex araneus* L. *Proceedings of the Zoological Society of London,* **140**, 99-114.

Shorten, M. (1957) Squirrels in England, Wales and Scotland, 1955. *Journal of Animal Ecology,* **26**, 287-294.

Shorten, M. (1962) Squirrels. *Ministry of Agriculture, Fisheries and Food, Bulletin no. 184.* London, H.M.S.O.

Shorten, M. & Courtier, F.A. (1955) A population study of the grey squirrel (*Sciurus carolinensis*) in May 1954. *Annals of Applied Biology,* **43**, 494-510.

Simpson, G.G. (1945) The principles of classification and a classification of mammals. *Bulletin of the American Museum of Natural History,* **85**, 1-350.

Skinner, J.O. (1905) The house sparrow (*Passer domesticus*). Report of the Smithsonian Institute, 1904, pp. 423-428.

Smith, A.T. (1974) The distribution and dispersal of pikas: consequences of insular population structure. *Ecology,* **55**, 1112-1119.

Smyth, M. (1966) Winter breeding in woodland mice, *Apodemus sylvaticus,* and voles, *Clethrionomys glareolus* and *Microtus agrestis,* near Oxford. *Journal of Animal Ecology,* **35**, 471-485.

Solomon, M.E. (1964) Analysis of processes involved in the natural control of insects. *Advances in Ecological Research,* **2**, 1-58.

Solomon, M.E. (1969) *Population Dynamics.* London: Arnold.

Southern, H.N. (1940) The ecology and population dynamics of the wild rabbit, *Oryctolagus cuniculus* (L.). *Annals of Applied Biology,* **27**, 509-526.

Southern, H.N. (1954) Tawny owls and their prey. *Ibis,* **96**, 384-410.

Southern, H.N. (1965) The trap-line index to small mammal populations. *Journal of Zoology, London,* **147**, 216-238.

Southern, H.N. (1970) The natural control of a population of Tawny owls (*Strix aluco*). *Journal of Zoology, London,* **162**, 197-285.

Southern, H.N. (1973) A yardstick for measuring populations of small rodents. *Mammal Review,* **3**, 1-10.

Southern, H.N. & Laurie, E.M.O. (1946) The house mouse (*Mus musculus*) in corn ricks. *Journal of Animal Ecology,* **15**, 134-149.

Southern, H.N. & Lowe, V.P.W. (1968) The pattern of distribution of prey and predation in tawny owl territories. *Journal of Animal Ecology,* **37**, 75-97.

Southwick, C.H. (1956) The abundance and distribution of harvest mice (*Micromys minutus*) in corn ricks near Oxford. *Proceedings of the Zoological Society of London,* **126**, 449-452.

Stebbings, R.E. (1966) A population study of bats of the genus *Plecotus. Journal of Zoology, London,* **150**, 53-75.

Stoddart, D.M. (1970) Individual range, dispersion and dispersal in a population of water voles (*Arvicola terrestris* (L.)). *Journal of Animal Ecology,* **39**, 403-425.

Stoddart, D.M. (1971) Breeding and survival in a population of water voles. *Journal of Animal Ecology,* **40**, 487-494.
Stout, I.J. & Sonenshine, D.E. (1974) Ecology of an opossum population in Virginia, 1963-69. *Acta theriologica,* **19**, 235-245.
Tamarin, R.H. & Krebs, C.J. (1969) *Microtus* population biology. II. Genetic changes at the transferrin locus in fluctuating populations of two vole species. *Evolution,* **23**, 188-211.
Tapper, S.C. (1976) Population fluctuations of the Field vole (*Microtus*): a background to the problems involved in predicting vole plagues. *Mammal Review,* **6**, 93-117.
Taylor, J.C. (1970) The influence of arboreal rodents on their habitat and man. *European and Mediterranean Plant Protection Organization Publications,* Series A, No. **58**, 217-223.
Trent, T.T. & Rongstad, O.J. (1974) Home range and survival of cottontail rabbits in southwestern Wisconsin. *Journal of Wildlife Management,* **38**, 459-472.
Twigg, G.I. (1975) Marking mammals. *Mammal Review,* **5**, 101-116.
Üttendörfer, O. (1939) *Die Ernährung der deutschen Raubvögel und Eulen, und ihr Bedeutung in der heimischen Natur.* Berlin: J. Neumann-Neudamm.
Varley, G.C., Gradwell, G.R. & Hassell, M.P. (1973) *Insect Population Ecology: an Analytical Approach.* Oxford: Blackwell Scientific Publications.
Vaughan, T.A. & Hanson, R.M. (1964) Experiments on interspecific competition between two species of Pocket Gophers. *American Midland Naturalist,* **72**, 444-452.
Wallace, A.R. (1902) *Island Life.* 3rd edn. London: Macmillan.
Watson, A. & Hewson, R. (1973) Population densities of Mountain hares (*Lepus timidus*) on western Scottish and Irish moors and on Scottish hills. *Journal of Zoology, London,* **170**, 151-159.
Watts, C.H.S. (1969) The regulation of wood mouse (*Apodemus sylvaticus*) numbers in Wytham Woods, Berkshire. *Journal of Animal Ecology,* **38**, 285-304.
Whittaker, J.B. (1971) Population changes in *Neophilaenus lineatus* (L.) (Homoptera: Cercopidae) in different parts of its range. *Journal of Animal Ecology,* **40**, 425-443.
Wijngaarden, A. van & Bruijns, M.F. Mörzer (1961) De Hermelijnen, *Mustela erminea* L. van Terschelling. *Lutra,* **3**, 35-42.
Windberg, L.A. & Keith, L.B. (1976) Snowshoe hare population response to artificial high densities. *Journal of Mammalogy,* **57**, 523-553.
Wodzicki, K.A. (1950) *Introduced Mammals of New Zealand.* Wellington: Department of Scientific and Industrial Research.
Wodzicki, K. & Roberts, H.S. (1960) Influence of population density on the adrenal and body weights of the wild rabbit *Oryctolagus cuniculus* L. in New Zealand. *New Zealand Journal of Science,* **3**, 103-120.
Wynne-Edwards, V.C. (1962) *Animal Dispersion in Relation to Social Behaviour.* Edinburgh: Oliver & Boyd.

Community structure and functional role of small mammals in ecosystems

4

G.F. HAYWARD
AND J. PHILLIPSON

4.1 Introduction

The title of this chapter suggests that small mammals, however defined, constitute a single component or state variable in each of a number of multi-component systems. Further, for those systems in which small mammals are identified as a single component the title implies that interactions occur between the 'small mammal' state variable and others i.e. functional relationships exist between the components of the system (system variables). In such circumstances it is generally assumed that any change in one state variable produces corresponding changes in others via functional relationships (transfer functions). On the basis of the above interpretation the major aims of this chapter are clear and comprise (i) characterization of the state variable 'small mammals' within ecosystems and (ii) evaluation of the transfer functions which link the component 'small mammals' to other system variables.

4.2 Characterization of 'small mammals' as a single state variable

The phrase 'small mammals' does not constitute a clearly defined taxonomic entity and is generally considered to refer to any assemblage of mammal species whose individual live weights do not exceed 5 kg when adult. A definition as broad as this means that members of the mammalian orders Insectivora, Rodentia, Lagomorpha, Carnivora, and Chiroptera can all be grouped under the single state variable 'small mammals'. Consideration of the general trophic status of these groups makes clear that it is not necessarily helpful, from the viewpoint of elucidating the functional role of small mammals in ecosystems, to group such mammals into one state variable on the basis of size-weight categories alone. Nevertheless, in certain circumstances, such grouping is permissible; a subject area under investigation could well be the rate of respiratory energy loss by 'small mammals'. In an ecosystem context the transfer function of interest would be that between energy input from food and respiratory energy output per unit area per unit time, a situation that could be adequately illustrated by reference to the pictorial model shown in Fig. 4.1. Although simple in concept, accurate

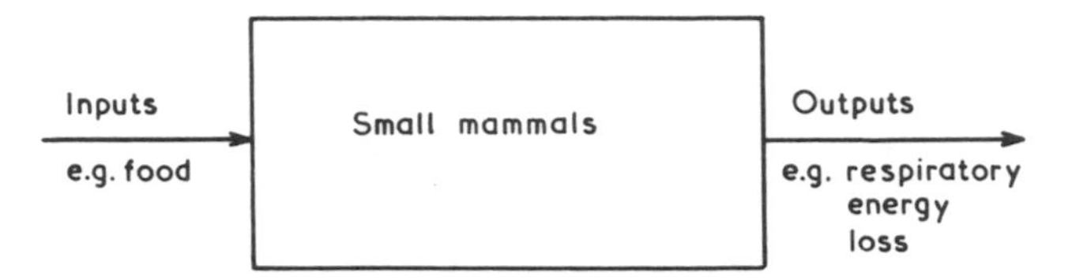

Fig. 4.1 Extreme form of a holological model.

quantification of this model for a specific ecosystem would require detailed knowledge over time of (1) small mammal species composition, (2) age or size class distribution of each species, (3) food habits and food intake by each age/size class, (4) assimilation efficiencies of each age/size class, (5) energy equivalents of the food materials assimilated by each age/size class, (6) energy equivalents of the respiratory outputs of each age/size class. Given such information one might well argue that the simple synthetic model of Fig. 4.1 should be expanded in such a manner that for all species each age/size class is represented as a state variable. Figure 4.2 illustrates such an expansion for a hypothetical system containing one species of herbivorous small mammal and one species of carnivorous small mammal which depends wholly on the herbivore as a food source. The two models outlined were chosen to represent extremes of a spectrum of models that could be constructed, Fig. 4.3 for example illustrates an intermediate stage.

The fact that a series of models can be constructed to represent a 'single' functional relationship between small mammals and their food supply highlights the problem of structuring a chapter on the functional role (a synthesis of all functional relationships?) of small mammals in ecosystems. From the examples given it is obvious that in the preparation of this account we were faced with a choice of adopting either a holological or a merological approach. The holological approach, as exemplified by Fig. 4.1, refers to a situation where the 'small mammal' component is treated as a black box, that is as a single state variable, the functioning of which is evaluated without reference to its parts. In contrast the merological approach, illustrated in Fig. 4.2, is such that parts of the earlier identified 'small mammal' component become state variables in their own right, these are dealt with in such a way that a picture of the functioning whole is built up from them.

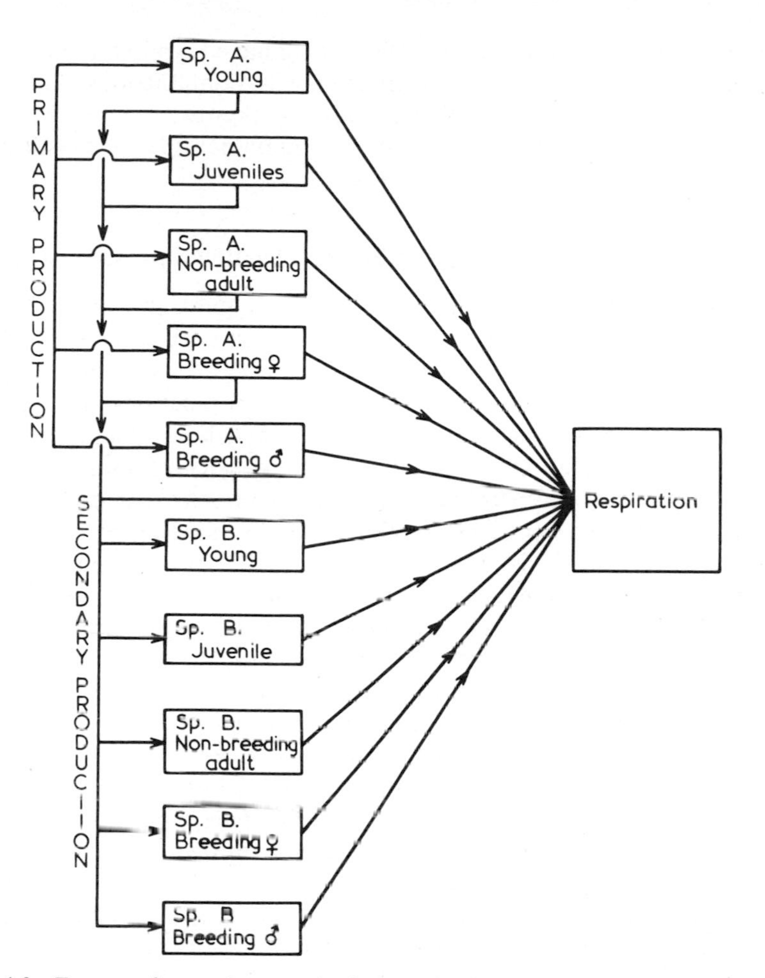

Fig. 4.2 Extreme form of a merological model for a two small mammal species community.

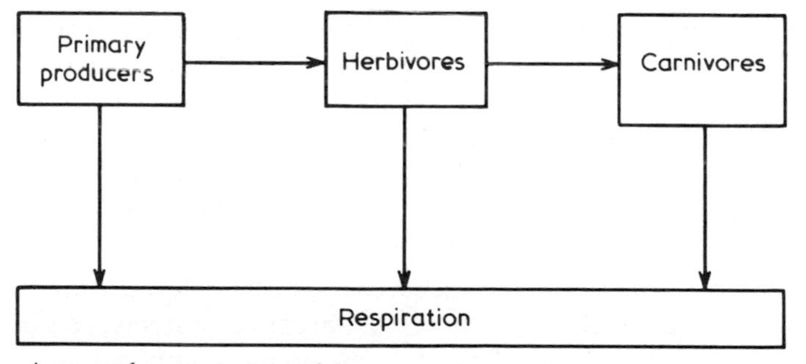

Fig. 4.3 A general ecosystem model.

Insofar as current knowledge has allowed we have favoured a merological approach for we believe that it is from the nexus of interactions so identified that the functional role of small mammals in ecosystems can best be evaluated. For the purposes of this account small mammals ideally constitute not one state variable but many.

4.3 Characterization of small mammals as different state variables

In an extreme form the merological approach requires that each and every life stage of all species be assigned the status of a state variable. Additionally, all transfer functions attributable to these system variables should be identified and evaluated. Studies of single populations are generally amenable to such treatment but in ecosystem modelling a major obstacle to progress at the same level is the sheer complexity of most ecosystems (Mann, 1969). For this account, and with our present state of knowledge, adoption of the extreme merological approach is impossible; indeed, in an ecosystem context it is arguably undesirable. Apart from the sheer complexity, and consequent production of models of unmanageable proportions, too much detail frequently obscures the identification of functional relationships; by contrast over-simplification due to excessive grouping of data leads to the loss of useful information. For all practical purposes some intermediate level of approach is desirable.

In this account the level of approach has been set by the available information; we have variously allocated small mammals to state variables according to a variety of criteria e.g. size, taxonomic position and trophic status. We make no apologies for the uneven treatment and trust that it will focus the reader's attention on the similarities and differences between the selected state variables and transfer functions in terms of (1) their structure, (2) their functional roles and (3) our state of knowledge about them.

4.4 The structure of small mammal faunas

There are many types of ecosystem and some knowledge of the diversity to be found in the structure of their small mammal faunas is essential before state variables can be defined and meaningful statements about transfer functions can be made. Such knowledge is particularly important for comparative purposes, especially in view of the general assumption that any change in one state variable produces corresponding changes in other via the functional relationships.

Features of faunal structure believed to be of relevance to community functioning, and which vary between different ecosystems, can be categorized under four major headings (1) taxonomic composition, (2) species diversity, (3) relative abundance and (4) biomass and density.

4.4.1 Taxonomic composition

On a world scale the taxonomic composition of small mammal faunas varies between zoogeographical regions. Among the non-chiropteran small mammal faunas the Rodentia and Insectivora are the most important groups in terms of numbers of species in the Ethiopian, Oriental, Nearctic and Palaearctic regions. In the Neotropical and Australasian regions the Rodentia and Marsupialia are numerically the most important. Because of the paucity of data on feeding habits, especially among tropical taxa, the extent to which these differences in taxonomic composition reflect differences in trophic structure is difficult to ascertain. It could be postulated that with additional knowledge about small mammal faunas few trophic differences of significance to the study of the functional role of small mammals in ecosystems will be found. The evidence available in support of such a postulate is slight, one example includes the marsupial Didelphidae which occupy the insectivore niche (sensu Elton, 1927) of lowland Neotropical communities; a niche occupied elsewhere by the Insectivora. A further example is Fleming's (1973a) analysis of the non-chiropteran small mammal faunas of Malaysia and Panama where few major trophic differences were discerned. To counter these examples are the suggestions by Wilson (1973) and Fleming (1973a) that there are important taxonomic and trophic differences between tropical bat faunas.

By comparison with other small mammals little is known about the trophic requirements of Chiroptera but in general insectivorous bats are probably trophically important in all zoogeographical regions. Frugivorous species are only of importance trophically in the Neotropical and Australasian regions, while nectarivorous forms assume some significance in Neotropical areas (Fleming, 1975).

In terms of ecosystem modelling the validity of using taxonomic criteria to define system variables at the level of small mammal major taxa must remain questionable until such time that additional information about food and feeding habits becomes available.

4.4.2 Species diversity

From a modelling viewpoint, the greater the number of small mammal species in an ecosystem the more likely it is that species will be grouped to reduce complexity in the number of system variables. As was indicated in section 4.3 oversimplification due to excessive grouping of data leads to a loss of useful information, it follows that differential species richness or diversity in ecosystems could affect analyses of the functional role of their small mammal components. The nature of the problem can best be recognized through consideration of small mammal species richness in different ecosystems.

Table 4.1 *Number of small mammal species per habitat in the northern hemisphere.*

Habitat	Location and elevation	No. of species	Source
Tundra plant formation	(1) Seward Peninsula, Alaska (65°N; *c.* 250 m)	12	Quay (1951)
White spruce-paper birch formation	(2) Fairbanks, Alaska (65°N; *ca.* 250 m)	9	Dice (1920a)
Mixed deciduous woodland	(3) Wytham Woods, Oxford (50°N; *c.* 130 m)	23	Unpubl.
Virgin hemlock-hardwood forest	(4) Charlevoix Co., Michigan (45°N; *ca.* 300 m)	23	Dice (1925) Burt (1954)
Deciduous beech-maple forest	(5) Ann Arbor or Warren Woods, Michigan (42°N; 250 m)	23	Dice (1920b) Davis (1925) Burt (1954)
Deciduous beech-maple forest	(6) Great Smoky Mountains, Tennessee (35°N; 1200 m)	22	Komarek and Komarek (1938)
Dry tropical forest	(7) Guanacaste Prov., Costa Rica (10°N; 45 m)	46	Fleming (1973a) Organization for Tropical Studies unpubl. data
Wet tropical forest	(8) Heredia Prov., Costa Rica (10°N; 100 m)	60	Fleming (1973a)
Second growth dry tropical forest	(9) Balboa, Panama Canal Zone (9°N; 50 m)	58	Fleming (1970, 1972); Fleming *et al.* (1972)
Moist tropical forest	(10) Cristobal, Panama Canal Zone (9°N; 5 m)	57	Fleming (1970, 1972); Fleming *et al.* (1972)
Deciduous tropical forest	(11) Mount Nimba, Liberia (7°N; *c.* 500 m)	49	Coe (1976)
Acacia-Nubica semi-arid savannah	(12) South Turkana, Kenya (2°N; *c.* 500 m)	32	Coe (1972)

Table 4.1 shows the number of species of small mammal found in twelve habitats in the northern hemisphere. Starting with the tundra as a base-line (12 species) the temperate forests are twice as rich in species whilst Panamanian rain forest has approximately five times the tundra total. The increase in species richness with decreasing latitude is particularly marked in forests and is mainly attributable to the large bat faunas of tropical rain forests. Similar trends do exist however in the rodent faunas of grasslands. For example, French *et al.* (1976) reported from 4–8 species of rodent in different North American grasslands while Dieterlen (1967) found at least 12 species of rodent inhabiting *Pennisetum* grasslands in Zaire. Similarly Bellier (1967) found 14 species of rodent in savanna regions of the Ivory Coast and Coe (1972) counted 14 rodent species in the *Acacia-Nubica* savanna of South Turkana, Kenya.

On the available, but scanty, evidence there appear to be consistent differences between habitats with respect to their small mammal species diversity. Tropical small mammal communities are typically richer in species than temperate ones and grasslands usually contain fewer co-existing species than forests. In support of this latter contention one can refer to accounts of Sheppe (1972) and Fleming (1975), the first author noted the potential co-existence of only five species of rodent on the seasonally inundated Kafue River floodplain in Zambia while the second pointed out that less than 10 rodent species are likely to co-exist in the grasslands of Central America. These values can be contrasted with those of tropical forests which often contain from 10–16 species of co-existing rodent (Fleming, 1975). Information about non-rodent small mammal communities is particularly scarce but it would appear that the number of sympatric species of either marsupial or insectivore is always less than the number of rodent species. The relationship of marsupial, insectivore, lagomorph and carnivore species richness to latitude or habitat type is difficult to ascertain on the basis of available data.

Estimates of the species richness of bat faunas are primarily for Neotropical regions. Fleming (1973a, 1975) and Fleming *et al.* (1972) indicate that from 27–40 or more species can be found in one habitat and can comprise at least 66 per cent of the total number of small mammal species. The relationship to other small mammals of the 32 species of bat reported by Liat (1966) from the forests around Kuala Lumpur, Malaysia is unknown but probably parallels neotropical situations.

Clearly, our knowledge of the species diversity of small mammal faunas in different parts of the world is far from complete. Further, the data available on the various major taxa of small mammals are uneven, most effort having been directed towards rodents.

In modelling terms it is obvious that the choice of state variables will be guided by the available information. Inevitably some models, through lack of data, will be reductionist in the extreme and follow the holological approach; the consequences of this being a loss of information regarding transfer functions and hence the functional roles of different small mammal species. For systems where considerable data are available, a more merological approach can be adopted, yet the greater the species diversity the greater the number of possible state variables and hence complexity. It is often desirable to reduce the complexity of such models and from the viewpoint of studying functional relationships this may be best achieved by grouping those species of similar trophic status which are not very abundant. By contrast the abundant and, by inference, 'key species' of the system can be treated as separate state variables.

Whether for modelling purposes or otherwise, consideration of the relative abundance of different small mammal species is of importance to the elucidation of their functional role in ecosystems.

4.4.3 Relative abundance

It is well known that in many animal communities a few species are very common whilst others are relatively rare, small mammal communities are no exception.

In the tropics both rodent and bat communities have been shown to exhibit high levels of numerical dominance by one or two species. Information from nine tropical localities was summarized by Fleming (1975) who demonstrated that the three most common species of rodent or bat account on average for 69 per cent (range 55–85 per cent) of their respective communities. Likewise Coe (1972) recorded that three genera (*Tatera, Taterillus* and *Gerbillus*) constituted 62 per cent of the rodent fauna of South Turkana, Kenya while two species of bat (*Pipistrellus nanus helios* and *Cardioderma cos*) comprised 85 per cent of the chiropteran fauna. Other tropical examples can be found in Hanney (1965), Fleming (1973b) and Handley (1967); the first named author studied rodents in Malawi and found that one or two species accounted for more than 65 per cent of the individuals caught, *Praomys jacksoni* formed 66 per cent of the rodents trapped in montane forest whereas *Mastomys natalensis* was dominant in cultivated areas. Fleming (1973b) similarly found that *Heteromys desmarestianus* and *Liomys salvini* accounted respectively for 94 and 96 per cent of rodents trapped in Costa Rican wet and dry tropical forests. Handley (1967) studied bats in Brazilian forests and found *Artibeus lituratus* and *A. jamaicensis* to be dominant in the canopy; these species only ranked third and fourth in abundance at ground level, the first and second places being occupied by *Carollia perspicillata* and *C. subrufa*.

There is not, to our knowledge, detailed information on relative abundance in bat communities from higher latitudes but examples of numerical

Table 4.2 *Wet weight biomass estimates for populations or communities of small mammals.*

Location and habitat	No. of species	Average density (no. ha^{-1})	Average biomass [g (wet weight) ha^{-1}]	Reference
Alaska: *Picea glauca* forest	5 (rodents)	4.60	253.14	Grodziński (1971a, b)
	1 (shrew)	4.00	15.04	
Sweden: *Picea abies* forest	4 (rodents)	13.25	344.88	Hansson (1971a)
	3 (shrews)	6.25	65.83	
Poland: *Querceto-Carpinetum* forest	5 (rodents)	8.00	183.23	
	2 (shrews)	5.25	24.50	Grodziński (1961)
U.S.A.: Bunchgrass grassland	5 (rodents)		299.00	French *et al.* (1976)
U.S.A.: Tallgrass prairie	5 (rodents) 1 (shrew)		813.00	French *et al.* (1976)
U.S.A.: southern Shortgrass prairie	7 (rodents) 1 (shrew)		255.00	French *et al.* (1976)
U.S.A.: desert	11 (rodents)*	1.57	61.36	Chew and Chew (1970)
	1 (lagomorph)	0.20	455.00	
Panama: dry tropical forest	11 (rodents)	18.90	4025.00	Fleming (1970, 1971, 1972)
	5 (marsupials)	3.50	1538.00	
Panama: wet tropical forest	9 (rodents)	11.30	6304.00	Fleming (1970, 1971, 1972)
	4 (marsupials)	2.10	1293.00	
Lake Kivu, Zaire: *Pennisetum* grassland	5 (rodents) >1? (insectivores)	370.00	14 000.00– 15 000.00	Dieterlen (1967)
South Turkana, Kenya: *Acacia-Nubica* semi-arid savannah	7 (rodent genera)	21.52	1334.10	Coe (1972)
	2 (insectivores)	3.48	200.00	

*6 resident species, 5 transient species

dominance by a single species of rodent are well documented. Chew and Chew (1970), for example, demonstrated that *Dipodomys merriami* comprised 66 per cent of all rodents in *Larrea tridentata* scrub desert and Hansson (1971a), working in a Swedish spruce plantation, showed that *Microtus agrestis* constituted 72 per cent of all rodents captured.

Studies of relative abundance can have considerable influence in governing which state variables are likely to be of greatest importance in the evaluation of the functional role of small mammals. All matters, including size, being equal it would seem reasonable to assign dominant species to individual state variables and group other species of similar trophic status into 'collective' state variables. However, if marked size differences occur between species of the community then relative abundance alone is not an ideal criterion and biomass too must be considered, as indeed must density if absolute relationships are to be explored.

4.4.4 Biomass and density

Table 4.2 gives small mammal biomass and density data for 11 ecosystems, as with species diversity both biomass and density tend to be highest at low latitudes. One probable reason to account for the fact that such high biomasses can be supported in the tropics is the relatively high primary production and year round availability of food. Nevertheless, despite the general tendency for biomass and density to increase with decreasing latitude some temperate ecosystems have a small mammal carrying capacity as high as tropical ones, for example, Ashby and Vincent (1976) recorded a biomass of 15000 g ha^{-1} for *Arvicola terrestris* in northern England. In Britain *A. terrestris* is restricted to 'linear' territories along water courses and the reported high biomass for this species is no doubt supported by the highly productive stream-side vegetation. A further example of sizeable biomass and density can be found in the work of Abaturov and Kuznetsov (1976) on the sousliks of steppe grasslands. In this instance the densities of *Spermophilus pigmaeus* (= *Citellus pygmaeus*) ranged from 33–66 ha^{-1} and correspond to an approximate biomass of 10000–20000 g ha^{-1}. Unlike tropical and waterside vegetation, annual primary production levels in the grassland steppes are low and hence high souslik biomasses are best explained in terms of the seasonal activity of this animal, a species which hibernates in winter.

Biomass, like relative abundance, is inequitably distributed in small mammal communities; moreover, high biomasses are not always coincident with high densities. It will be recalled that in the scrub desert ecosystem studied by Chew and Chew (1970) the numerically dominant species was *Dipodomys merriami*, a much rarer species which constituted a mere one per cent of the total number of small mammals was *Lepus californicus* and yet each of these species accounted for approximately 40 per cent of the total small mammal biomass. Similarly, in Panamanian tropical forest *Dasyprocta punctata,* with an individual live weight of approximately 4000 g, occurs at densities of less than one per hectare and yet accounts for between 29 and 52 per cent of biomass (Fleming, 1975).

In section 4.4.2, reference was made to the concept of 'key species' and their designation as single state variables on the grounds of their numerical abundance. It should be obvious by now that numerical or relative abundance alone are not particularly good criteria to choose when identifying which groups of animals are 'key species' in the functioning of ecosystems. Taxonomic composition, species diversity, relative abundance, biomass and density, as well as trophic status should all be taken into account when selecting system variables of potential importance to the evaluation of small mammal functional relationships within ecosystems. Parameters as diverse as species diversity and biomass are not easy to combine within a single index but it is clear that indices of 'ecological diversity' could be of great help to both theoreticians and field workers.

4.4.5 Ecological diversity

Ecological diversity emphasizes the kinds of animals found in ecosystems and is defined as the distribution of species in various classes of body size, feeding adaptation and food habit in a manner analogous to the way 'species diversity' refers to the distribution of individuals among species. Fleming (1973a) stressed the importance of considering ecological diversity in relation to community structure and, without doubt, indices of 'ecological diversity' could be most useful in the selection of appropriate system variables for modelling purposes.

Figure 4.4 shows the percentage composition of small mammal communities in seven habitats according to taxonomic composition, body size, spatial adaptation and feeding strategy. On the basis of the frequency distributions it was shown statistically that there are interhabitat differences between the small mammal communities but the degree of differentiation varies according to which criterion is used. Differences between communities with respect to taxonomic composition were referred to in section 4.4.1, here we concentrate on body size, adaptations and food habits.

Body size – no significant differences were found between body size distributions in the seven communities ($\chi^2 = 6.4$, d.f. = 6, $P > 0.1$). Nevertheless, there appears to be a trend towards smaller body size with decreasing latitude.

Adaptations – significant differences in the distributions of spatial adaptations between communities were demonstrated ($\chi^2 = 64.03$, d.f. = 12, $P = < 0.001$). Aerial species dominate the small mammal faunas

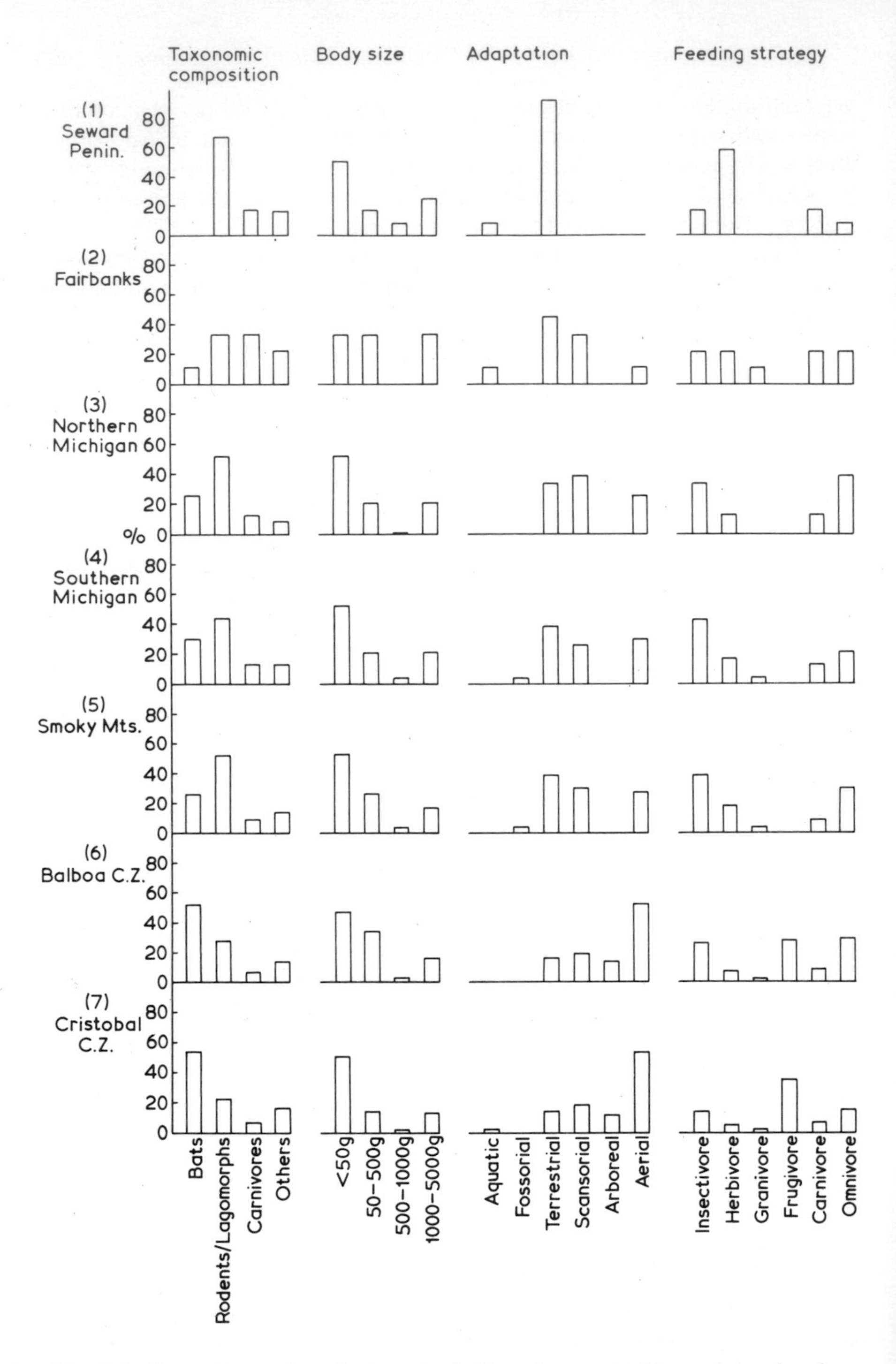

Fig. 4.4 Percentages of species in seven habitats by mammalian order and various classes of body sizes, adaptations and food habits (Based on Fleming, 1973a.)

of the tropics while ground dwelling (≡ terrestrial) and scansorial species are more important at higher latitudes.

Food habits – no significant differences were detected between the communities ($\chi^2 = 11.42$, d.f. = 8, $P > 0.1$) but it is possible that the non-significance could be an artefact caused by the grouping of data to obtain expected values of five per cell in the association analysis. Within the food habit category two trends are apparent: (i) a proportional decrease in the number of herbivorous species in tropical habitats and (ii) the occurrence of frugivorous species only in the tropics. This latter, and seemingly clear, distinction is partly the result of an hological approach in that fruit production in temperate latitudes is seasonal and hence temperate 'frugivores' are paradoxically categorized as omnivores.

Comparisons of the percentage composition of small mammal communities according to the chosen criteria indicate that the structure of such communities varies with habitat type. The question arises as to whether each small mammal community possesses a structural uniqueness such that general principles regarding the functional role of small mammals in ecosystems cannot be identified. We do not believe that this is the case and are of the opinion that it is possible to accept a form of reductionism such that small mammal communities from different habitats can be grouped by reference to a combination of criteria. This approach is easily illustrated by reference to the small mammal communities of the single tundra and six forest ecosystems shown in Fig. 4.4.

Following Fleming (1973a), and using body size, adaptation and food habit categories, we have calculated a series of ecological diversity indices for each of the seven communities. We used the Shannon – Wiener formulation $H' = -\Sigma_{p_i} \log_e p_i$ where p_i is the proportion of species found in each arbitrarily defined class within a particular category. As species become more evenly distributed between the classes the index of diversity increases and the best measure of diversity within each category is equitability, or E, where $E = H'/H_{max}$ when H_{max} is the natural logarithm of the number of classes.

Figure 4.5 is a three dimensional representation of the indices of diversity of the seven small mammal communities under investigation. The communities of the seven ecosystems separate into three groups: (i) the tundra community which is characterized by high body size diversity and low adaptation and food habit diversities, (ii) the taigan and temperate forest communities which exhibit lower body size diversity but higher adaptation and food habit diversities when compared with the tundra community and (iii) the tropical rain forest communities which, by comparison with temperate forest communities, have low indices of body size diversity and high adaptation and food habit ones. It would seem that within forests indices of ecological diversity, other than body size, increase as latitude

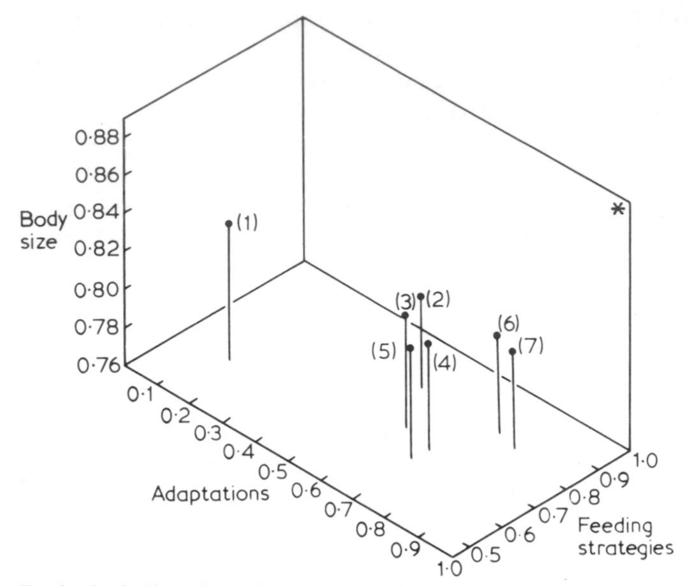

Fig. 4.5 Ecological diversity of the mammal communities in seven North and Central American habitats. Numbers refer to habitats shown in Fig. 4.4 (Based on Fleming, 1973a.) *The point of maximum diversity in all respects.

decreases, i.e. there is a more even distribution of species between category classes as one moves from high to low latitudes. In the case of body size the reverse appears to be true.

Available information on the small mammal communities of grasslands suggests that apart from food habit diversity there are few differences between ecosystems. When compared with forest communities the body size diversity of grassland rodents is low and relatively constant. Low indices of diversity with respect to spatial adaptations in grasslands are clearly correlated with a lack of spatial heterogeneity in the habitat, most small mammals being either fossorial or ground dwelling. However, food habit diversity indices for grassland mammals are generally quite high although there are exceptions, one example being the *Acacia-Nubica* savanna studied by Coe (1972), a habitat where most species of small mammal possessed catholic food habits.

Indices of ecological diversity, when used in combination as in Fig. 4.5, draw attention to both similarities and differences in small mammal community structure; the uniqueness of each community is not overemphasized and the construction of realistic models to illustrate general principles becomes a distinct possibility. Figure 4.6 provides examples of pictorial models for the three types of small mammal community identified in Fig. 4.5; these are based on a study by Fleming (1973a). The emphasis on vertical distribution in the models can be justified on the grounds that highly

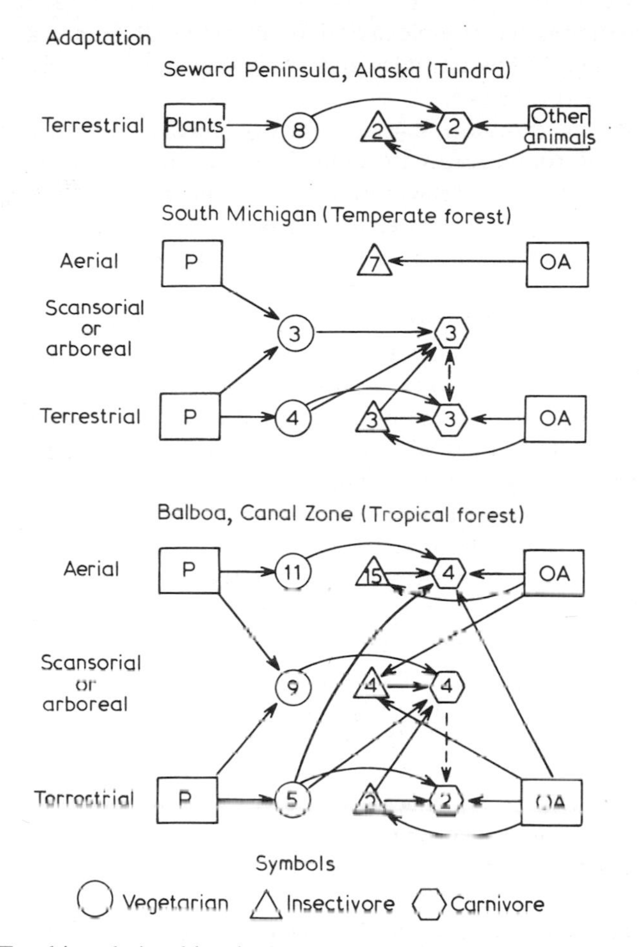

Fig. 4.6 Trophic relationships between mammals in three North and Central American habitats. Arrows indicate the direction of energy flow. Numbers within the symbols indicate number of species in each class. The vegetarian class includes herbivores, frugivores and granivores. Omnivores have been equally partitioned among other feeding classes. (Based on Fleming, 1973a.)

significant differences in ecological diversity with respect to spatial adaptations were demonstrated. In this connection it is of interest to note that a quantitative approach has confirmed earlier, but non-statistically based, recommendations by Elton and Miller (1954) that the fauna of a specific ecosystem is best recorded by reference to habitat structure.

Within each ecosystem model it is clear that subdivision of the small mammal community of each horizontal stratum into state variables which reflect trophic status aids conceptualization of their functional relationships. The models are qualitative rather than quantitative, nevertheless they illustrate the increasing complexity of small mammal faunas as one progresses from high to low latitudes and imply that this is closely associated with habitat heterogeneity.

Taxonomic composition, species diversity, relative abundance, spatial adaptations and trophic status are all useful parameters in the production of acceptable qualitative representations of small mammal communities. However, additional data such as food processing information, body size, biomass and density are necessary if the dimensions of the chosen state variables and transfer functions are to be defined. In the absence of such information it is not possible to write, except in general terms, of the functional role of small mammals in ecosystems. To date insufficient data have accumulated for a detailed account to be made of the structure and function of small mammals in any single ecosystem. The small mammals of Wytham Woods, U.K. have been studied more or less continuously over the last 30 years and Fig. 4.7 is a summary of our knowledge about their structure and function. The picture is incomplete but illustrates an advance over the conceptual models of Fig. 4.6.

Most mammalogists with an interest in the functional role of small mammals have, for mainly logistical reasons, restricted their activities to studies of single species populations. Furthermore they have concentrated on only one or two aspects of function, e.g. consumption as a percentage of primary production (e.g. Drożdż, 1966), the cycling of specific plant nutrients (*see* Gentry *et al.*, 1975), or respiratory energy loss (e.g. Górecki, 1968). Despite many years of investigation, small mammal studies have not yet reached a stage at which it is possible to collate the findings within ecosystem models and then write, other than in qualitative manner, about the functional role of small mammals in ecosystems. In summary, we have the means to construct conceptual frameworks for potentially useful models but insufficient data to fully quantify them. Inevitably subsequent sections of this account will be concerned with either general ecological characteristics of small mammals or particular functional roles of single species populations.

4.5 Some ecological characteristics of small mammals

Small mammals are highly adaptable animals and may be found in practically all habitats ranging from polar regions to the tropics. Despite accepted feeding strategies such as herbivore, granivore, insectivore, the feeding categories are not hard and fast and indeed most small mammals, notably rodents, are omnivorous. Herbivorous small mammals have high digestibility coefficients (assimilation/consumption) relative to ungulates for example, and this characteristic combined with the high ratio of respiration

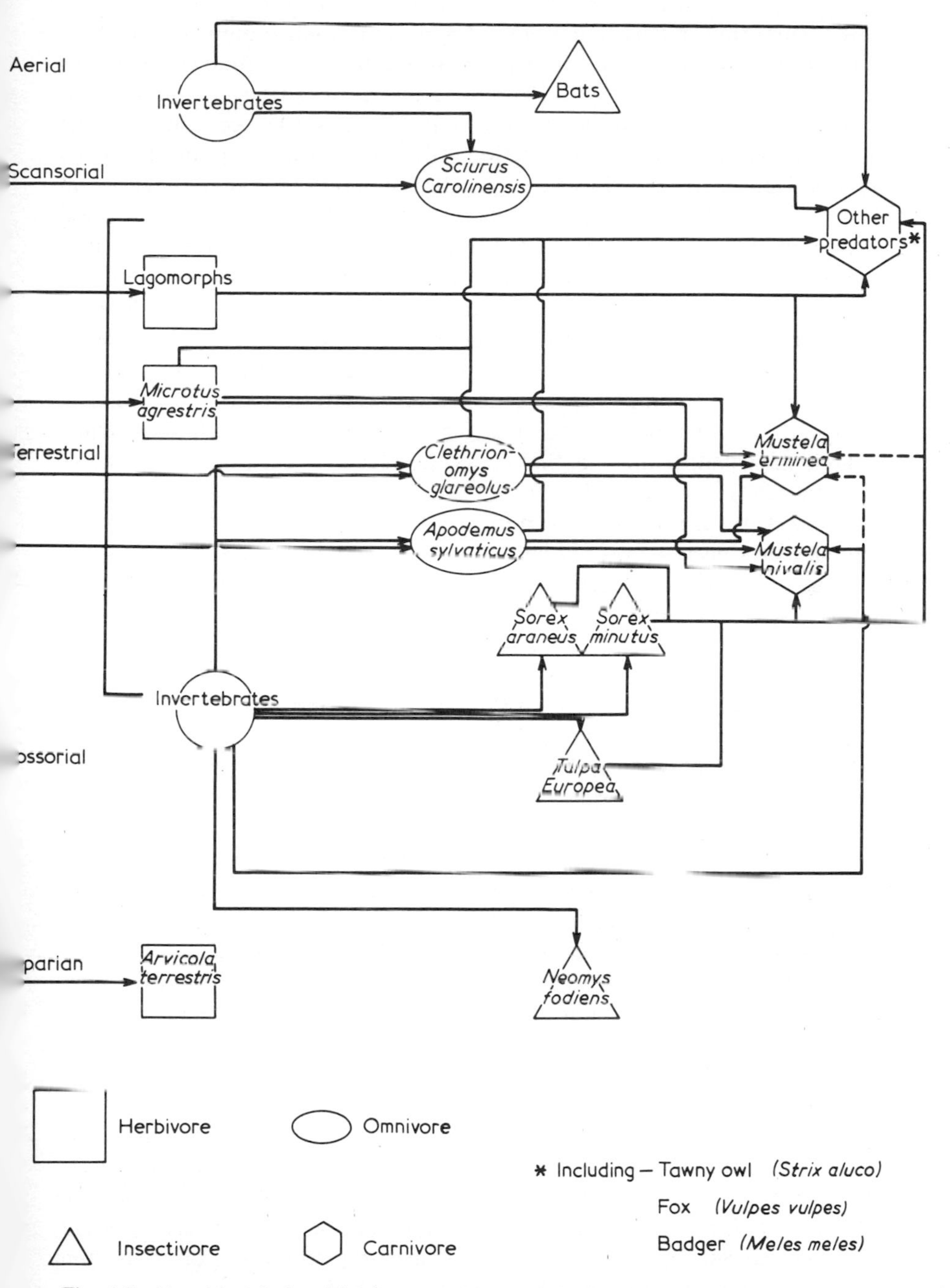

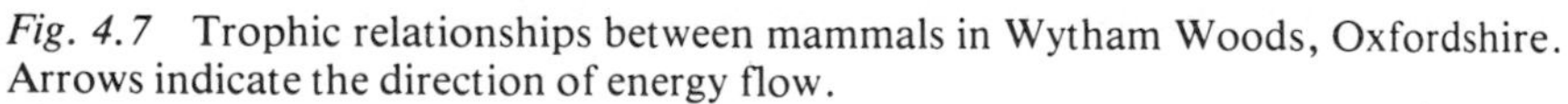

Fig. 4.7 Trophic relationships between mammals in Wytham Woods, Oxfordshire. Arrows indicate the direction of energy flow.

to consumption means that weight for weight small mammals are more efficient in effecting the mineralization of organic matter than either insects or ungulates (Golley *et al.*, 1975a). Energetic efficiency of production (production/assimilation) is low and approximates 2 per cent on average; within ecosystems this low production efficiency is usually counterbalanced by relatively high reproductive rates. Reproduction and survival show an inverse correlation and Golley *et al.* (1975a) have argued that small mammals are characterized by either high reproduction with low survival or low reproduction with high survival. These are, of course, extremes of an *r* and *K* selection continuum and intermediate states do exist (*see* Fleming, Chapter 1). Typically, small mammals of direct economic importance to man via population irruptions show high reproduction and low survival.

Turnover of biomass (production/biomass) reflects the relationship between reproduction and survival; small mammals, especially rodents, exhibit high values and this highlights the fact that even at relatively low densities they can be an important source of food for many predators. Many rodents are omnivorous and as a taxon form a link between many primary producers and secondary consumers. This is especially the case in tundra and desert ecosystems where rodents are important 'key' vertebrate species (*see* Golley *et al.*, 1975a).

Using demographic and bioenergetic features Golley *et al.* (1975a) characterized small mammals as ecosystem components having (i) high turnover rate, (ii) very high energy costs of production, and (iii) small dependence on climatic conditions due to endothermy (with the exception of hibernating species). Such characteristics influence both the adaptability of small mammals and their input-output efficiencies in terms of energy throughput and mineral cycling. In general, the output of any component or state variable per unit time can be defined as the product of mass times turnover. It would appear that small mammals are more flexible in their level of standing crop than in turnover.

Standing crop is a result of reproduction and survival and because turnover is determined by survival it follows, as pointed out by Golley *et al.* (1975a), that the variation in ecosystem output of small mammals is largely influenced by variation in the reproductive processes of populations. Ecological parameters of reproduction such as length of breeding period, intervals between successive pregnancies in the same female, and number of pregnant females are more variable than the physiological parameters of litter size or pregnancy duration. One must not ignore, however, the possibility that variability in ecological parameters may be food dependent (Shvarts, 1975). Another factor which should not be forgotten is dispersion; as pointed out by Lidicker (1975) dispersion can have an important influence on estimates of standing crop and reproduction.

As we have seen the energy costs of production by small mammals are high and are equivalent on average to approximately 98 per cent of the assimilated energy. Clearly, small mammals as individuals are poor accumulators of both energy and matter, the corollary being that as individuals they are excellent processors of both. In later sections we shall learn that a considerable amount of information has been collected about small mammals as processors of energy but that little is known about the role they play in processing and cycling matter. With regard to this latter aspect the studies of Gentry *et al.* (1975) suggest that specific differences among small mammals of like trophic status are small compared to those found among plants. On the present evidence it would seem that small mammals as a whole are relatively homogeneous in terms of mineral cycling.

Trophic status, digestibility coefficients, ratios of respiration to consumption, production efficiencies, reproductive strategies, standing crops and biomass turnover all play their part in governing the role of small mammals within ecosystems. In part they set the choice and dimensions of the various small mammal system variables but more importantly in this context they influence the values attributable to different transfer functions.

4.6 Evaluation of transfer functions which link the component 'small mammals' to other system variables

Transfer functions are representations of the functional relationships which exist between system variables. We have seen that dimensions of system variables can be defined in a variety of units, e.g. species number, density, or biomass in weight or energy equivalents; it is not surprising therefore to learn that transfer functions can also be expressed variously in matter or energy units. The choice of units depends entirely on which aspect of function one is currently interested in, as does the selection of particular transfer functions for investigation.

Figure 4.8, based on Pitelka's (1957) study of lemmings in a tundra ecosystem, illustrates some of the numerous functional relationships that can exist between a small mammal state variable and other system variables. In this example the population of herbivorous lemmings interacts with vegetation not only via consumption but through a variety of other activities e.g. girdling and trampling. These activities, along with others such as burrowing and defaecation, not only affect the system variable 'vegetation' and its associated transfer functions but also have an impact on the system variable 'soil' and its transfer functions. Soil and vegetation constitute resources for the lemmings and in their turn are a resource which is exploited by parasites, predators and scavengers. Although constructed with lemmings in mind the

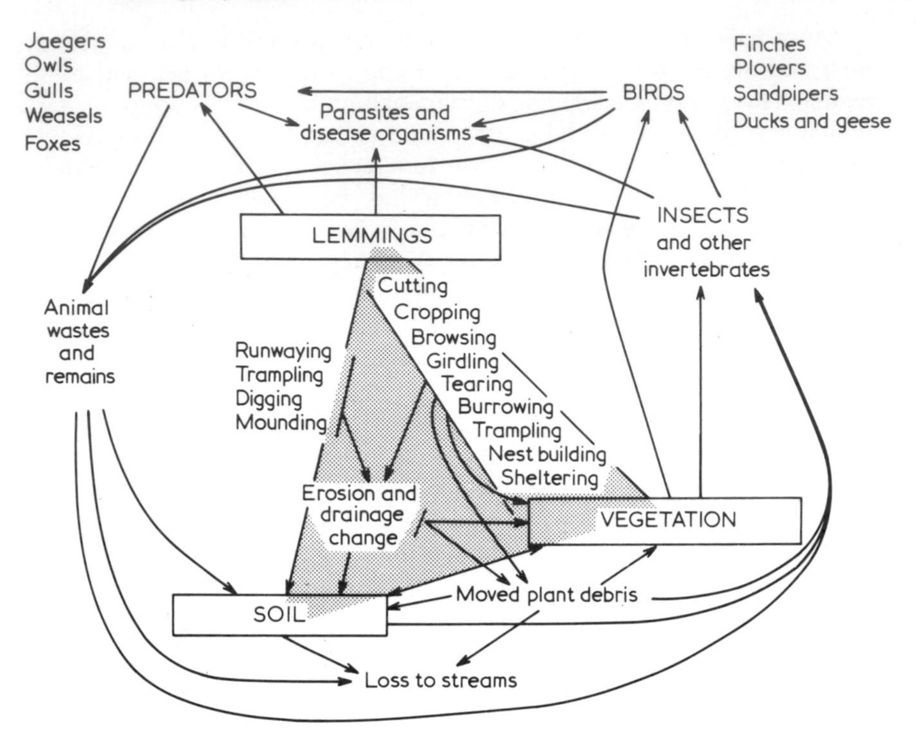

Fig. 4.8 Diagram of food web and lemming activities in coastal tundra of northern Alaska. (From Pitelka, 1957.)

diagram would, with minor modifications, be applicable to other small mammal species of similar trophic status. As Batzli (1975) has pointed out small mammals influence at least three major components of any ecosystem, e.g. the soil, the vegetation and predators. We agree with Batzli (op. cit.) when he states, 'Any evaluation of the significance of small mammals must be concerned with the degree to which the characteristics of these important components are interdependent. Insofar as these relationships can be stated quantitatively our understanding of ecosystem dynamics will be enhanced.'

The most obvious relationships between biotic state variables are those connected with feeding, the impact of which on an ecosystem may be direct or indirect. Transfer functions representing the direct impact of consumption can be evaluated in terms of biomass or energy units. More difficult to comprehend and evaluate are the indirect impacts associated with feedback from the effects of consumption and other activities. Figure 4.9, after Golley *et al.* (1975b) illustrates the distinction between direct and indirect influences in a three compartmental system.

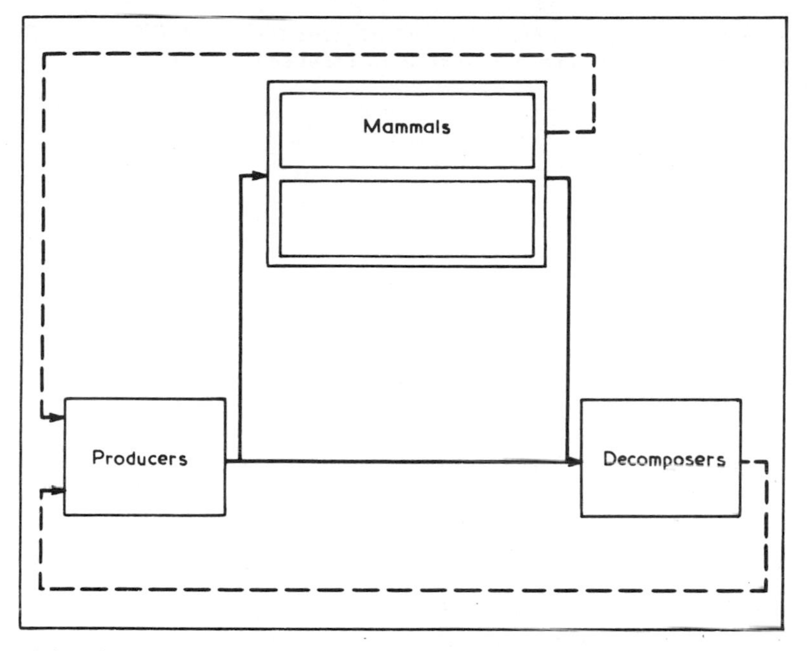

Fig. 4.9 Diagram of a generalized ecological system with three compartments. Mammals are one subcompartment of the consumers. Feedback loops are indicated by dashed lines. (From Golley *et al.*, 1975).

4.6.1 Consumption: its direct impact and the functional relationships between vegetation and primary consumers

It is important to recognize that not all primary production in an ecosystem is suitable for, or available to, small herbivores. Grodziński (1968) defined available primary production (A.P.P.) as 'that food which is easy to find and is being chosen and eaten by these animals (small mammals).' Table 4.3 gives data on A.P.P. for a number of systems and shows that the percentage of net primary production (N.P.P.) that can be considered available as small mammal food varies with both the ecosystem and the food habits of the species under consideration. Clearly, the proportion of N.P.P. available as food to rodents is greatest in grasslands and agrocoenoses, intermediate in desert ecosystems and lowest in forest ecosystems. Exceptions do arise however, for example in forests A.P.P. need not necessarily be a small percentage of N.P.P.; nor, as Odum *et al.* (1962) or Pearson (1964) have maintained, need only a fraction of the food available to herbivore species be available to granivores. An example of this can be found in the study by Phillipson *et al.* (1975) of a beech woodland area of Wytham Woods. It was shown that tree fruits, a preferred food of the woodmouse *Apodemus sylvaticus* and grey squirrel *Sciurus carolinensis*, contributed differentially

Table 4.3 *The percentage of net primary production (N.P.P.) available (A.P.P. to different species of rodent in different ecosystems.*
H = herbivore *G = granivore* *O = omnivore*

Ecosystem	Species and food habits	A.P.P./N.P.P. %	Source
Fagetum carpaticum forest	*Clethrionomys glareolus* [H]	4.40	Grodziński *et al.* (1970)
	Apodemus flavicollis [G(O)]	2.40	Grodziński *et al.* (1970)
Picea glauca forest	*Microtus oeconomus* [H]	6.57	Grodziński 1971a, b
	Clethrionomys rutilus [O]	5.34	Grodziński 1971a, b
	Tamiasciurus hudsonicus [G] *Glaucomys sabrinus* [G]	5.16	Grodziński 1971a, b
Tilio-Carpinetum forest	All rodent species	5.16	Medwecka-Kornas *et al.* (1973)
Forest plantation	*Microtus oeconomus* [H]	55.74	Gebcyzńska (1970)
Water meadow	*Arvicola terrestris* [H]	95.12	Ashby and Vincent (1976)
Poa grassland	*Microtus pennsylvanicus* [H]	100.00	Golley (1960)
Agrocoenosis	*Microtus arvalis* [H]	100.00	Trojan (1969)
Agrocoenosis	*Cricetus cricetus* [G(H)]	61.70	Górecki (1977)
Desert	*Lepus californicus* [H] *Spermophilus audubonii* [H]	39.46	Chew and Chew (1970)
	Dipodomys merriami [G]	12.40	Chew and Chew (1970)
	Peromyscus spp. [O]	16.85	Chew and Chew (1970)

to N.P.P. from year to year. In a good fruiting year (1970–71) the weight of beech mast produced was equivalent to 51.7 per cent of the total above ground litter input (= N.P.P. above ground minus stem increment) whereas in a poor year (1969–70) it equalled 2.2 per cent. Clearly, such variations in A.P.P. with time would affect the dimensions of system variables associated with vegetation and presumably also those other variables which are linked to them via transfer functions.

One cannot overemphasize the dynamic nature of ecosystem structure and function, or the need for long term studies.

Without doubt, the amount of primary production available to small mammals varies in both space and time; a question to which we can address ourselves is whether the impact small mammals have on this primary production also varies spatially and temporally. Table 4.4 provides data on annual consumption by small mammals as a percentage of available primary production in a number of ecosystems. In general, consumption is usually less than 6 per cent of the available primary production, and since

Table 4.4 *Energy consumption of small mammals as the percentage of available primary production consumed yearly in ecosystems.*

Ecosystem	Available primary production 10^3 mJ ha^{-1} $year^{-1}$	Percentage consumption	Reference
Grassland			
Old field, Michigan	196.46	1.60	Golley (1960)
Grass field	169.71	1.30	Trojan (1965)
Northern shortgrass	–	0.50	French *et al.* (1976)
Midgrass	–	1.67	French *et al.* (1976)
Tall grass	–	5.00	French *et al.* (1976)
Southern shortgrass	–	5.20	French *et al.* (1976)
Serengeti	300.00	4.40	Phillipson (1973)
Serengeti	–	5.60	Sinclair (1975)
Deserts			
Desert shrub	10.03	5.50	Chew and Chew (1970)
Desert grassland	–	8.73	French *et al.* (1976)
Forests			
Pine lichen	4.18	1.90	Ryszkowski (1969)
Vaccinium-pine (40 years)	10.03	0.90 1.20	Ryszkowski (1969)
Vaccinium-pine (140 years)	29.26	0.60	Ryszkowski (1969)
Oak	–	0.73	Phillipson (1973)
Oak-pine	54.34	0.60–0.80	Ryszkowski (1969)
Oak-hornbeam	8.78	4.60	Grodziński (1961)
Mixed forest	67.72	0.60	Ryszkowski (1969)
Taiga *Picea glauca* forest	5.43–6.81	3.00–3.80	Grodziński (1971a)
Circaeo-Alnetum	26.24	2.20	Aulak (1973)
Agrocoenosis and Plantations			
Rye field	170.13	0.50	Trojan (1969)
Alfalfa	166.36	0.80	Trojan (1969)
Alfalfa	162.18	3.00–3.80	Ryszkowski *et al.* (1973)
	153.41	0.50	Gorecki (1977)
Augustów Forest	28.01	3.10	Gebczyńska (1970)
Spruce plantation	141.70	1.50	Hansson (1971a)

A.P.P. constitutes only a small fraction of N.P.P. the role of small mammals as energy and material processors in most ecosystems is, despite their individual excellence, neligible. Their small contribution to energy and material processing is not unusual among mammals for, as Phillipson (1975) has suggested, consumption by large mammals in the semi-arid savanna of Tsavo National Park (East) ranges between 5 and 15 per cent respectively in good and bad years of primary production. In the same area small mammal annual consumption, despite an estimated threefold change in their standing crop, was calculated to approximate 5 per cent of above ground net primary production in all years. This example provides support for the suggestion made in section 4.5. that small mammals, because of early maturity and high reproductive rates, are extremely flexible in their level of standing crop. Among mammals they are at the *r* end of the *r* and *K* selection continuum and, as opportunists, are able to respond relatively rapidly to increased food supplies. Because small mammals are capable of responding so rapidly to food supply some explanation is needed to account for the fact that these animals consume only 6 per cent or less of the available primary production in any one year.

Annual consumption, when expressed as a percentage of A.P.P., is an average of values over time and does not give any indication of the variations which occur with season. In mildly seasonal ecosystems, primary production will be spread fairly evenly throughout the year and in such a way that herbivore food supplies remain relatively stable from month to month; under such circumstances periods of relative food abundance and scarcity rarely occur and consumption as a percentage of A.P.P. remains fairly constant with time. For seasonal environments much of the annual primary production occurs over a limited period of time and herbivore food supplies range from being relatively abundant to relatively scarce, one might expect variation therefore in the seasonal values for consumption as a percentage of A.P.P. High consumption values as a percentage of A.P.P. certainly occur during periods of food scarcity, Kalela *et al.* (1971) and Kalela and Koponen (1971) reported that in a winter of high lemming densities 30–70 per cent of the moss in alpine habitats of Finnish Lapland was consumed. Further, Batzli (1975) calculated that to sustain the observed densities of lemmings during a peak year, 109 per cent (!) of the monocot stems in a given habitat would have to be eaten. Given such high values for consumption during periods of food scarcity it is clear that an annual value of 6 per cent or less can only be achieved if consumption values during periods of food abundance are very low. It must be concluded that under most conditions, despite high reproduction rates, increases in small mammal population densities are not rapid enough to take full advantage of the seasonal increase in food supplies. Irruptions, mainly in agrocoenoses and plantations, constitute the exceptions to this rule. Because of the fluctuating nature of food supplies in highly seasonal ecosystems it can be postulated that annual consumption as a percentage of A.P.P. could well be less than that recorded for mildly seasonal ecosystems. Some indications to this effect can be seen in Table 4.4 where percentage consumption in tropical and semi-tropical systems is generally higher than that recorded for temperate ones. True or not, what is abundantly clear is that in most cases the direct impact on vegetation by small mammal consumption is relatively slight. In man-managed systems the economic consequences can be great; for example three species of rat (*Rattus exulans, R. rattus* and *R. norvegicus*) are responsible for losses averaging 4.5 million U.S. dollars per year to the Hawaiian sugar cane industry (Hood *et al.*, 1970; Teshima, 1970). Kanervo and Myllymäki (1970) estimated that between 1954 and 1966 *Microtus agrestis* caused 2 million dollars of damage annually to orchards and ornamental plants. In Sweden in 1962 *M. agrestis* alone caused 4–8 million dollars of damage to orchards (Stenmark, 1963). Clearly, the impact of small mammals on an ecosystem is a far more subtle process than simply consumption alone. Indeed, consumption is a very poor index of impact since its consequences may exceed the actual amount of primary production eaten (Spitz, 1968; Ryszkowski *et al.*, 1973; Abaturov *et al.*, 1975). Zlotin (1975) estimates that the impact of small mammals in grassland due to consumption does not exceed 5 per cent of the total impact. In considering the impact and role of small mammals in ecosystems it is the feedback from the effects of consumption on other ecosystem components (*see* Fig. 4.9) that is of much greater importance than the consumption itself.

4.6.2 Consumption: its indirect impact and the functional relationships between primary consumers and primary producers

The immediate and most direct impact of small mammals as consumers of vegetation is a reduction of primary producer standing crop but in the long term the indirect effects of grazing may be any of a number of alternatives, for example stimulation of production, alteration of the species composition of plant communities, and changes in plant stature and reproduction. As noted by Batzli (1975), which of these alternatives becomes most apparent is largely dependent on the species of small mammal involved and its population density.

An immediate problem that presents itself when discussing the indirect impact of small mammal consumption is how to quantify it for modelling purposes. Changes in species composition can be reflected in state variables while plant stature is more difficult; on the other hand increased plant production or changes in reproduction may be accommodated in appropriate transfer functions.

As has been already indicated, small mammal impact in terms of the percentage of N.P.P. or A.P.P. consumed is negligible for most systems but as Petrusewicz and Macfadyen (1970) pointed out the best measure of 'ecological pressure' on vegetation is the amount or percentage of material removed (MR) rather than consumption (*see* Fig. 4.10). Unfortunately, estimates of MR are difficult to obtain but this does not detract from the fact that there may be considerable differences between MR and consumption for grazing small mammals. Indeed, for these animals consumption may not even be a reliable index of the degree of reduction of standing crop. Koshkina (1962) observed that high populations of *Lemmus lemmus* under

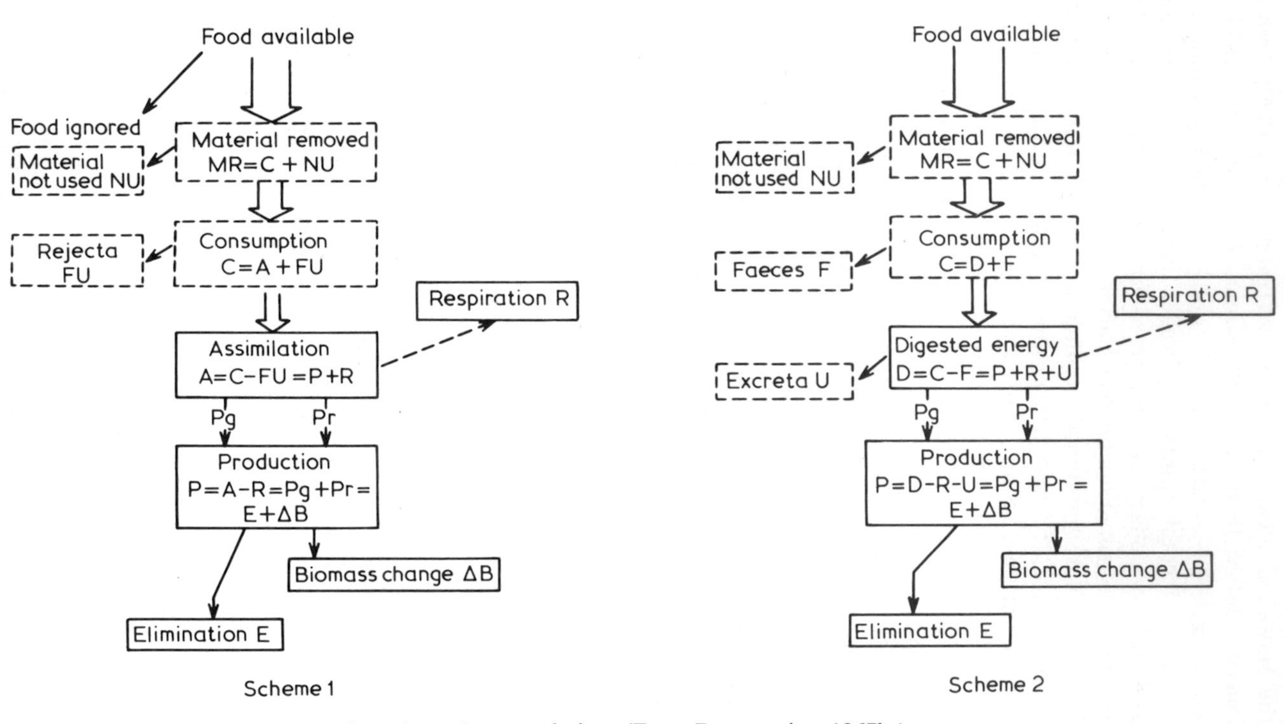

Fig. 4.10 Schemes of energy flow through a population. (From Petrusewicz, 1967b.)

snow clip virtually all the monocot stems present while Pitelka (1957) and Tikhomirov (1959) reported a similar situation with high densities of *L. obensis* and *L. trimucronatus*. Not all of the clipped stems are ingested and therefore constitute material removed but not consumed. The quantitative effects of clipping in tundra ecosystems have been estimated by Dennis and Tiezsen (1972) and Bunnel (1973) to be equivalent to a reduction in the standing crop of green stem bases from 15–30 g m^{-2} in September to 10 g m^{-2} by January.

In a summer following peak lemming densities the monocot standing crop may be depressed 50 per cent below that found in exclosures (Batzli, 1975). Where soil has been overturned to gain access to rhizomes production may be depressed by as much as 90 per cent (Pitelka and Schultz, 1964); further, Peiper (1963) observed that mosses are similarly devastated by high lemming densities. A decrease in yield equivalent to a destruction of 5 kg of plant material per animal was noted by De Vos (1969) in an enclosure containing 15 kangaroo rats per hectare.

Light grazing by small mammals may actually stimulate plant production. For example Smirnov and Tokmakova (1971, 1972) showed that moderate grazing by *Microtus oeconomus* and *M. middendorffi* increase the production of *Eriophorum* and *Carex* by stimulating new shoot growth. However, above the optimal density of 30–50 voles ha^{-1} grazing caused a decline in productivity.

Small mammal foraging may also substantially alter the species composition of plant communities. In many instances, specific plant species compositions are found associated with rodent colonies. For example, Kucheruk (1963) found that *Marmota* spp. colonies were characterized by *Hyosciamus niger, Chenopodium foliosum* and *Sisymbrium* spp.; *Alticola strelzowi* colonies by *Crossilaria acicularis,* and *Lugurus* spp. by *Carduus unicinatus*. Furthermore, it may take several years for the plant composition to return to its initial composition after the colonies have been abandoned, in the case of *Spalax micropthalamus*, for instance, the period is 4–6 years (Skvorcova and Utehin, 1969) while the period is 20 years for *Citellus* colonies (Formosov *et al.*, 1954).

Small mammals may alter the vegetation of an area by selective feeding. In a series of exclosure experiments, Batzli and Pitelka (1970) showed that grazing by *Microtus* kept the habitat open and increased plant species diversity. If voles were excluded then grasses, the preferred food type, increased and became dominant. Koford (1960) also asserts that the foraging activities of prairie dogs tend to make vegetation composition more heterogeneous.

Both Tikhomirov (1959) and Pjastalova (1972) suggest that microtine activity enhances the dominance of monocots over mosses in the tundra. The end result of this, they conclude, at least in the Yamal and Taimyr peninsulas of the USSR, is the creation of sedge and grass meadows.

Batzli (1975) comments that results from exclosure experiments conducted at Barrow, Alaska over the last 18 years would seem to indicate that the effect of microtine activity on plant species composition is dependent upon habitat. Thus in upland areas when grazing is prevented lichens and moss develop at the expense of monocots whilst in lowland areas mosses may develop at the expense of monocots as shown by the large amounts of standing dead material surrounding grazed areas. However, in some areas, especially in wet *Eriophorum* sites, monocot growth may be more vigorous within the exclosure. Despite the fact that none of these trends have been quantified, Batzli (1975) concludes that in the long term microtine activity will disrupt the ground layer of mosses and lichens in such a way that a monocot dominated vegetation results.

Rodents may also prevent succession by maintaining a sub-climax type of vegetation (Formosov, 1928). Thus, prior to the advent of myxomatosis in Britain rabbits and sheep maintained short grass communities in the lowlands by grazing down tree seedlings. Koford (1958) pointed out that the activities of prairie dogs (*Cynomys* spp.) maintain and extend short grass associations while Tikhomirov (1959) and Carl (1971) record that ground squirrel living in tundra habitats graze down *Dryas* and shrubs thereby allowing invasion of herbs and grasses, in particular *Calamagrostis* and *Poa*.

Small mammal activity may also result in a series of secondary successions because digging and the formation of runways creates vegetation free patches. Tikhomirov (1959) records that at times of peak lemming densities runways may cover 20 per cent of the surface and holes can occur at a density of 1 m^{-2} in heavily utilized areas. The result of these disturbances is the production of sites that give rise to early successional associations and hence a mosaic of vegetation types across the habitat.

Bond (1945) comments that under some conditions prairie rodents and lagomorphs may assist in the recovery of a deteriorated habitat via differential pressure on plant species typical of early successional stages. For example, (i) in Oklahoma, prairie dogs in association with domestic stock control the shrub component of the community and appear to be able to maintain indefinitely a condition resembling short or mixed grass prairies (Osborn, 1942), (ii) in enclosures on the San Joaquin experimental range in California *Citellus beecheyi*, over a five year period had the effect of substantially decreasing the abundance of *Erodium botrys* and *Lupinus bicolor* while increasing *Bromus mollis,* a grass of a higher successional stage than the two forbs (Horn and Fitch, 1942), (iii) Vorhies and Taylor (1940) noted that on ranges that are not overgrazed by domestic stock, the wood rat (*Neotoma albigula*) by feeding preferentially on woody plants and weedy herbs, may to some extent aid the climax perennial grasses to re-establish themselves.

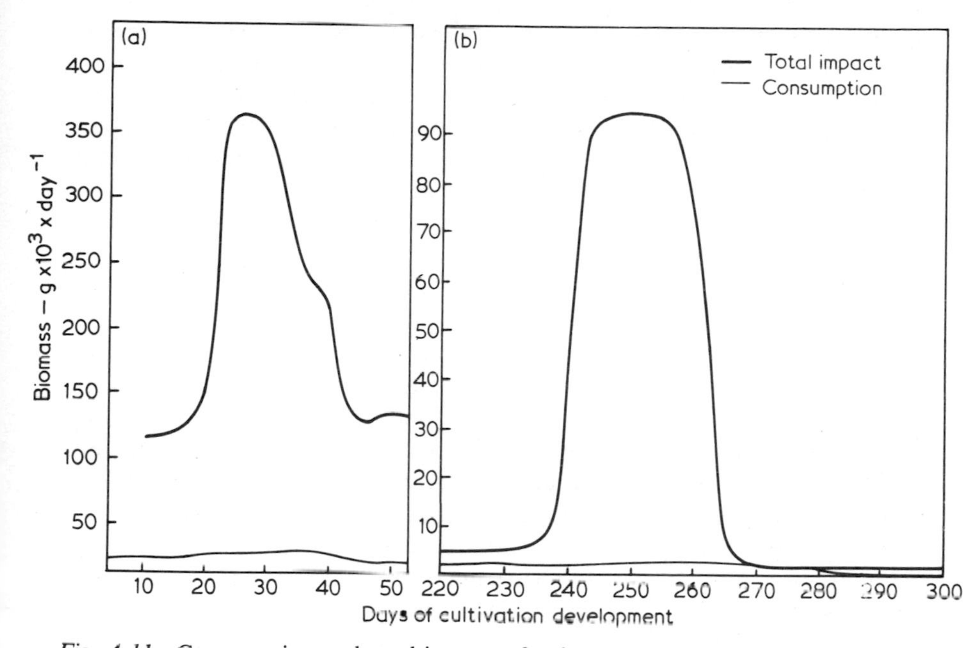

Fig. 4.11 Consumption and total impact of vole population in outbreak years on alfalfa (a) and winter wheat (b) crops. (From Grodziński *et al.*, 1977.)

Grazing by the desert rodent *Rhombomys opimus* can either destroy the surface vegetation or alter the species composition in favour of invading species which include a number of annual grasses and Cruciferae (Naumov, 1975). Suppression of flowering of *Eriophorum angustifolium* and the tundra grass *Dupontia fischeri* has been attributed to grazing by microtines (Tikhomirov, 1959). Batzli (1975) comments that the flowering of *Eriophorum* in north Alaska varies widely though Pitelka (after Batzli, 1975) concluded that flowering was heaviest during the summer before a lemming peak. Tast and Kalela (1971) however found no differences in the flowering of *Eriophorum* inside and outside exclosures though, as Batzli (1975) points out they give no data for a summer following a lemming peak.

Small mammal grazing may alter the stature of a plant without actually affecting production. For example Spitz (1968) found that *Microtus arvalis* had little impact on alfalfa production but strongly influenced the size of stems and the leaf/stem ratio. Grodziński *et al.* (1977) also record a similar effect by the same species on alfalfa crops in Poland. They demonstrated that the total impact of *M. arvalis* was far in excess of that indicated by consumption of A.P.P. alone. Figure 4.11a shows changes in consumption and total impact during a 42 day period of alfalfa growth. Though maximum daily consumption represented only 0.18 per cent of the potential alfalfa crop other effects such as loss of yield, weed succession and crop quality represented some 2.5 per cent of the potential crop, 13 times greater than that indicated by consumption during the period of highest alfalfa damage. In other periods these impacts exceeded consumption by a factor of 4–8. Fig. 4.11b illustrates a similar phenomenon for winter wheat. In both these cases there was a reduction in the quality of the crop rather than quantity, a factor of considerable importance in agricultural production.

Tahon (1969) referred to 'the non-related criteria between food consumption by small mammals and waste caused to vegetation' and stressed that for man at least it is loss of potential yield rather than the amount of material consumed that it critical. This is particularly important for trees where grazing by small mammals may substantially reduce future yield.

Small mammals will preferentially graze tree seedlings because of their high nutritional value. Dinesman (1961) found that young trees injured by *Microtus socialis* were reduced in height though only a few of the older trees died as a result (Table 4.5). Buchalczyk *et al.* (1970) estimated damage done to *Populus nigra, Salix purpurea* and *Alnus glutinosa* by *Microtus oeconomus* at 48–67 per cent of all the trees surveyed though only 6–7 per cent of the trees died as a direct result of microtine activity. By contrast, tree seedlings are destroyed in a much higher proportion than older trees; for example, Golley *et al.* (1975a) ascribe the destruction of 78 per cent of seedlings in a deciduous forest to *Apodemus flavicollis*.

Table 4.5 *Effects of injury by* Microtus arvalis *on the height of different species of tree. (Based on Dinesman, 1961).*

Tree	Average height when uninjured (cm)	Average height when injured (cm)
Maple (*Acer negundo*)	123	106
Elm (*Ulmus pinnato-ramosa*)	123	105
(*Elaeagnus angustifolia*)	118	103

Sviridenko (1940) placed trees into three categories according to rodent damage:

(i) 80–100 per cent of seedlings destroyed: oaks, elms, maples, lime trees
(ii) 50–60 per cent of seedlings destroyed: ashes, rowan
(iii) 0–20 per cent of seedlings destroyed: hazels, bird cherries.

There is, in general, a lack of data about the ratio of tree biomass consumed to biomass destroyed though Pivovarova (after Dinesman, 1961) compares the destruction brought about by rodents at peak densities to that caused by insects and fungi (Table 4.6).

Tahon (1969) emphasizes that for commercial trees and shrubs, especially orchard trees, the effects of grazing may be particularly disastrous in terms

Table 4.6 *Destructive activity of rodents at peak density compared with that of fungi and insects (Provovarova (after Dinesman, 1961)).*

Species of seedling	Number of seedlings ha^{-1}		Percentage injured by rodents	Percentage injured by fungi and/or insects
	Beginning of spring	End of autumn		
Hornbeam (*Carpinus*)	20000	4600	32	18
Maple (*Acer*)	7200	1660	25	44
Oak (*Quercus*)	8800	2800	41	27

of long term production. He pointed out that the disappearance of a pear tree due to consumption of 3 per cent of its biomass by *Arvicola terrestris* results in a loss of twenty years of production; a loss of 500 kg of economic yield for a consumption of only 2 kg of tissue. In the case of beech plantations a consumption of a mere 2 kg of tissue can result in a loss of 6 tonnes of economic production.

The problem is somewhat different in a mature beech forest where populations of rodents have no great influence on the production of wood. Consumption of dead leaves, twigs, fruit and some roots (Watts, 1968) causes 'no fundamental effect on future growth' (Tahon, 1969). Rodents may not cause appreciable harm to mature beech trees but there is a strong feeling among timber growers in Britain that the damage to pole stage beech caused by the bark stripping activities of the introduced grey squirrel *Sciurus carolinensis* will be reflected in timber quality when these trees reach the saw mills.

A number of options are open to plants in their response to small mammal grazing. Firstly, plants may increase production if grazing is only light and interferes with plant maturation (Golley *et al.*, 1975b). Secondly, there may be a switch between plant parts or in species composition towards those plants or parts which are less favoured food (Golley *et al.*, 1975b). Thirdly, palatability may change with time with the production of plant toxins or protective spines. Although a complex subject Whittaker and Feeny (1971) concluded that palatability is a compromise between selection of plant materials for nutrients and avoidance of toxins with body size, plus the form of the digestive tract influencing the relationship of the animal to toxic compounds. Very little is known about the effects of plant toxins on small mammals though it may be of considerable economic importance. For instance, Golley *et al.* (1975b) suggest that devastation of agricultural crops by rodents may be due to the selection by man against toxins in food crops.

There is considerable circumstantial evidence to support the contention that seed predation by vertebrates is an important regulating factor for at least some plant species (Janzen, 1971a). The impact of seed predation by small mammals on plant reproduction has been most fully documented for forest trees. High seed losses are common among forest trees (Smith and Aldous, 1947; Tevis, 1953; Shaw, 1968) and these losses are thought to influence forest regeneration. Small mammals are certainly capable of consuming large quantities of seed. Radvanyi (1966) found that small mammals consumed 47 per cent of the *Picea glauca* seed in a forest where the density of mice was only 15 ha^{-1}. Abbot (1961), using white pine seed in feeding hoppers, showed that *Peromyscus maniculatus* and *Clethrionomys gapperi* were respectively capable of consuming 260 and 232 seeds $animal^{-1}$ day^{-1}, while Mathies *et al.* (1972) working in hickory forests in Tennessee estimated that *Blarina brevicauda* and *P. maniculatus* respectively consumed 64 and 17 per cent of pine seeds supplied in hoppers. Using an exclosure technique, Gashweiler (1970) estimated that, depending on tree species, between 22 and 44 per cent of the conifer seeds were destroyed by mice and shrews in Oregon. Further, Bramble and Goddard (1942) demonstrated that seeds protected from small mammals had an increased probability of survival and germination. Finally, Zhiyakov (1977) found that small mammals consume 34–60 per cent of the Schenk's spruce seed that fall in Northern Shan province, USSR.

Smith and Aldous (1947), Spencer (1955), Ashby (1959) and Gashweiler (1970) all concluded that such seed losses would adversely affect forest regeneration. Dinesman (1961) summarizing the work of other authors, states that *Apodemus flavicollis* consumes all the maple seeds available in a woodland while all the elm and hazel survive. In Kankaz forests this mouse consumes all the beech seed. Zhukov (1949) demonstrated that the number of oak seedlings in a forest was dependent on rodent density (Table 4.7).

Table 4.7 *Consumption of oak seedlings at different densities of rodents (After Zhukov, 1949 and Golley* et al., *1975a).*

Crop of acorns	Number of seedlings ha^{-1}	
	Without rodents	At peak density of rodents
Very poor	280	138
Poor	1880	920
Moderate	9870	4830
Good	108600	52900

Unlike the abovementioned authors, Tanton (1965) thinks the impact of seed predation on forest regeneration has been overemphasized. It is certainly true that most tree species are long lived and hence experimental studies to evaluate the effect of seed loss due to small mammals are, because of the time factor, difficult. However, the same is not true for short-lived species and there is some evidence to indicate that rodent seed predation is an important mortality factor. In Californian annual grass populations for example, Marshall and Jain (1970) found that seed predation on *Avena fatua* varied, with site, between 0 and 65 per cent. Batzli and Pitelka (1970) failed to detect significant seed losses in grasslands during *Microtus californicus* population peaks but calculated, on bioenergetic grounds, that at least 35 per cent of the available seeds should be consumed. The reason for this discrepancy was, they suggest, due to early summer storage of seed by the vole, something their experiments would not have detected. Pearson (1964) made similar estimates of the energy requirements of peak populations of small mammals and concluded that in his study area only 7 per cent of the annual seed crop would escape predation.

Borchert and Jain (1978), using exclosures and plots sown with known quantities of seed, demonstrated that a mouse population with an approximate density of 296 ha^{-1} consumed 75 per cent of *Avena fatua* seed, 44 per cent of *Hordeum leparinum* seed and 37 per cent of *Bromus diandrus* seed.

The abundance of certain members of a plant community whose seeds represent the favoured food of small mammals can be strongly affected by seed predation. Thus Soholt (1973) found that in the Mojave desert there was a reduction in the numbers of *Erodium cicutarium* which, owing to its high nutritional value, is the favoured food of kangaroo rats; in effect this species consumed 95 per cent of the seed. Janzen (1971a) used this type of argument when suggesting that the relatively low abundance of hickory and beech in eastern forests of the U.S.A. was due to the preference by squirrels for the fruits of these species. It is of interest that in the study by Borchert and Jain (1978) *Avena fatua* seeds were chosen twice as frequently as any other species studied, even though *Avena* represented only 4 per cent of the total plant community cover. They concluded that the 'cumulative effect of predation over the years could be to change the relative abundance of a species in the plant community to a low equilibrium level.'

Small mammals could, therefore, regulate annual grass populations in three ways; but cutting mature plants, eating seedlings, or consuming the seed crop. Batzli and Pitelka (1970) found that a population of *Microtus californicus* which exceeded 1500 voles ha^{-1} caused a substantial reduction in the densities of *Avena fatua, Lolium multiflorum* and *Bromus diandrus*. Losses of seedlings due to small mammal consumption tend to represent a component of overall high seedling mortality and not necessarily an increased pressure on population numbers (Borchert and Jain, 1978). In practice, reduction of seed numbers by small mammals may increase the probability of a seed becoming a mature plant by reducing intra- and interspecific seedling competition.

Seed losses can occur in two stages. Firstly, heavy cropping of plants during reproduction can suppress flowering (e.g. Tikhomirov, 1959) and reduce the number of viable seeds below carrying capacity; for example, Batzli and Pitelka (1970) reported a 70 per cent seed loss following heavy *Microtus* clipping. Secondly, seeds may suffer post-dispersal losses due to predation.

Using population data from a number of authors (Pearson, 1963; De Long, 1967; Batzli and Pitelka, 1971) combined with information on seasonal and annual fluctuations in plant numbers (Talbot *et al.*, 1939; Heady, 1976; Bartolome, 1976), Borchert and Jain (1978) concluded that for most years 'rodent seed predation could regulate populations of California annual grasses either by influencing plant numbers in relation to the carrying capacity or by changing the outcomes of interspecific plant competition for each of these species.'

Janzen (1971a) stated ' ... seed predation [is] the cost of reliable dispersal and [is] directly analogous to a juicy fruit or complex exploding pod.' He also pointed out that it is difficult to separate the act of seed predation from dispersal for a given food type because the end result of seed handling by small mammals may either be death (seed eaten) or successful germination (seed cached but forgotten). In general, the importance of seed predators is that they help to determine the 'seed shadow' of a plant which in turn determines where the offspring, if any, of a particular plant will be located. Such effects may ultimately help determine the density of a particular species and also the plant species composition of the habitat.

Small mammals are not only important as seed predators but also as seed dispersers. In tropical ecosystems the importance of this role of small mammals has been demonstrated by Janzen (1971b, 1972) and Wilson and Janzen (1972). In Costa Rica, the vine *Dioclea megacarpa* suffers some 'pre-dispersal mortality' due to milk stage predation of seeds (0 – 43 per cent but usually less than 10 per cent of the seeds are eaten) by *Sciurus variegatoides* but this animal is also one of the plant's main dispersal agents. The squirrel, along with other mammals such as agouti and paca, carries mature fallen seeds away from the vine thereby reducing predation by the bruchid beetle *Caryedes brasiliensis* and the larvae of the noctuid moths.

A similar situation obtains in the case of tropical trees such as the palm *Scheelea robusta* and the Panama tree *Sterculia apetala*. Seeds of these trees, when not dispersed by mammals, suffer intense seed predation by either two species of bruchid beetle (*Scheelea*) or the cotton stainer bug *Dysdercus fasciatus* (*Sterculia*). The further a seed is removed from a conspecific plant the greater is its probability of germinating (Janzen, 1970).

Few quantitative data are available on the effects of seed caching by small mammals. Certainly as Sviridenko (1957) has demonstrated, large quantities of seed may be stored (Table 4.8). West (1968) concluded that seed caches may result in clumps of seedlings and that 50 per cent of the bitter bush and 15 per cent of the ponderosa pines in Oregon resulted from rodent

Table 4.8 *Storage of seeds and other foods by rodents (after Sviridenko, 1957).*

Species	Food stored	Amount stored
Sciurus vulgaris	Acorns	650 seeds
Eutamias sibiricus	Various seeds	up to 8 kg
Citellus undulatus	Various seeds and grasses	up to 6 kg
Glis glis	Nuts	up to 12–15 kg
Spalax leucodon	Potatoes	up to 12 kg
Microtus arvalis	Herbs, grasses	up to 1.4 kg
M. socialis	Seeds	up to 1.2 kg
M. nivalis	Grass	up to 2.0–2.3 kg
Arvicola terrestris	Roots	up to 1 kg
Apodemus flavicollis	Acorns	up to 4 kg
Cricetus raddei	Seeds, herbs, fruit	up to 7 kg
*Cricetus cricetus**	Seeds	3477.76 mJ ha^{-1} $year^{-1}$

*From Górecki (1977).

seed caches. Golley *et al.* (1975b) presumed that the pattern of tree spacing, location of seeds in sites favourable to germination, and the improvement of the chances of survival must be of some advantage to plants though the exact nature of this advantage still remains to be identified and quantified.

4.6.3 Activities other than consumption: their impact and the functional relationships between primary consumers and abiotic state variables

Reference has already been made to the indirect impact of burrowing and trampling on vegetation but these activities clearly have a direct impact on the abiotic state variable soil.

Small mammals can dislodge prodigious quantities of earth which, apart from altering the drainage characteristics of soil, modifies the microtopography. Voronov (1953) made quantitative measurements on the burrowing activity and structure of nests for a variety of mammals; Table 4.9 summarizes his data and the relative importance of individual *Talpa europaea* and *Arvicola terrestris* in disturbing soil can be clearly seen. Kuznetzov (1970) established that 11–12 moles (*T. europaea*) could dislodge 0.15 m^3 of soil ha^{-1} day^{-1} in deciduous forest while Zimina *et al.* (1970) estimated that marmots (*Marmota* spp.) could move 18–20 m^3 of earth ha^{-1} $year^{-1}$. In a laboratory experiment, Dufour (1971) showed that *Apodemus sylvaticus* could dislodge from 1–3 kg of earth in two hours, an amount corresponding to 50–150 times the animal's own body eight. Hodashova (1950), working with *Microtus socialis*, showed that this species

Table 4.9 *Burrowing activity of small mammals (Voronov, 1953).*

	Length of corridor (cm)	Volume of nest (cm^3)	Amount of removed soil (m^3)
Talpa europaea	6000	1096	0.0770
Apodemus sylvaticus	100	351.2	0.0012
Apodemus flavicollis	100	2280	0.0043
Clethrionomys glareolus	95	283	0.0011
Arvicola terrestris	1500	4886	0.0530
Microtus arvalis	40	3299	0.0037

can disturb 2.1 m^3 of surface earth ha^{-1} at peak density and that this soil may cover 0.004 per cent of the animal's territory. For *Pitymys subterraneus*, Novikov and Petrov (1953) estimated the amount of soil brought to the surface to be 12 000 kg ha^{-1} $year^{-1}$. Golley *et al.* (1975b) point out that voles at a density of 200 – 400 ha^{-1} probably dislodge 10 m^3 of earth ha^{-1} $year^{-1}$ and that this activity is restricted to the top 40 cm of soil; a simple calculation indicates that this is equivalent to 0.25 per cent of 'available' soil, i.e. a soil turnover time of 400 years. Fossorial rodents such as *Ellobius* spp., *Myospalax* spp. and *Spalax* spp. move between 13 and 15 m^3 of earth m^{-2} $year^{-1}$.

Such activities influence microtopography and hence surface water runoff. Lemming burrows, for example, are often found to be associated with stone polygons and Batzli (1975) has suggested that these voles may influence indirectly the formation of such polygons on the well-drained soils of northern Alaska. Lemming runways are frequently found in association with frost cracks, particularly in polygon troughs, and although it is debatable whether these animals are merely selecting a suitable habitat or actually altering microtopography they must at least be altering the drainage and thermal regime of the soil (Pitelka, 1957; Tikhomirov, 1959).

Because of water entering the holes, the moisture content of the soil close to the burrows is higher than that at the same depth in undisturbed soil. For example, Kazmarczyk (quoted in Golley *et al.*, 1975b) compared the percentage difference in moisture due to the burrows of *Microtus arvalis* in alfalfa fields with that of unburrowed soil at the same depth:

Days after rain	Percentage difference in burrowed soil
1	+ 1.5
4	+ 1.1
20	+ 0.8

Small mammal activities also have an effect on soil density; Abaturov and Zubkova (1969) found that in *Citellus pygmaeus* colonies the relative density of solonetz soil was 1.29 compared with 1.45 for unaffected soil of the same type.

Soil disturbance affects the distribution of chemicals; Abaturov *et al.* (1969) reported that on the dry steppe *Citellus pygmaeus* dislodges 1500 kg of soil ha^{-1} $year^{-1}$ from depths of 40–200 cm. In this area of steppe the water table is low and there is a corresponding lack of capillary movement, hence little sousliks are the main agents for renewal of minerals at the upper surface of the soil. The raised soil was found to contain 14 kg ha^{-1} of SO_4; 5 kg ha^{-1} of Na; 3 kg ha^{-1} of Cl; 77 ha^{-1} of Ca; 122 kg ha^{-1} of Al; 49 kg ha^{-1} of Fe; and 18 kg $ha^{-1}$of S. Later estimates by Abaturov (1972) for spruce forests indicate that moles dislodge 19 000 kg of soil ha^{-1} $year^{-1}$ from depths of 10–40 cm and that this amount of material contains 47 kg of Si; 47 kg of Fe; 139 kg of Al; and 30 kg of Ca and Mg.

Burrowing clearly leads to a mixing of the soil horizons and Kucheruk (1963) noted that in soil burrowed by rodents organic matter increased below the A horizon. In some instances, for example, *Spalax* colonies, the humus horizon can extend to a depth of 20 cm; this is accompanied by an increase in the Ca and Mg content which alters the pH.

It is obvious that small mammal activity can affect the distribution of minerals within soils and it is of interest to note that their activities may also influence the rates of mineralization of organic matter; Hodoshova (1970), for example, reported an increase in decomposition rates in the presence of rodents. Burrowing does alter the crumb structure of soils and this alone will affect decomposition by changing the surface area of organic material available to bacteria and fungi. However, moisture and pH changes brought about by small mammal activity will also influence decomposition processes.

In addition to influencing physical changes in the soil small mammals also contribute to its chemical nature, particularly its organic content and the rate of mineral cycling, which in turn may well affect plant stature. Zimina *et al.* (1970) noted that plants growing within colonies of *Marmota* spp. were taller than those growing outside. Similarly, Formosov and

Table 4.10 *Effect of* Meriones tamariscinus *colony on the growth of* Xanthium strimarium *L. (Kucheruk, 1963).*

	Within colony			Outside colony		
	Mean	Minimum	Maximum	Mean	Minimum	Maximum
Number of stems	5.5	1.0	14	2.2	1.0	7.0
Height (cm)	193.5	160.0	220	116.5	95.0	145.0
Maximum diameter of stem (cm)	3.3	1.8	5.0	2.0	1.2	2.7

Voronov (1939) reported that within colonies of *Lagurus lagurus* the plants *Stipia, Festuca* and *Artemisia* appeared to be larger. The stature of *Xanthium strimarium* was shown by Kucheruk (1963) to be strongly influenced by the presence of *Meriones tamariscinus* (Table 4.10). It can be assumed that rodent activity enhances the availability of plant nutrients. Small mammal faeces and urine probably have an important role to play and Kucheruk (1963), following Richards, has pointed out that the faeces of rodents may influence the growth of *Azotobacter* in the soil. Batzli (1975) stated that the deposition of faeces and urine, combined with fast rates of decomposition, could lead to rapid accumulation of nutrients at the soil surface. Górecki (1977) calculated that hamsters must return annually some 15 kg of faeces and urine to each hectare of cultivated land while Kucheruk (1963) estimates that the steppe rodents of Eurasia produced up to 2500 kg ha^{-1} $year^{-1}$. Urine, and some faeces, are very rich in nitrogen (Drożdż, 1968) and are thus an important component of the nitrogen cycle. Ulehla *et al.* (1974) found that lucerne around the burrows of hamsters had a greater biomass per unit area than elsewhere, a fact they attributed to the 'particularly intense fertilization' in the vicinity of the burrows.

Besides urine and faeces, the non-ingested products of clipping add organic material to the soil sooner than expected and accelerate the rate of mineral cycling. Stark (1973) described the effect of *Tamiasciurus* on Jeffrey pine as an increased litter fall of 18 g m^{-2} $year^{-1}$ which added an extra 137 g Ca; 284 g Na; and 18 kg K per 0.25 ha to the litter. It was suggested by Golley *et al.* (1975b) that this may be an important effect in a number of forest ecosystems.

Clipping is particularly imprtant in tundra ecosystems where it may affect the thermal regime of the soil. According to Pitelka and Schultz (1964) the clipped material that is left on the soil surface is practically all inedible and the reduction in stature of the surface vegetation seems to result in a reduction of its insulating properties; the end result being a deeper thaw in unclipped areas. In low lying regions, the spring melt water moves the clippings about and windrows may form at pond edges as the water recedes; this leads again to a reduction in the depth of thaw, also a decrease in shoot growth and a possible production of peaty mounds.

Clipping, by adding organic material to the soil, plays an important part in mineral cycling; which in turn appears to be correlated with lemming cycles. Nutrient concentrations in the vegetation have been shown to be high in those summers when lemming populations are at a peak; it has been suggested that some of the nutrients which reach the soil surface via clippings, faeces and urine are rapidly released, leached into the soil, and quickly reabsorbed by plants (Peiper, 1963; Schultz, 1969; Bunnel, 1973). Concentrations of nutrients in the vegetation are low in the summer following a lemming peak and it would appear that the surface litter nutrients have

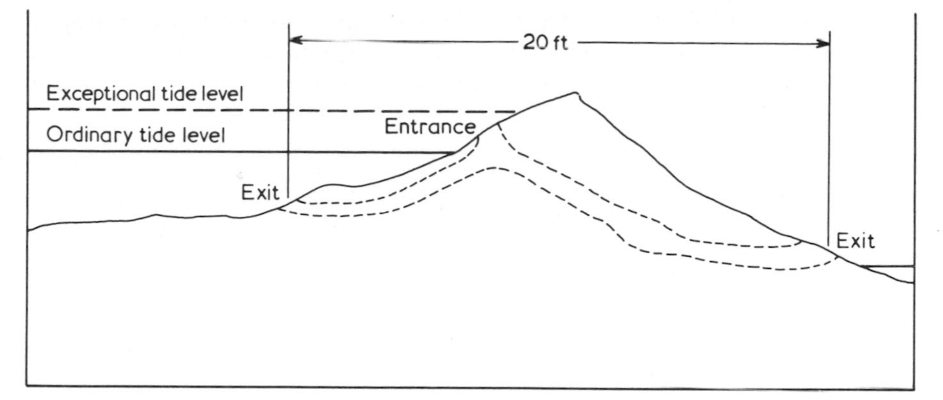

Fig. 4.12 Section of river wall showing tunnel made by coypus, at Haddiscoe in Norfolk.

not been readily released during the winter and spring because of the low temperatures caused by lack of insulation from the vegetation. Some two to three years is required for decomposition to occur under ameliorated cover conditions such that the vegetation again becomes nutrient rich and capable once more of supporting high lemming densities. Some evidence to support this hypothesis has been provided by Peiper (1963) and Schultz (1969) who have noted that the nutrient levels of vegetation taken from lemming free exclosures do not show cyclic fluctuations. Although cyclic patterns in plant nutrient content require further explanation there is clearly some link between nutrient flux and lemmings in tundra regions; it may be that similar phenomena exist in other ecosystems.

Thus far, we have dealt mainly with the useful role of small mammals in relation to abiotic state variables there are, of course, destructive aspects, particularly in man-made systems. One such example can be found in the U.K. where the introduced South American rodent, *Myocastor coypus*, burrows into the dyke banks and river walls of East Anglia. The burrows are some 20 – 50 cm in diameter and can extend up to 5 m; the entrance at the river end is at water level and there can be an opening at the landward side. Figure 4.12 is based on a diagram from a Coypu Strategy Group Report (1978) and illustrates how coypu activity can breach flood defences with consequent economic losses to agriculture in periods of exceptional high tides.

Because of the diverse nature of the functional relationships that exist between small mammal and abiotic state variables, and because of the difficulty of quantifying the transfer functions, very few attempts have been made to produce quantitative models of these relationships. Woodmansee *et al.* (1978), however, have produced a model for the nitrogen budget of a short grass prairie ecosystem. It is a synthesis of data from several sites in the U.S.A. and shows that the role of small mammals in effecting nitrogen flux is greater than that of other vertebrate groups (birds, snakes, antelopes) but nevertheless is much smaller than that of either above ground or soil invertebrates.

4.6.4 Small mammals as secondary consumers: their impact and functional relationships with other consumer state variables

The role of small mammals as secondary consumers within ecosystems has not been investigated as fully as their role as primary consumers. A lack of information about bat feeding, particularly with respect to insectivorous forms, has already been noted and not much more is known about the Rodentia, Insectivora or Carnivora.

Many rodents eat material of both plant and animal origin, strictly speaking they are omnivores and yet most studies on the impact of their feeding activities have concentrated on herbivory and granivory as opposed to carnivory. The work of Drożdż (1966) was exceptional in this respect and examined the total amount of food available in a beech forest to *Clethrionomys glareolus* and *Apodemus flavicollis*. The secondary production available to *C. glareolus* amounted to a mere 2.6 per cent of the total available food supply, the equivalent figure for *A. flavicollis* was 4.3 per cent. In the light of these values it is not surprising that most effort has been directed towards elucidating the functional relationships between rodents and the primary producer state variable.

Insects and other invertebrates fall into the category of items that might be eaten by insectivorous small mammals. In the case of the mole, *Talpa europaea*, numerous examinations of stomach contents have been carried out by various authors, but quantitative estimates from these studies are very difficult since digestion rates vary according to food type (Godfrey and Crowcroft, 1960). Bolton (1969) studied mole stomach contents with season and found that in his alder-birch woodland study area the earthworm *Allolobophora rosea* occurred commonly in the diet; despite total earthworm densities of 102 m^{-2} he expressed the view that large *A. rosea* ($\approx$ 20 m^{-2}) were the major object of continuous and high predation.

Shrews are voracious insectivores and Varley and Gradwell (1968) have shown them to be a density dependent factor in the population dynamics of the winter moth, *Opheroptera brumata*, in Wytham Woods, Oxford. Pernetta (1973, 1976a, 1976b) made a detailed study of the two species of shrew (*Sorex araneus* and *S. minutus*) in Wytham and showed that together they consumed an average of 217.36 MJ ha^{-1} $year^{-1}$. Unfortunately we do not have any idea as to what percentage of the total available food supply this figure represents.

There can be little doubt that insectivorous small mammals have a considerable impact on invertebrate populations but in the absence of adequate field data it is not possible to determine whether, as Holling's (1959) experiments with *Sorex* and *Blarina* suggest to be a possibility, they ever regulate prey populations.

Small mammal insectivores, as well as having a direct impact on their prey, could influence indirectly the primary producer state variable. Destruction of large parts of the larval population of larch sawflies by shrews (Buckner, 1964, 1966) or indeed the consumption of any invertebrate herbivores, may affect the primary production of our ecosystem by reducing plant losses due to invertebrates. Conversely, small mammals may destroy those invertebrates which are themselves predators or parasites of phytophagous invertebrates. For example, Obtrel *et al.* (1978) estimated that *Apodemus sylvaticus* destroyed up to 50 per cent of the eonymphae of *Diprion pini* (Hymenoptera: Diprionidae) overwintering as cocoons in Moravia.

Southern (this volume, Chapter 3) has already provided examples of the intensity of predation by small carnivores on small mammals (e.g. MacLean *et al.*, 1974; Pearson, 1966, 1971; Fitzgerald, 1977) and these will not be discussed further. We have, however, calculated a provisional figure for the impact of the weasel *Mustela nivalis* on small mammal prey in Wytham Woods, Oxford. Using data on weasel energetics from Moors (1974, 1977) and Hayward (unpubl.) and population data from King (1975) we estimate that between September 1968 and September 1969 weasels consumed 14.2 per cent of the small mammal production. During this one year, out of thirty recorded, the small mammals were present at a medium level of density. The result of weasel predation must be a reduction in the impact of small mammals on the rest of the ecosystem although the extent of this reduction will no doubt vary according to the demographic states of the small mammal populations (*see* Chapter 3).

4.6.5 Small mammals as prey items: their impact and functional relationships with other consumer state variables

Small mammals serve as a food supply for a number of predators. Southern (this volume, Chapter 3) has referred to the relationships that exist between small mammals and their predators and the possible role of small mammals in predator-prey cycles. He has also pinpointed the effect small mammals have on the reproductive success of the tawny owl (*Strix aluco*). The interactions between small mammals and their predators are complex, Ryszkowski *et al.* (1973) have provided one such example in discussing the reciprocal role of hares and the vole *Microtus arvalis* in the diet of foxes. A similar situation occurs in Wytham Woods where reciprocal relationships exist between tits (Paridae) and voles in the diet of weasels (Dunn, 1977).

In general, as pointed out by Golley *et al.* (1975b), interactions between the state variable 'small mammals' and other consumer state variables should be interpreted in terms of transfer functions within the consumer complex. Such evaluations are important because the internal stability of the consumer complex is essential to system stability.

4.7 Energy budgets

In section 4.6 we laid stress on the fact that the dimensions of system variables and transfer functions can be expressed in a variety of units; further, we made the point that the choice of units depends entirely on which aspect of structure or function one is currently interested in. For certain purposes units of energy have advantages over species number, density, or biomass in units of weight. Calories, joules or watts are either expressions of the amount of bound energy potentially available for work or the amount of chemically bound energy that has been transformed to heat during work processes. Energy equivalents of those plant parts which form the food items of small mammal herbivores are clearly a measure of the energy potentially available to these animals, while the energy content of the mammals themselves is obviously a measure of the energy potentially available to predators. Energy expended in the course of work, be it food searching, digestion, or burrowing, is obviously a measure of the animals' overall activities and, as such, reflects the part played by small mammals in utilizing the available food energy. Energy units have universal applicability and it is not surprising that attention has been directed towards the elucidation of energy budgets as one means of expressing the functional role of small mammals in ecosystems.

The basic equations used in energy budgets studies, which are illustrated in Fig. 4.10 and appear in Petrusewicz and Macfadyen (1970), are:

$$\begin{aligned} MR &= C + NU \\ C &= A + FU = D + F \\ A &= P + R \\ D &= P + R + U \end{aligned}$$

where MR is material removed, C is consumption, NU is material not utilized, A is assimilation, D is digestion, F is faeces, U is urine, FU is rejecta (faeces + urine), P is production and R is respiration.

The distinction between assimilated energy (A) and digested energy (D) becomes necessary for logistic reasons; in some studies faeces (F) and urine (U) are not, or cannot easily be, separated and hence the employment of the formulation $C = A + FU$. In other studies where F and U are, or can be, measured separately the preferred equation is $C = D + F$. The parameters MR, C, F and U, although not referred to under these symbols, were dealt with in earlier sections and will not be considered in detail here, emphasis is placed on assimilation (A), digestion (D), production (P) and respiration (R). The first two are measures of how much food energy is actually processed while the last two describe the fate of this processed energy.

4.7.1 Assimilation and digestion

The coefficients assimilation/consumption (A/C) and digestion/consumption (D/C) are measures of the efficiency with which animals utilize consumed energy. Grodziński and Wunder (1975) listed the assimilation and digestibility coefficients of a large number of species of small mammal but for the purposes of this account their summary, according to major feeding habits, has been used. For each of the categories grazing herbivore, omnivore, granivore, insectivore and carnivore Grodziński and Wunder (loc. cit.) computed the average assimilation and digestibility coefficients; in some instances only the digestibility coefficient was known and so they calculated the assimilation coefficient as being 2–3 per cent less (the generally accepted level of urine energy content as a proportion of FU). The values are listed in Table 4.11 and shown diagrammatically in Fig. 4.13.

Table 4.11 *Digestibility and assimilation of natural foods in mammals, as percentage of energy or organic matter consumed (After Grodziński and Wunder, 1975).*

Feeding type	Number of species (number of data used)	Digestibility coefficient (%)	Assimilation coefficient (%)
Grazing herbivore	14 (14)	67	65
Omnivore	10 (12)	77	75
Granivore	6 (8)	90	88
Insectivore	9 (9)	–	85
Shrews	6 (6)	–	90
Carnivore	2 (2)	90	–

Grazing herbivores such as voles and lagomorphs are the least efficient in utilizing consumed food and exhibit digestibility and assimilation efficiencies of the order of 65–67 per cent. The coefficients are higher for omnivorous species, for example mice and squirrels, and average 75–77 per cent; while for granivorous rodents, insectivores and carnivores the coefficients reach 90 per cent or more.

Digestibility and assimilation depend upon a number of factors but primarily on the quality of the food itself. Bulky food of high fibre content is generally less digestible than concentrated food and data based on diets of

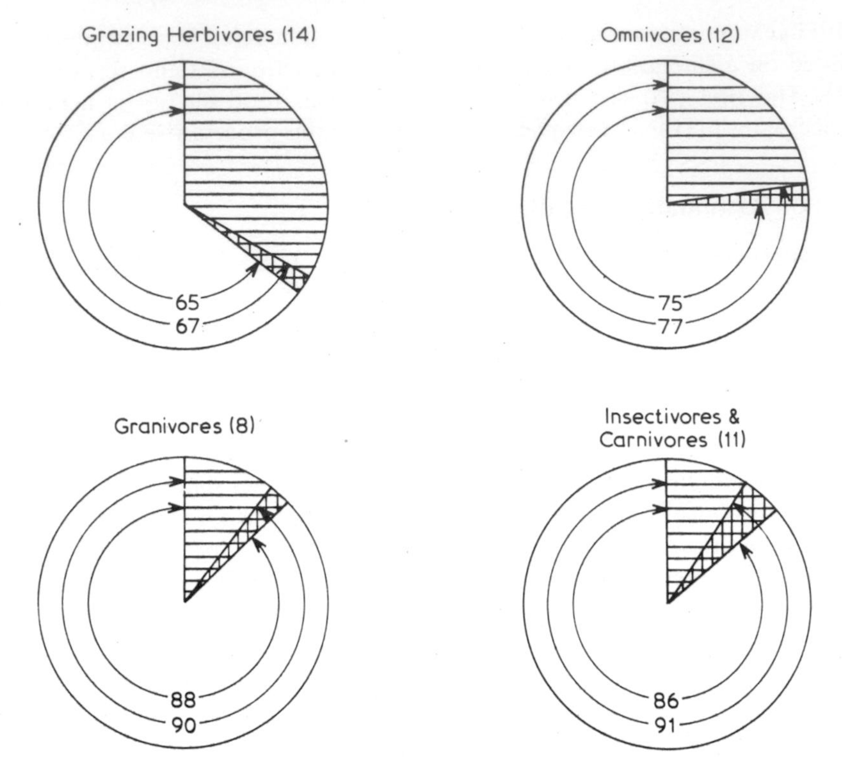

Fig. 4.13 Utilization of food energy by small mammals. The top number on each scheme is the coefficient of assimilation and the bottom number the coefficient of digestibility. Both numbers are given as percentages. The lined areas represent energy lost with faeces and the cross-hatched areas energy lost through urine. (From Grodziński & Wunder, 1975.)

concentrated artificial food are of limited ecological value. Chemical composition is also important (Van Soest, 1966), only a fraction of the cellulose and hemicellulose content of ingested food is utilized and lignin frequently remains undigested. According to Grodziński and Wunder (1975), digestibility is a function of the lignin content of the food. Further, diets which are low in protein but have a high mineral content have a low digestibility, whilst age and physiological condition may also play a significant role.

4.7.2 Production

Production is the amount of consumed or assimilated food which is utilized for body growth and repair of tissues, also for the provision of fat reserves, eggs, sperm, embryos and secretions such as milk. It refers to the total quantity of new material produced during a given time interval, irrespective of whether all the material produced remains to the end of the period. Production is clearly a measure of the amount of new material made potentially available to predators.

With respect to small mammals it is of some interest to consider what proportion of their food appears as production, hence the quotients production/consumption (P/C) and production/assimilation (P/A) are worthy of study. Duncan and Klekowski (1975), who refer to the two quotients as K_1 and K_2 respectively, have pointed out that in terms of the individual these values vary with age. In the early stages of a life cycle K_1 (= (= P/C) can be as high as 0.65, that is, 65 per cent of the consumed food is used for production, while in the final stages it can be zero. Similarly K_2 (= P/A) for embryonic growth may reach 0.80 while that for senescent adults is zero. Winberg (1962) has made the point that the coefficient K_2 is less affected than K_1 by the nature of available food, its digestibility and other external environmental factors, and is thus a more stable index of energy balance. Table 4.12 shows some K_2 values calculated by Duncan and

Table 4.12 *Some efficiencies and ratios for small mammals. [Selected from Grodziński* et al. *(1975), page 139].*

Population	Annual population production (kcal m^{-2} year $^{-1}$	K_2 = P/A x 100	100 – K_2/K_2 = R/P
Microtus pennsylvanicus	0.517	2.95	32.9
Mustola rixosa allegheniensis	0.013	2.34	41.8
Clethrionomys rutilus dawsoni	0.00024	1.89	51.9
Tamiasciurus hudsoni preblei	0.00020	1.37	72.0
Microtus oeconomus macfarlani	0.00024	1.82	54.0
Glaucomys sabrinus yukonensis	0.00002	1.18	84.0
Pitymys subterraneus	0.0955	1.57	60.0
Peromyscus polionotus	0.12	1.79	54.7

Klekowski (1975) for various small mammal populations; the range of values exhibited by individual quotients is damped, the population coefficients range from 1.18 to 2.95, indicating that of the food energy assimilated by small mammals less than 3 per cent appears as production. These low values can be contrasted with those for invertebrate populations which frequently exceed 50 per cent. On a unit area basis, small mammals show poor productivity compared with invertebrates; nevertheless, they form an important source of food for many predators.

In certain circumstances, depending on the period of time over which studies are made and the nature of the equation P = A — R, production can be negative. For example, the little souslik (*Citellus pygmaeus*) has no

Table 4.13 *Comparison of net production of small rodents in different ecosystems (P = total net production due to reproduction).*

Ecosystem	Dominant species	g ha^{-1} $year^{-1}$	mJ ha^{-1} $year^{-1}$	Pr/P %	Source
FOREST TYPE ECOSYSTEMS					
Taiga forest (Spruce), Alaska	*Clethrionomys rutilus* (Pall)	730	4.5980		Grodziński (1971a)
	Tamiascurus hudsonicus	530	3.3440		Grodziński (1971a)
	Microtus oeconomus (Pall)	270	1.6720		Grodziński (1971a)
	5 small mammal species	1667 (400–5200)	10.4500 (2.5100–12.6500)		Grodziński (1971a)
Picetum myrtillosum Vitoska Mountain, Bulgaria	*C. glareolus*	825 (729–1338)	4.9810 (4.4400–5.5900)		Petrusewicz *et al.* (1972)
Picea abies plantation Bjorn Störp, Sweden	*M. agrestis* *C. glareolus* *Apodemus sylvaticus*	3147 (2320–3973)	19.7500 (14.5600–24.9400)		Hansson (1971a)
Fagetum carpaticum Ojców, Poland	*C. glareolus*	1111	6.9800		Grodziński *et al.* (1970)
	A. flavicollis	(469–2083)	(2.8500–12.6500)		Grodziński *et al.* (1970)
Tilio-carpinetum Mazurian lake region, Poland	*C. glareolus*	3017 (2857–3265)	18.3600 (17.3800–19.8600)		Petrusewicz *et al.* (1971)
Tilio-carpinetum Niepolomicka forest, Poland	*C. glareolus*	947	5.7600		Bobek (1973)
	A. flavicollis	366	2.2170		Bobek (1973)
	All rodents	1360	8.2967		Bobek (1973)
Tropical rain forest, Panama		4000	25.3700		Gliwicz (1973)

Table 4.13 *(cont.)*

Ecosystem	Dominant species	g ha^{-1} $year^{-1}$	mJ ha^{-1} $year^{-1}$	Pr/P %	Source
GRASSLAND TYPE ECOSYSTEMS ECOSYSTEMS					
Agrocoenose, Tinew, Poland	*M. arvalis*	22752	139.86	92	Trojan (1969)
Agrocoenose, Bulgaria	*Cricetus cricetus* (Linn.)		46.82		Górecki (1977)
Old field community (Poa compressa) Okenos, Michigan	*M. pennsylvanicus*	3774	21.63		Golley (1960)
Old field community Savanna river, S. Carolina	*Peromyscus polionotus* (Wagn.)	600	5.02		Odum *et al.* (1962)
Alpine meadows, Bieszczady, Poland	*Pitmys subterraneus* *M. agrestis*	650	4.00		Grodziński *et al.* (1966)
Bunchgrass: *Agropyron spicatum* *Stipa comata*, U.S.A.	5 rodent species		3.52		French *et al.* (1976)
Midgrass: *Agropyron smithii* *Stipa viridula*, U.S.A.	4 rodent species		0.83		French *et al.* (1976)
Tall grass: *Andropogon gerardi* *Panicum virgatum*, U.S.A.	5 rodent species 1 shrew		13.50		French *et al.* (1976)
Northern shortgrass: *Boutelona gracilis* *Buchloe dactyloides* U.S.A.	4 rodent species		1.93		French *et al.* (1976)
Southern shortgrass: *Boutelona gracilis* *Aristida longiseta* U.S.A.	8 rodent species		4.18		French *et al.* (1976)
Grass desert: *Boutelona eriopoda* *Sporobolus flexnosus* U.S.A.	8 rodent species		7.61		French *et al.* (1976)
Shrub desert, *Larrea tridentata*	12 rodent species	3593	24.11	90	Chew and Chew (1970)

energy intake during the winter months of hibernation ($A = P + R = 0$); however, respiration occurs at a low level throughout this period and production is thus negative. According to Abaturov and Kuznetsov (1976) the decrease in individual weight of *C. pygmaeus* during hibernation can be as much as 62 per cent of the initial value. It is clear that productivity calculations must include weight losses when $A < R$.

It is also of importance to note that production of new tissue can result from two major processes, production due to normal growth (P_g) and production due to reproduction (P_r), i.e.

$$P = P_g + P_r$$

In most instances for which we have data, production due to reproduction represents a considerable percentage of total production and in precocial small mammals may constitute more than 60 per cent of production (Petrusewicz *et al.*, 1968; and Table 4.13). Correct estimation of production due to reproduction is therefore of importance to energy budget evaluation.

In productivity investigations, many small errors in parameter measurement become compounded by the use of composite formulae. Petrusewicz and Hansson (1975) review details of the methods employed in production estimation and here we concentrate only on the possible sources of error in such estimates.

The accuracy of a production estimate depends upon two factors, the precision of the empirically determined parameters and the validity of the formula employed. Two types of parameter may be recognized: physiological and ecological (Petrusewicz and Hansson, 1975). Physiological parameters, for example gestation, litter size and growth may vary with different conditions but the range of variations is limited by the specific properties of the parameters. Thus the weight of a particular age class may vary geographically (Shvarts, 1969); growth rate may vary with season (Shvarts *et al.*, 1964), the phase of the population cycle (Krebs, 1966), population density (Grodziński *et al.*, 1966; Bobek, 1971) and with food conditions (Ashby, 1967; Hansson, 1971b). However, the degree of variation is restricted; for example, adult *C. glareolus* vary in weight between 18 and 27 g and adult hares (*Lepus europaeus*) between 3.7 and 5.5 kg. Ecological parameters are not subject to the same or similar constraints; depending upon conditions population density may range from zero to thousands of animals per hectare whilst the pregnancy ratio can vary from 0 to 100 per cent.

Of particular concern is the problem of density estimation (reviewed by Smith *et al.*, 1975). Since this is a basic parameter in all population productivity studies, correct estimation is essential but often difficult. It is only the trappable part of the population that is measured with any accuracy while the non-trappable part (nestlings and small juveniles which die between trap rounds) may be repsonsible for as much as 81 per cent of the total population production (Petrusewicz *et al.*, 1968).

Other sources of error result from the inevitable use of approximation at times when census information is reduced or unreliable. When approximations are combined with the variation found in physiological and ecological parameters to compute production by means of composite formulae the margin of error may be large. Table 4.14 shows the results of calculations

Table 4.14 *Total coefficient of variation of composite estimates at various levels of the individual coefficients of variation (γ = number of newborn).*

Estimate	Coefficient of variation of individual factors 5%	20%	50%
$\gamma_r = \frac{\bar{N}.S.\bar{f}.T.\bar{L}.}{tp}$	11	45	112
$P = N.\bar{N}_t \cdot \theta_N$	9	35	87
$P = \bar{N}.W^+ . \theta_N$	10	40	100

by Petrusewicz and Hansson (1975) who computed total coefficients of variation for various factors when individual factor variation was 5, 20 and 50 per cent. It is clear that composite formulae which include two or more stochastically variable elements inevitably result in errors of production estimates unless large and frequent samples are obtained to reduce the confidence limits of individual factors. However, as Grodziński (1975) pointed out, a simple analysis of sensitivity indicated that errors associated with estimating net production itself will have little influence on the whole energy balance. Nevertheless, it is obvious that for some purposes, e.g. energy flow to a higher trophic level in an ecosystem, an accurate calculation of net production is important.

Despite advances in the methodology of production estimation (cf. Petrusewicz and Macfadyen, 1970; Grodziński *et al.*, 1975; Petrusewicz and Hansson, 1975) it is clear that in modelling terms the accurate quantification of transfer functions associated with production is far from easy.

4.7.3 Respiration

Respiration is that part of assimilation (the sum of production and respiration) which is transformed to heat energy and lost in life processes. Because small mammal population production amounts to only 2–3 per cent of

Table 4.15 *Components of metabolic measurements. From Gessaman (1973)*

Metabolic measurement	Components			
	Basal metabolism	Thermoregulatory metabolism	Specific dynamic effect	Activity metabolism locomotion
Basal metabolic rate (BMR)	+	–	–	–
Standard metabolic rate (SMR)	+	–	–	–
Fasting metabolic rate (FMR)	+	–	–	+
Resting metabolic rate (RMR)	+	+	+?	–
Average daily metabolic rate (ADMR)	+	+	+	+

assimilation population respiration must be equivalent to the remaining 97–98 per cent, a level that can be accounted for on grounds of the high respiratory cost of endothermy. In the light of this high value it is not surprising that the sensitivity analysis referred to by Grodziński (1975) pinpointed the estimation of respiration as one of the major factors influencing the accuracy of computations of energy flow.

For most ecological purposes respiration (= metabolic rate) should ideally be measured under natural field conditions. A number of methods have been devised for the determination of metabolic rate in the field; for example, the deuterium method (Mullen, 1973) and biotelemetric systems (Morhardt and Morhardt, 1971); but the majority of workers have relied on a variety of laboratory based methods. A resumé of the main methods, shown in Table 4.15, is given below.

(i) Basal metabolic rate (BMR) – measurements are made in small cages to minimize the activity of experimental animals and at temperatures within the thermoneutral zone. No nest is provided and the animal is deprived of food for 3–4 hours previously so that it is in the post-absorptive state. Measurements are then made over a period of about two hours (e.g. Brody, 1945).

(ii) Standard metabolic rate (SMR) – a measure which is essentially identical to BMR except that the determinations are made at various but stated temperatures (e.g. Packard, 1968).

Basal and/or standard metabolic rate measurements provide the most satisfactory means for making strictly physiological comparisons between species since they are measured under the most constant conditions. Hart (1971) and Kalabukhov (1969) summarize data of this kind and Kleiber (1961) reviewed the relationship between BMR and body size.

Basal metabolic rate may be estimated as a function of body weight by use of the refined Brody-Proctor equation (Morrison *et al.*, 1959).

$$BMR = 3.8\ W^{-0.50}$$

where BMR is expressed as ml O_2 g^{-1} h^{-1} and W in g. Clearly, employment of an appropriate oxycalorific equivalent will convert the volumetric units of oxygen into energy equivalents.

Basal and standard metabolic rate measurements are difficult to apply in ecological investigations; for example, it is not possible to state whether an animal in the field is post-absorptive or not, similarly it may only infrequently be in a state of total inactivity. These disadvantages have resulted in the appearance of the less rigorous fasting metabolic rate (FMR) and resting metabolic rate (RMR).

(iii) Fasting metabolic rate (FMR) – seems to have come into use because of the difficulty of preventing activity during BMR determinations, but all other BMR constraints apply (e.g. Golley, 1960; Wiegert, 1961).

(iv) Resting metabolic rate (RMR) – the measurements are made at various temperatures with the animals provided with food but kept in small chambers so that activity is minimized (e.g. Górecki, 1968).

The problems associated with FMR and RMR relate to their applicability in a field situation. Golley (1960) for instance, felt it necessary to double his FMR measurement for *Microtus pennsylvanicus* in order to obtain a reasonable estimtae of metabolic rate in the field.

(v) Average daily metabolic rate (ADMR) – this is the least controlled of all the various metabolic rate measurements. It is primarily an ecological measurement, the purpose being to ensure that the animals have an environment as close to natural as possible. Animals are placed in large cages and are provided with food and nests. Measurements are made over 24 hour periods so as to include variations in metabolic rate with time. The temperature of measurement is usually 20°C though some workers; for example, Campbell (1974), Moors (1977) and Grodziński *et al.*, (1977), simulated ambient temperatures.

Average daily metabolic rate (ADMR) may be divided into two major components; active metabolic rate (AMR) which is the energy expended during activity, and nesting metabolic rate (NMR) the comparable measurement in the nest. The measurements over 24 hours encompass both these types of activity, the importance of this can be seen in the studies of Campbell (1974) on *Apodemus sylvaticus*. He found that although the majority of animals conformed to a set pattern of circadian activity certain 'rogue' animals showed a different pattern. This difference in activity did not affect the calculations of total daily energy expenditure shown by individuals but would have done so if measurements had only been taken over short periods.

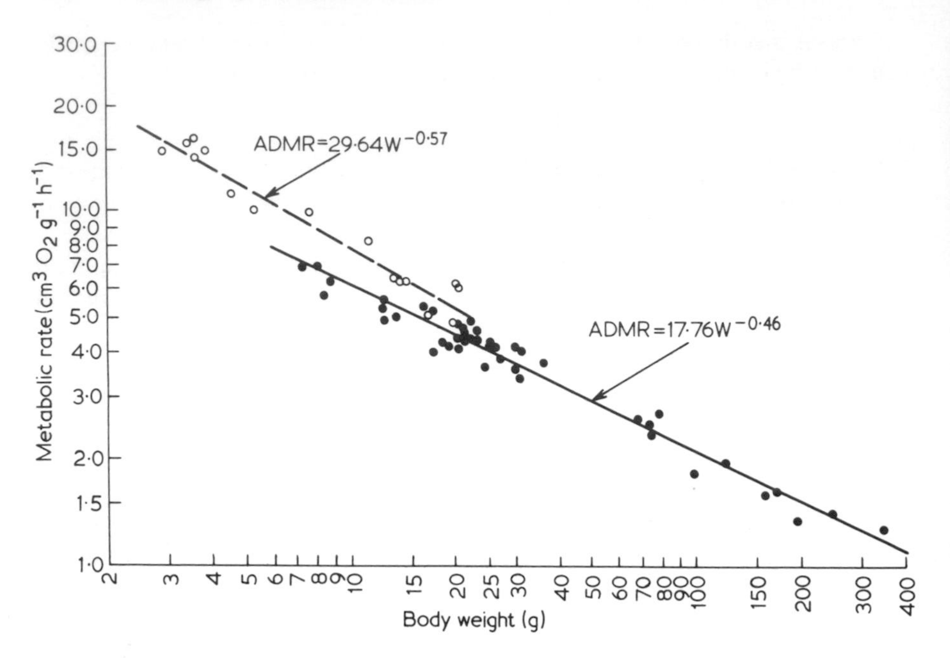

Fig. 4.14 Relationship between average daily metabolic rate (ADMR in cm^3 O_2 g^{-1} h^{-1}) and body weight (W in g) for insectivores (broken line) and rodents (solid line). (From Grodziński, 1975).

Analyses of the relationship between ADMR and body size in voles, mice and squirrels show that it is intraspecifically allometric (Grodziński and Wunder, 1975). However, the exponents of the intraspecific relationships, when expressed as metabolism per whole animal, have been found to be close to 0.50 and not 0.75 as is the case with BMR (Hansson and Grodziński, 1970; Grodziński, 1971a, b; Drożdż *et al.*, 1971; Górecki, 1971; Campbell, 1974). Two general interspecific functions of ADMR against body weight were calculated by French *et al.* (in Grodziński and Wunder, 1975), one for small rodents and one for insectivores (*see* Fig. 4.14). In all, 72 data points were used and included 36 species of rodent (ranging from a 370 g *Cricetus cricetus* to a 7 g *Perognathus*) and 8 species of insectivore with a range of weights between 21 g (*Blarina brevicauda*) and 3 g (*Sorex minutus*). The regression lines of the two derived equations possess significantly different intercepts but the exponent was close to 0.50 in both (-0.57 and -0.46 respectively). Forcing the regressions into an average slope of 0.50 the following equations can be used to predict the ADMR of animals at 20°C.

$$\text{Rodent ADMR} = 19.94\ W^{-0.50}$$
$$\text{Insectivore ADMR} = 26.80\ W^{-0.50}$$

where ADMR is in ml O_2 g^{-1} h^{-1} and W is in g.

For ecological purposes it is clear that ADMR is the most suitable measure for inclusion in energy budgets. It covers different kinds of activity and providing sufficient animals are subjected to experiment over the seasons can include the energy costs of thermoregulation, aestivation, hibernation and reproduction.

Thermoregulatory costs certainly vary with species, behaviour and season. When a nest is not present, huddling can reduce metabolic rate per unit weight by between 5 and 45 per cent (Ponugaeva, 1961); Campbell (1974) suggested that this is a result of the heat production of individual animals being used to maintain the temperatures of adjacent forms. In the presence of a nest even heat losses from a solitary individual are not wasted, heat is retained by the nest materials and this leads to higher temperatures and a lowering of the individual's metabolic rate.

The specific effects of huddling and nest insulation on metabolic rate are generally small when taken as a percentage change but are important to the consideration of daily energy budgets. Most small mammals spend much of their time huddled in the nest (Frank, 1953) but, in addition to sleeping, they display activities such as grooming and feeding (Bashenina, 1968). The effects of these phenomena are manifested in a daily energy budget by modifying the temperature to which the animals are exposed.

The energy costs of seasonal bouts of torpor (aestivation and hibernation) also vary with time. In continuously active animals variation in energy dissipation is relatively independent of season and does not usually exceed 30 per cent (Górecki, 1977). On the other hand the energy expenditure of seasonally active animals can vary markedly with season. Gebczyński *et al.* (1972) studying three species of Gliridae showed that *Glis glis* used 88.7–91.8 per cent of its yearly energy requirement during a six month period of activity; *Muscardinus avellanarius* dissipated about 92 per cent of its energy in the same period while *Dryomys nitedula* utilized 92.5 per cent of its yearly energy budget during the active part of its year. Similarly, Górecki (1977) demonstrated that a hamster (*Cricetus cricetus*) population respired 83 per cent of its annual expenditure when out of hibernation. The deposition of adipose tissue, particularly in the Gliridae, is energetically expensive (King and Farner, 1961; Blaxter, 1966) and this increases the total energy expenses during the active period. The hamster, being more dependent on stored food supplies than stored lipids, expends somewhat less energy than glirids in the active period of the year.

The energetic differences between continuously and periodically active animals will clearly reflect the difference in impact that these different types of small mammal have on ecosystems.

Reproduction may increase individual metabolic costs by 50–70 per cent (Table 4.16). Available data on reproductive costs (Kacmarski, 1966; Trojan and Wojceichowska, 1967; Migula, 1969; Campbell, 1974; Górecki, 1977)

Table 4.16 *Additional energy requirements of reproducing females during gestation and lactation (values recalculated from references). (After Grodziński and Wunder, 1975.)*

Rodent species	Percentage increase		
	Assimilation	Respiration	Reference
Clethrionomys glareolus	57.8	49.4	Kaczmarski (1966)
Mus musculus	77.7	64.7	Myrcha *et al.* (1969)
Microtus arvalis	82.5	69.3	Migula (1969)
	80.5	69.4	Trojan and Wojceichowska (1969)

*Respiration represents assimilation minus production of litters including placentae and foetal membrane.

show that for voles, mice and hamsters at least there is a slight increase in energy requirements during pregnancy and a much larger increase during lactation (Fig.4.15). As Grodziński and Wunder (1975) point out, the magnitude of this effect on a population energy budget depends on two factors: (i) the number of reproducing females and (ii) the length of the breeding season.

Any factor which modifies the rate of energy expenditure in an individual will ultimately influence the energy budget of a population and hence the impact (particularly in terms of consumption) of that population on the ecosystem. As we have seen, estimates of active daily metabolic rate (ADMR) from carefully designed experiments in the laboratory cover many of the activities which modify rates of energy expenditure; nevertheless to obtain realistic estimates for field situations it is necessary to relate time budgets of laboratory activity to time budgets of field activity.

The errors associated with the estimation of production by small mammals, some 2–3 per cent of total energy flux, were referred to in section 4.7.2 and it is of interest to note that measures of metabolic rate can be used to predict production. Grodziński (1975) refers to analyses of production and respiration in 44 populations of small mammals from Europe and North America. He distinguishes two separate regression relationships, one representing rodents and the other insectivores, which can be used to estimate production from the respiration of their populations:

$$\text{Rodent} \quad \log P = -2.2627 + 1.1365 \log R$$
$$\text{Insectivore} \quad \log P = -0.4723 + 0.6081 \log R$$

where both P and R are expressed in kcal ha^{-1} $year^{-1}$.

4.7.4 Energy budget models

From section 4.7 it is clear that a generalized formula for an energy budget model is

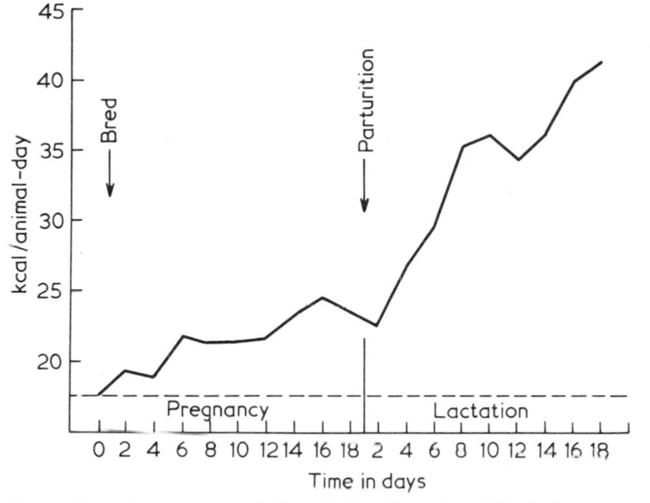

Fig. 4.15 Cost of maintenance of female bank voles (*Clethrionomys glareolus*) during pregnancy and lactation. The level of energy requirements in non-reproducing females of the same size is shown by the broken line. (From Kaczmarski, 1966.)

$$MR = NU + P + R + FU$$

where P + R + FU – C, P + R = A and A represents total energy flow through either an individual or a population.

In terms of total energy flow through small mammal populations respiratory energy losses (R) are large compared with the energy of production (P), the R/P quotient being between 32.9 and 84.0 (Table 4.12). Energy of respiration is thus a major component of energy flow and accordingly has been afforded considerable emphasis.

Gesaman (1973) reviewed a number of energy flow models. One of the earliest referred to was that of Golley (1960) where daily energy expenditure was

$$M_{adults} = FMR \times 2 \times 24\,h$$
$$M_{young} = 7.11\,kJ\,24\,h^{-1}*$$
(*changed to kJ from original kcal)

Another early model (Pearson, 1960) attempted to account for various ecological parameters and included effects such as huddling, nest insulation, and a factor for activity. McNab (1963) modelled R for a deer mouse with the general intention of producing a model with general applicability to other mammals. Metabolism was considered to be a function of temperature, the time spent at that temperature, and whether the animal was active. For the purposes of the model the day was divided into a sinusoidal function of high temperature, low activity (day) and low temperature, high activity (night). The energy costs of activity were then introduced as maximum and minimum limits following the sine wave function. In addition, and independently, the thermal conductance of each animal was used in the model.

For all practical purposes two general types of model are in current use.

(i) The resting metabolic rate (RMR) model – RMR was first used to compute energy flow by Wiegert (1961) and Grodziński (1961). Subsequently, Trojan (1969) modified this approach to produce a more complex and ecologically realistic model. The most complete RMR model to date is that of Chew and Chew (1970). The model has two important limitations: (a) there is no simple way of estimating the cost of activity and (b) the necessity of feeding in metabolic data specific to the animal being studied.

Wunder (1975) produced a generalized model to estimate R in small mammals taking into account body size, air temperature and the degree of running activity

$$R = \alpha M_B + M_{TR} + M_A$$

where α is the coefficient to modify metabolism for the posture associated with activity (Schmidt-Nielsen, 1972), M_B is basal metabolism, M_A metabolism due to activity, and M_{TR} metabolism associated with temperature regulation below thermoneutrality. The extended form of the model is

$$R = \alpha(3.8\,W^{-0.25}) + 1.05\,W^{-0.50}[(38 - 4\,W^{+0.25}) - T_A] + (8.46\,W^{-0.40})\,V$$

where α is as before, W is weight in g, T_A is ambient temperature and V is velocity of running in km h^{-1}. This model coupled with estimates of ambient conditions in the field and estimates of time budgets can be used in the estimation of energy flux through an individual over time.

(ii) The average daily metabolic rate (ADMR) model – early models of this type referred only to the respiratory costs of an animal of mean body weight (Grodziński, 1966; Grodziński and Górecki, 1967). More recent models have been based on intraspecific allometric functions and describe the relationship between ADMR and body size (e.g. Grodziński *et al.*, 1970; Hansson and Grodziński, 1970; Grodziński, 1971a; Drozdz *et al.*, 1971; Moors, 1977).

Weiner (quoted in Grodziński and Wunder, 1975) developed a daily energy budget (DEB) for small mammal populations based on ADMR and produced the equation

$$DEB = ADMR + f\,[TC\,(t_k - t_a)] + CR\,(R/P.TR)$$

where f = fraction of the day spent out of the nest
TC = thermal conductance
t_k = lower critical temperature, assumed to be 20°C
T_a = ambient temperature

CR = coefficient of respiration increase in a reproducing female
RP = fraction of reproducing females in the population
TR = duration of the breeding period as a fraction of one year.

Models of this type allow adjustments to be made to specific conditions in the determination of DEBs when the essential physiological and ecological parameters are known (Grodziński and Wunder, 1975). Clearly, this form of model, like RMR models, affords the opportunity to produce reasonably realistic estimates of energy flow through field populations. However, the models as outlined here relate to respiratory costs only and adjustments to allow for production should be included if total energy flow is the desired estimate. Similarly, the addition of estimates of the energy equivalents of rejecta (FU) allow, by summation, the estimation of consumption, i.e. C = R + P = FU.

Energy budgets for individual animals are often presented as daily or annual energy flow:

$$\text{DEB} = \sum_{i=1}^{24} R + \sum_{i=1}^{24} P \quad \text{(daily energy budget)}$$

$$\text{YEB} = \sum_{i=1}^{365} \text{DEB} \quad \text{(yearly energy budget)}$$

Because of the influence of changing physiological and ecological factors the DEB will vary from day to day; however, the production of a YEB requires a cumulative total of DEBs. Obviously, DEBs must be calculated for different times of the year for it is important to know amongst other things (i) whether the animal exhibits torpor and if so for how long and under what conditions (ii) how ambient conditions vary in the field and their effect on metabolism (iii) whether behaviour changes with season (huddling, nest utilization, etc.) and (iv) the duration of pregnancy and lactation.

Calculation of a YEB from daily DEBs can be both time consuming and tedious. A compromise solution is the construction of a YEB from monthly or seasonal DEBs. Campbell (1974) adopted the month by month approach in his study of *Apodemus sylvaticus*, as did Górecki (1977) for a *Cricetus cricetus* population. Muul (1968) and Gebczynski *et al.* (1972) used a seasonal approach for squirrels and dormice respectively while Randolph (1973) presented a yearly model for *Blarina brevicauda*. It has been suggested by Grodziński and Górecki (1967) that at a minimum the DEBs from at least two seasons (summer and winter) should be considered when constructing YEBs.

In terms of population energy budgets, even a $\text{DEB}_{\text{pop.}}$ is the result of all the DEBs shown by different types of individual within the population:

$$DEB_{pop.} = \sum_{i=1}^{N} DEB$$

To derive a $DEB_{pop.}$ it is necessary to know the size and age structure of the population, the DEBs of different types of individual, the levels of activity and ambient conditions, the fraction of the population that is reproducing, and the length of the breeding season. Then,

$$YEB_{pop.} = \sum_{i=1}^{365} DEB_{pop.}$$

Computations of this sort have been used to estimate energy flow through small mammal populations in different ecosystems (e.g. Chew and Chew, 1970; Grodziński *et al.*, 1970; Grodziński, 1971a; Hansson, 1971a).

An assessment of the probable magnitude of errors associated with the estimation of parameter values for energy budget purposes has been made

Table 4.17 *Effects of errors in estimates on energy budgets (from Grodziński and Wunder, 1975.)*

Type of error	Size of error	Error as % R
(A) In R components		
(1) Estimating level of activity. (Largest error when level is low but estimated high)	0.9–2.1 km/h^{-1}	2–10
(2) Amount of time running	–	5–15
(3) Temperature when out of nest. (Depends on time out of nest: our limits 4–8 h)	10°C	6–12
(4) Time out of nest. (Low error if little time spent running, high error if much time running)	4–8 h	6–12
(5) Temperature of nest. (Depends on amount of time in nest)	5°C	12–17
(B) In P estimates		
Depends directly on the degree of error in estimating number of animals breeding and breeding period, but will affect only P		
(C) In N estimates		
These errors will affect DEB in direct relation to the magnitude of their error and will affect both P and R		

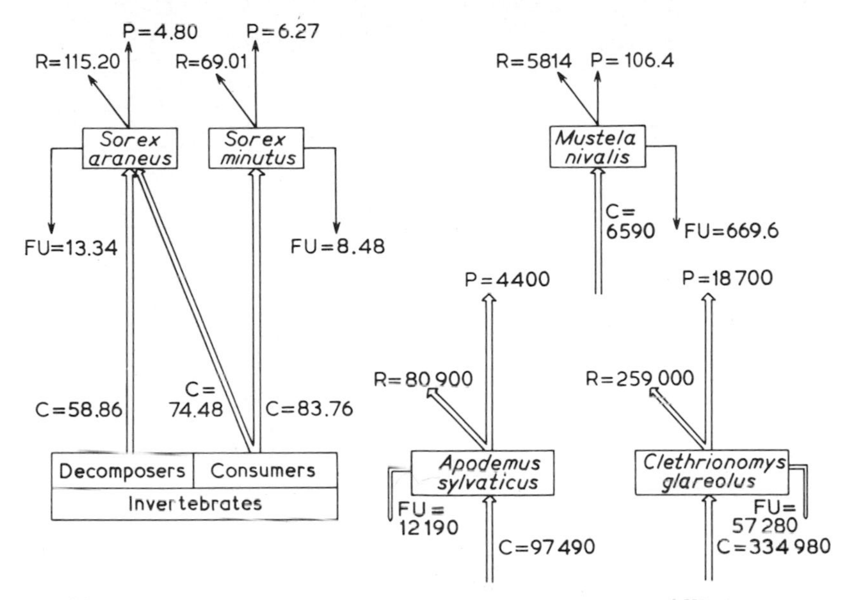

Fig. 4.16 Energy flow through the small mammal community of Wytham Woods, Oxfordshire. All figures in kJ ha^{-1} year^{-1} except for shrews which are in MJ ha^{-1} year^{-1}.

by Grodziński and Wunder (1975) and is shown in Table 4.17. The most likely source of large errors in energy budget evaluation is population estimation.

4.7.5 Energy budgets in an ecosystem context

An energy budget, when presented in the form of an energy flow diagram with its appropriate state variables and transfer functions, will provide a representation of certain aspects of the impact of small mammals within an ecosystem. Figures 4.16 and 4.17 show two such energy flow diagrams, one for Wytham Woods, Oxford and the other for the Niepolomice Forest, Poland. In these diagrams, two features are immediately apparent. First, they are incomplete in the merological sense because a number of variables have been combined into a single variable; for example, in the Wytham model the prey items of shrews are represented by the state variable 'invertebrates' divided into the two trophic categories 'detritivores' and 'consumers'. This is a gross oversimplification in that not only are the terms inadequate descriptions of true trophic status but also because the two species of shrew exploit different species of invertebrate to varying degrees. For a truly merological, and hence realistic model, we should at least have shown the shrew prey items as state variables referable to species. This could not be done because of the lack of adequate information. It was for

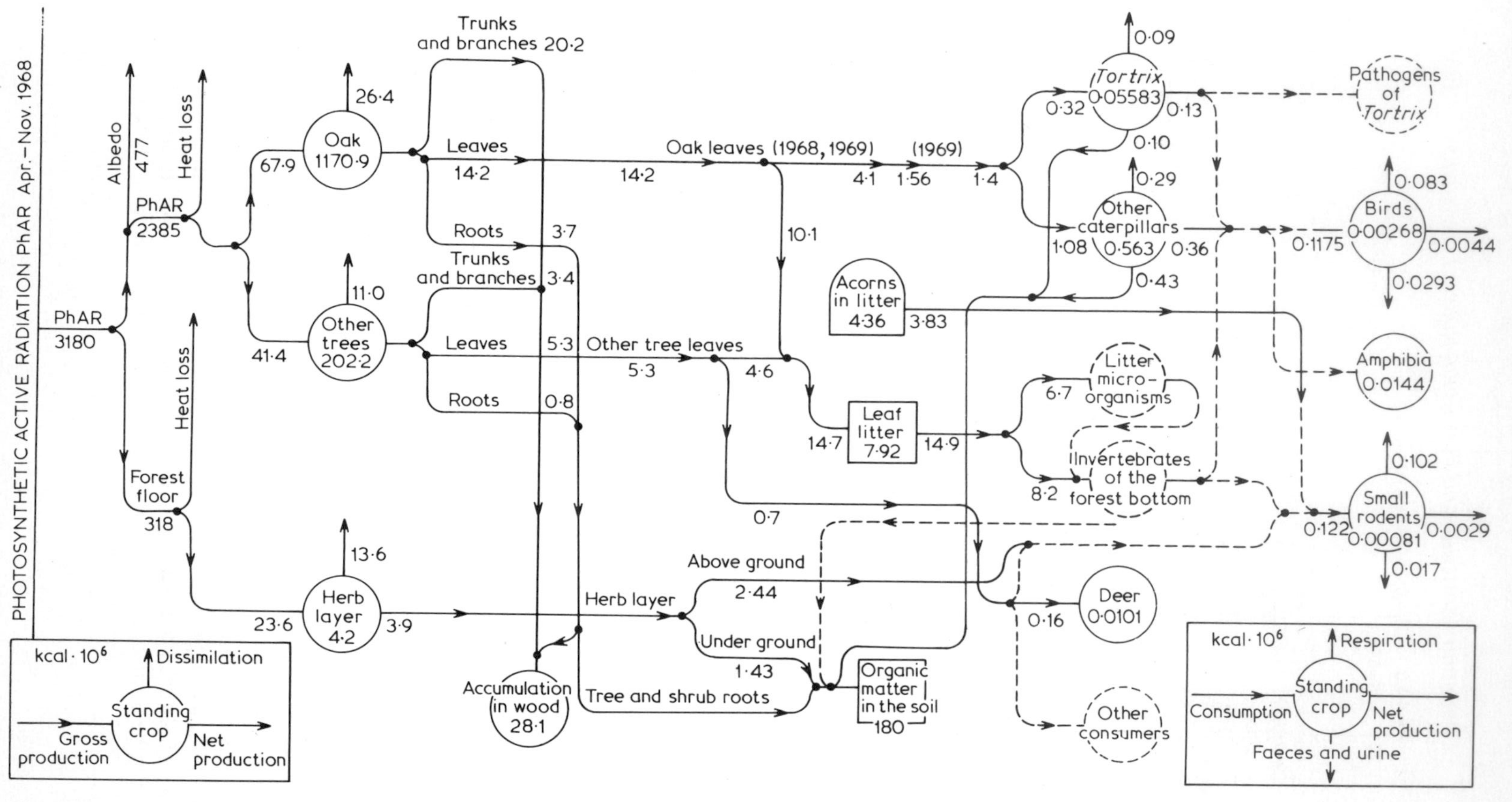

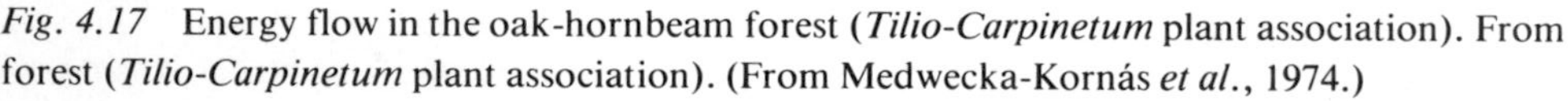

Fig. 4.17 Energy flow in the oak-hornbeam forest (*Tilio-Carpinetum* plant association). From forest (*Tilio-Carpinetum* plant association). (From Medwecka-Kornás *et al.*, 1974.)

identical reasons that rodents (which differentially exploit edible items) and their food resources could not be particularized in the model.

The second major feature to note from these diagrams is that only direct impact via consumption is included in the models. It has already been stressed that consumption *per se* represents a negligible part of the impact of small mammals on ecosystems and that the greatest impact lies in the indirect feedbacks such as burrowing and trampling. As yet, this latter type of transfer function has been but poorly evaluated and certainly not in terms of energy units.

Given the fact that small mammal ecology has been the subject of intensive study for many years (e.g. 30 years in Wytham Woods) and that energy flow diagrams of studied ecosystems are still incomplete it should be apparent that the data base necessary to construct adequate energy budgets in an ecosystem context for small mammals is enormous. However, the collection of data for such a purpose is not an impossible task, it is more a question of channelling research interests along the appropriate paths. Until such time as sufficient data are available for realistic simulation and modelling we can only hypothesize as to the true roles of small mammals within ecosystems. Progress is being made in this direction, albeit slowly.

4.8 The role of small mammals in ecosystems

In the introduction to this chapter we stressed the need to employ a merological approach in attempting to appreciate the role of small mammals in the ecosystem and in the preceding sections we have tried to convey some idea of the great variety of influences that small mammals exert on ecosystems.

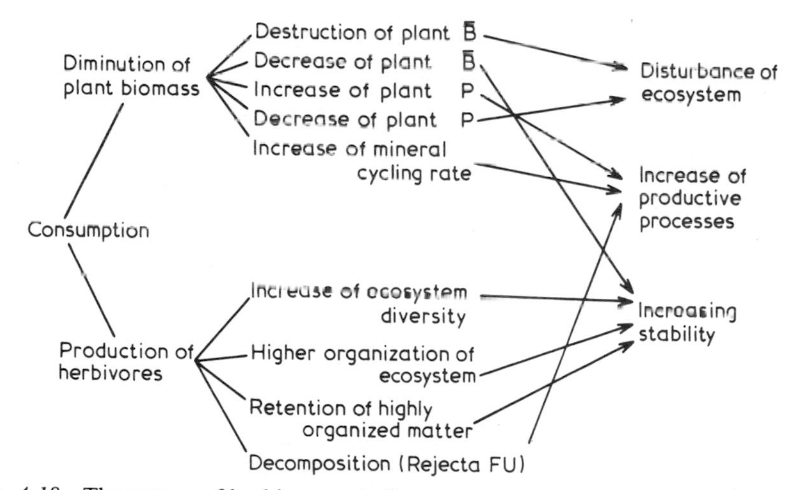

Fig. 4.18 The nature of herbivorous influence on ecosystems. (From Petrusewicz & Grodziński, 1975.)

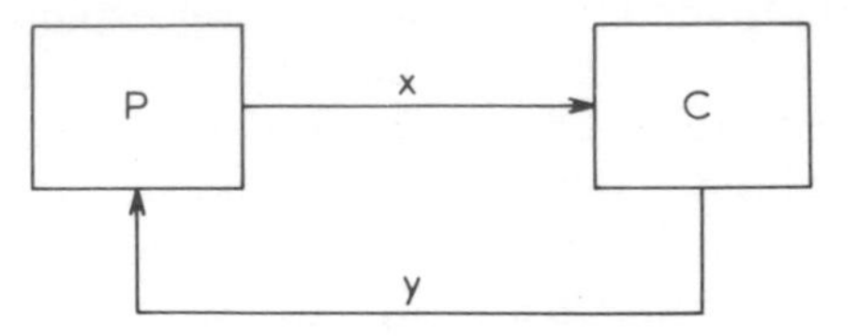

Fig. 4.19 Block diagram with primary producers, P, coupled to small mammals, C, directly through feeding, x, and through a feedback, y. (From Golley, 1973.)

Further, that these impacts are different for different types of small mammal (granivore, herbivore, fossorial etc.) and that the magnitude of the effects depends upon the nature of the ecosystem the small mammals are inhabiting. It is our task in this section to attempt to synthesize the available material and information on small mammals into some kind of general model or models of their role in all ecosystems.

The first step in the generation of such a model must be an attempt to generalize the activities of small mammals in ecosystems. Thus the effects of small mammal consumption can be shown diagrammatically as in Fig. 4.18 and further effects such as the impact of small mammals on the substrate should also be borne in mind.

Having generalized the impact of small mammals on the ecosystem, the problem becomes one of specification of the key role of the small mammal component within the ecosystem, if indeed such a role exists.

Clearly all ecosystems rely either directly or indirectly upon the producers to provide the energy required to maintain themselves. The primary producers are tied directly to the consumer component of the ecosystem through consumption of primary production and by the feedbacks from the effects of consumption (Fig. 4.19). Similarly detritivores and decomposers are tied both directly to the producers and indirectly through the feedback of nutrients (Fig. 4.20). Of the limiting factors acting upon the producer component of ecosystems, only nutrient supply is not under 'partial direct control of the vegetation' (Golley, 1973). Factors such as light, CO_2 levels, and temperature are, to a certain extent, influenced by the vegetation but nutrient supply is controlled by the twin processes of weathering and

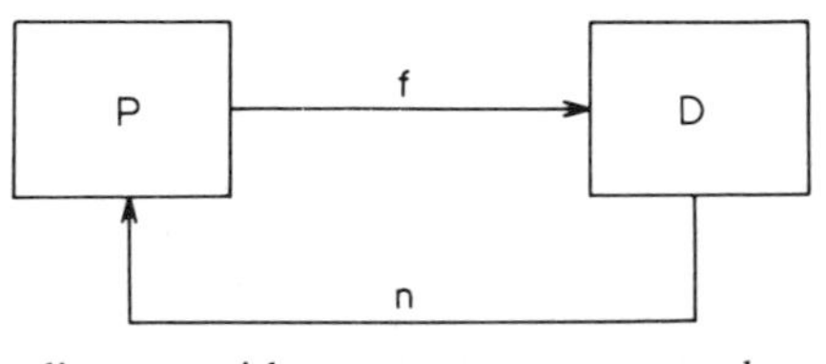

Fig. 4.20 Ecosystem diagram with two components: producers, P, and decomposers, D, coupled by a direct flow, f, and a feedback of nutrients, n. (From Golley, 1973.)

organic decay. Consequently, nutrient cycling and nutrient levels are primarily affected by the detritivore and decomposer components of the ecosystem since, without such a component, nutrients would become increasingly locked up in dead organic matter. Thus while it is possible to envisage an ecosystem consisting purely of producers and decomposers an ecosystem containing just producers and consumers is highly unlikely.

As Petrusewicz and Grodziński (1975) point out an ecosystem consisting merely of a producer component and decomposers would be a poor one and have a low level of complexity. In order to achieve complexity in an ecosystem a consumer component must also be present, since, firstly, the turnover and response times of producers and decomposers differ as do their responses to the environment and secondly excess production produced during optimum environmental conditions might be beyond the capacity of the decomposer component and will be 'shunted' into storage (Golley, 1973) (Fig. 4.21). Such storage alters the rate of flow of material to the decomposers and the timing of that flow. There is a systems advantage in storage since by reducing the effects of the environment on the producers it leads to ecosystem stability but the relative immobility of nutrients locked up in the storage tissues may eventually result in a reduced rate of nutrient cycling as storage becomes excessive, thereby having an adverse effect on production. Thus it is conceivable that the storage shunt could ultimately induce oscillations and instability into the ecosystem. In a complex system a means of 'short-circuiting' or regulating the storage shunt, either through destruction of storage tissues or by reducing excessive production directly, is required, ensuring that storage does not become excessive (Golley, 1973) (Fig. 4.22).

Such a regulator would be required to have the following attributes: (1) the ability to manipulate the population dynamics of the species so as to respond to changes in production over a short time period; (2) there must be an adequate variety of response to cope with the variation in production of a large number of different producers (Ashby's law of requisite variety);

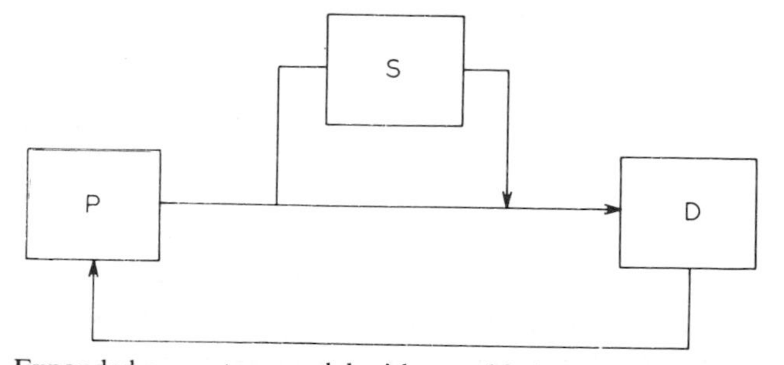

Fig. 4.21 Expanded ecosystem model with an added storage component, S, which may consist of root, rhizome, wood, litter or soil organic matter. (From Golley, 1973.)

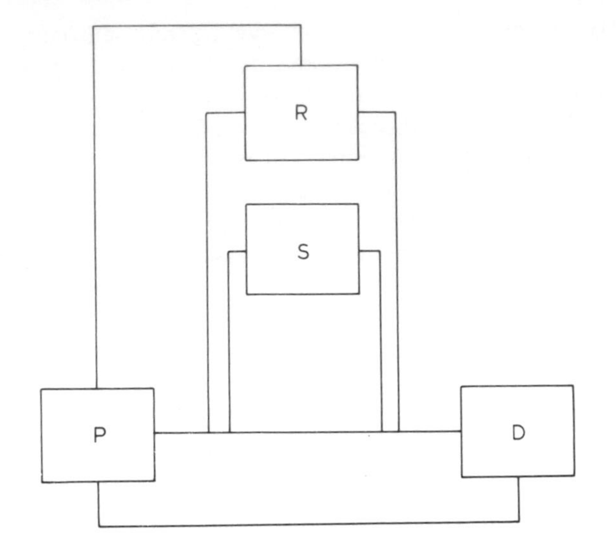

Fig. 4.22 Expanded ecosystem model with a regulator component, R. The regulators compete with storage for primary production to the decomposers and have a direct effect on the primary producers through feeding on photosynthetic tissues. (From Golley, 1973.)

and (3) the regulators must be coupled to both producers and decomposers.

Small mammals fill all these requisites for a hypothetical ecosystem regulator. Firstly their response time is short with an average gestation-lactation period of only 60 days plus large litter size (usually more than two and sometimes greater than 10) small mammals can effectively double population numbers in three months. Secondly small mammals do exhibit a great variety of response in their feeding habits, many being opportunistic omnivores though exhibiting definite food preference (Drożdż, 1966). Thirdly small mammals are limited to both the producer and decomposer components of the ecosystem in such a way as to maintain and regulate the flow of material between producer and decomposer. Some 80 – 90 per cent of the material ingested by small mammals is assimilated so 10 – 20 per cent is immediately made available to detritivores and decomposers plus that material which is clipped but not consumed albeit in a more compacted form. Of the assimilated energy less than 10 per cent is transferred into secondary production but this production, including the mineral content of the small mammal body, is not long lived because of predation or death.

Small mammals fulfill the requisites of an ecosystem regulator but do they actually regulate the storage shunt or reduce excess production. Clearly they cannot do this purely through consumption, except perhaps in those ecosystems subject to rodent outbreaks (e.g. the tundra), but as we have

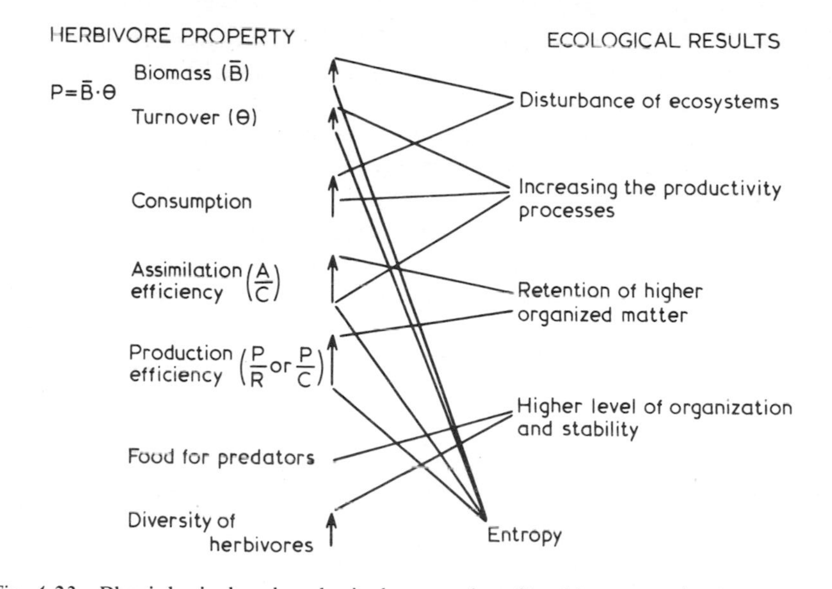

Fig. 4.23 Physiological and ecological properties of herbivorous animals in relation to their impact on ecosystems. (From Petrusewicz & Grodziński, 1975.)

stressed consumption alone is a poor index of small mammal impact. In the case of small mammals, the consequences of consumption on the producer component are far more important since these alter plant production characteristics by destroying the plant, suppressing successful reproduction (through suppression of flowering and through seed predation) and growth and by attacking those species or parts of plants which will contribute most to the future production of the plant community, for example seedlings. Obviously small mammals do not regulate the flow of material from producer to decomposer alone, this is a function of the consumer complex as a whole, but small mammals certainly display the correct characteristics to play an important role in such a regulatory complex.

Two other points should also be borne in mind. The extent of the impact and the importance of small mammals in their regulator role will vary between ecosystems. Thus Golley *et al.* (1975b) conclude that small mammals will exert the greatest pressure in grassland ecosystems and have the least impact in intensive agricultural systems and forests.

Finally, the ecological and physiological nature of a small mammal species will affect its importance as a regulator and maintainer of ecosystem stability. This is true not only of small mammals but of all herbivores and the relationships between ecological/physical properties of herbivores and their effect on the ecosystem is shown in Fig. 4.23.

In conclusion then despite the diversity of impacts of small mammals on ecosystems, the differences in the extent of these impacts in different ecosystems and the varying physiological/ecological characteristics of different species of small mammals can be adjudged to play a key role in regulating and maintaining ecosystems. This regulating effect is achieved largely as a result of the indirect consequences of consumption rather than as a direct effect of consumption itself. Though a great deal of data is now available on small mammal energetics, due largely to the efforts of IBP, it is only now that information on the all-important extraenergetic impacts of small mammals is being elucidated. Consequently, though the role of small mammals as ecosystem regulators is known in a semiquantitative way, there still remains a great deal of detail to be quantified before a truly comprehensive understanding of the functional role of small mammals in ecosystems can be realized.

4.9 References

Abbot, H.G. (1961) White pine seed consumption in small mammals. *Journal of Forestry,* **59**, 197-201.

Abaturov, B.D. (1972) The role of burrowing animals in the transport of mineral substances in the soil. *Pedobiologia,* **12**, 261-266.

Abaturov, B.D. & Kuznetsov, G.V. (1976) Formation of secondary biological production by little sousliks. *Zoologicheskii Zhurnal*, **55**, 1526-1537.

Abaturov, B.C., Devyatykh, V.A. & Zubkova, L.V. (1969) Rol voynschchei deyatelnosti suslikov (*Citellus pygmaeus* Pall.) v peremeshchenii mineralnykh veshchestr v poly pustynnykh poch vakh Zavolzhya. *Pochvovedenie,* **12**, 93-99.

Abaturov, B.D., Rakova, M.V. & Seredneva, T.A. (1975) Vozdestvie malyh suslikov na produktivnost rastitelnosti polupustyni severnogo Prikaspija. In *Rol živõtnych v funkcjovanii ekosystem.* eds. R.J. Zlotin, J.A. Isakov & K.S. Hodaśova. *Izdaniya Nauka,* 15-18. Moskva: MOIP.

Abaturov, B.D. & Zubkovo, L.V. (1969) Vliyanie malykh suslikov (*Citellus pygmaeus* Pall.) na vodno-fizycheskie svoista soloncovykh pochv polupustynii zavolzhya. *Pochvovedenie,* **10**, 53-69.

Ashby, K.R. (1959) Prevention of regeneration of woodland by field mice and voles. *Quarterly Journal of Forestry,* **53**, 228-236.

Ashby, K.R. (1967) Studies on the ecology of field mice and voles (*Apodemus sylvaticus, Clethrionomys glareolus* and *Microtus agrestis*) in Houghall Wood, Durham. *Journal of Zoology, London,* **152,** 389-513.

Ashby, K.R. & Vincent, M.A. (1976) Individual and population energy budgets of the water vole. *Acta Theriologica,* **21(34)**, 499-512.

Aulak, W. (1973) Production and energy requirements in a population of the bank vole in a deciduous forest of *Circaeo-Alnetum* type. *Acta Theriologica,* **18**, 167-190.

Bartolome, J.W. (1976) *Germination and establishment of plants in California annual grassland.* Ph.D. Thesis, University of California, Berkeley.

Basherlina, N.V. (1968) Basic bioenergetic indices in small non-hibernating mammals in natural populations. In *Energy flow through small mammal populations.* ed. Petrusewicz, K. and Ryszkowski, L. P.W.N. Polish Scientific Publishers, Warszawa 1969/1970.

Batzli, G.O. (1975) The role of small mammals in arctic ecosystems. In *Small Mammals: their productivity and population dynamics.* eds. Golley, F.B., Petrusewicz, K. and Ryszkowski, L. Cambridge University Press.

Batzli, G.O. & Pitelka, F.A. (1970) Influence of meadow mouse populations on California grassland. *Ecology,* **51**, 1027-1039.

Batzli, G.O. & Pitelka, F.A. (1971) Conditions and diet of cycling populations of the California vole, *Microtus californicus*. *Journal of Mammalogy,* **52**, 141-163.

Bellier, L. (1967) Recherches écologiques dans la Savane de Lamto (Cote d'Ivoire); densities et biomass des petits mammifères. *Terre et Vie,* **114**, 319-329.

Blaxter, K.L. (1966) Prezemiany energetyczne u przezuwaczy. *Pańmost owa Wydawnictwo Rolnické i Lesne,* 1-374 Warszawa.

Bobek, B. (1971) Influences of population density upon rodent production in a deciduous forest. *Annales Zoologici Fennici,* **8**, 137-144.

Bobek, B. (1973) Net production of small rodents in a deciduous forest. *Acta Theriologica,* **18**, 403-434.

Bolton, P.J. (1969) *Studies in the general ecology, physiology and bioenergetics of woodland Lumbricidae*. Ph.D. Thesis, University of Durham.

Bond, R.M. (1945) Range rodents and plant succession. *Transactions of North American Wildlife Conference,* **10**, 229-234.

Borchert, M.I. & Jain, S.K. (1978) The effect of rodent seed predation on four species of California annual grasses. *Oecologia (Berlin),* **33**, 101-113.

Bramble, W.C. & Goddard, M.K. (1942) Effect of animal coaction and seedbed conditions on regeneration of pitch pine in the barrens of central Pennsylvania. *Ecology,* **23**, 330-335.

Brody, S. (1945) *Bioenergetics and growth*. New York: Hafner Publishing Co. Inc.

Buchalczyk, T., Gebczyńska, Z. & Pucek, Z. (1970) Numbers of *Microtus Oeconomus* (Pallas, 1776) and its noxiousness in forest plantations. *EPPO Publications, Ser. A,* **58**, 95-99.

Buckner, C.H. (1964) Metabolism, food capacity and feeding behaviour in four species of shrew. *Canadian Journal of Zoology,* **42**, 259-179.

Buckner, C.H. (1966) The role of vertebrate predators in the biological control of forest insects. *Annual Review of Entomology,* **11**, 449-470.

Bunnel, F.L. (1973) *Computer simulation of nutrient and lemming cycles in an arctic tundra wet meadow ecosystem*. Ph.D. Thesis, University of California, Berkeley.

Burt, W.H. (1954) *The Mammals of Michigan*. Ann Arbor: University of Michigan Press.

Campbell, I. (1974) *The bioenergetics of small mammals, particularly* Apodemus sylvaticus *(L.), in Wytham Woods, Oxfordshire*. D.Phil. Thesis, University of Oxford.

Carl, E.A. (1971) Population control in Arctic ground squirrels. *Ecology,* **52**, 395-413.

Chew, R.M. & Chew, A.E. (1970) Energy relationships of the mammals of a desert shrub (*Larrea tridentata*) community. *Ecological Monographs,* **40**, 1-21.

Coe, M. (1972) The South Turkana expedition. Scientific papers IX. Ecological studies of the small mammals of South Turkana. *The Geographical Journal,* **138**, 316-338.

Coe, M. (1976) Mammalian ecological studies on Mount Nimba, Liberia. *Mammalia,* **39**, 523-587.

Davis, C.M. Jr. (1925) The mammal fauna of a wooded lot in Southern Michigan. *Papers of the Michigan Academy of Science,* **5**, 425-428.

De Long, K.T. (1967) Population ecology of the feral house mouse. *Ecology,* **48**, 611-634.

Dennis, J.G. & Tiezsen, L. (1972) Seasonal course of dry matter and chlorophyll by species at Barrow, Alaska. *Proceedings of the 1972 U.S. IBP Tundra Biome Symposium*. ed. J. Brown & S. Bowen, pp. 16-21.

De Vos, A. (1969) Production of wild herbivorous mammals. *Advances in Ecological Research,* **6**, 137-183.

Dice, L.R. (1920a) The land vertebrate associations of interior alaska. Occas. Pap. Mus. Zool. Univ. Mich. 85, 1-24.

Dice, L.R. (1920a) The land vertebrate associations of interior Alaska. *Occasional Papers of the Museum of Zoology of the University of Michigan*, **85**, 1-24. water voles (*Arvicola terrestris* L.) from Southern Moravia. *Annales Zoologici*

Dice, L.R. (1925) A survey of the mammals of Charlevoix County, Michigan, and vicinity. *Occasional Papers of the Museum of Zoology of the University of Michigan,* **159**, 1-33.

Dieterlen, F. (1967) Ökologische Populationsstudien an Muriden des Kivugebietes (Congo), Teil 1. *Zoologische Jahrbuch Systematische,* **94**, 369-426.

Dinesman, L.G. (1961) Vliyanie dikikh mlekopitayushchikh na formirovanie drevostoev. *Izvestiya Akademii Nauk SSSR, Seriya Biologicheskaya, Moscow.*

Drożdż, A. (1966) Food habits and food supply of rodents in the beech forest. *Acta Theriologica,* **11**, 363-384.

Drożdż, A. (1968) Digestibility and assimilation of natural foods in small rodents. *Acta Theriologica,* **13**, 367-389.

Drożdż, A., Górecki, A., Grodziński, W. & Pelikan, J. (1971) Bioenergetics of water voles (*Arvicola terrestris* L.) from Southern Moravia. *Annales Zoologici Fennici,* **8**, 97-103.

Dufour, B. (1971) Donées quantatives sur la construction du terrier chez *apodemus sylvaticus* L. *Revue Suisse de Zoologie,* **78**, 568-571.

Duncan, A. & Klekowski, R.Z. (1975) Parameters of an energy budget. In *Methods for Ecological Bioenergetics*. eds. Grodziński, W., Klekowski, R.A. & Duncan, A. IBP Handbook No. 24. Oxford: Blackwell Scientific Publications.

Dunn, E. (1977) Predation by weasels (*Mustela nivalis*) on breeding tits (*Parus* spp.) in relation to the density of tits and rodents. *Journal of Animal Ecology,* **46**, 634-652.

Elton, C. (1927) *Animal Ecology*. London: Sidgwick and Jackson.

Elton, C.S. & Miller, R.S. (1954) The ecological survey of animal communities with a practical system of classifying habitats by structural characters. *Journal of Ecology,* **42**, 460-463.

Fitzgerald, B.M. (1977) Weasel predation on a cyclic population of the Montane vole (*Microtus montanus*) in California. *Journal of Animal Ecology,* **46**, 369-397.

Fleming, T.H. (1970) Notes on the rodent faunas of two Panamanian forests. *Journal of Mammalogy,* **51**, 473-490.

Fleming, T.H. (1971) Population ecology of three species of neotropical rodents. *Miscellaneous Publications of the Museum of Zoology of the University of Michigan*, **143**, 1-77.

Fleming, T.H. (1972) Aspects of the population dynamics of three species of opossum in the Panama canal zone. *Journal of Mammalogy,* **53**, 619-623.

Fleming, T.H. (1973a) Numbers of mammal species in North and Central American forest communities. *Ecology,* **54**, 555-564.

Fleming, T.H. (1973b) The number of rodent species in two Costa Rican forests. *Journal of Mammalogy,* **54**, 518-521.

Fleming, T.H. (1975) The role of small mammals in tropical ecosystems. In *Small mammals: their productivity and population dynamics*. eds. Golley, F.B., Petrusewicz, K. & Ryszkowski, L. Cambridge University Press.

Fleming, T.H., Hooper, E.T. & Wilson, D.E. (1972) Three Central American bat communities' structure, reproductive cycles, and movement patterns. *Ecology,* **53**, 555-569.

Formosov, A.N. (1928) Mammalia in the Steppe biocenose. *Ecology,* **9**, 449-460.

Formosov, A.N., Hodashova, K.S. & Golov, B.A. (1954) Vlianie gryzunov na rastitelnost pastbishch i senosokov glinistykh polupustyn mezhdurchya Volga, Ural. *Izvestiya Akademii Nauk, U.S.S.R.,* 331-340. (In Russian).

Formosov, A.N. & Voronov, A.G. (1939) Deyatel 'nost' gryzunov na pastbishchakh i senokosnykh ugod'yakh Zapadnogo Kazakhstana i ee khozyaistvennoe znachenie. *Uchenye Zapiski Moskovskogo Gosudarstven nogo Universitet, Zoologiya,* **20**, 25-63.

Frank, F. (1953) Zur Entstehung übernormalen Populationsdichten im Massenweschel de Feldmaus, *Microtus arvalis* (Pallas). *Zoologische Jahrbücher Systematik,* **81**, 610-624.

French, N.R., Grant, W.E., Grodziński, W. & Swift, D.M. (1976) Small mammal energetics in grassland ecosystems. *Ecological Monographs,* **46**, 201-220.

Gashweiler, J.S. (1970) Further study of conifer seed survival in a western Oregon clear cut. *Ecology,* **51**, 849-854.

Gebczyńska, Z. (1970) Bioenergetics of a root vole population. *Acta Theriologica,* **15**, 33-66.

Gebczyński, M., Górecki, A., & Drożdż, A. (1972) Metabolism, food assimilation and bioenergetics of three species of Dormice (Gliridae). *Acta Theriologica,* **17**, 271-294.

Gentry, J.B., Briese, L.A., Kaufman, D.W., Smith, M.H. & Wiener, J.G. (1975) Elemental flow and standing crops for small mammal populations. In *Small Mammals: their productivity and population dynamics.* eds. Golley, F.B., Petrusewicz, K. & Ryszkowski, L. Cambridge University Press.

Gesaman, J.A. (1973) Methods of estimating the energy cost of free existence. In *Ecological Energetics of Homeotherms – a view compatible with ecological modelling.* ed. Gessaman, J.A. Utah State University Press.

Gliwicz, J. (1973) A short characteristic of a population of *Proechimys semispinosus* (Tomes 1860) – a rodent species of the tropical rain forest. *Bulletin de l'Académie Polonaise des Sciences, Biology Series,* **21**, 413-418.

Godfrey, G. & Crowcroft, P. (1960) *The life of the mole* (Talpa europaea *Linnaeus*). p. 152. London: Museum Press.

Golley, F.B. (1960) Energy dynamics of a food chain of an old-field community. *Ecological Monographs,* **30**, 187-206.

Golley, F.B. (1973) Impact of small mammals on primary production. In *Ecological Energetics of Homeotherms.* ed. Gessaman, J.A. pp. 142-147. Logan, Utah: Utah State University Press.

Golley, F.B., Petrusewicz, K. & Ryszkowski, L. (eds.) (1975a) *Small Mammals: their productivity and population dynamics.* Cambridge University Press.

Golley, F.B., Ryszkowski, L. & Sokur, I.T. (1975b) The role of small mammals in temperate forests, grasslands and cultivated fields. In *Small Mammals: their productivity and population dynamics.* eds. Golley, F.B., Petrusewicz, K., & Ryszkowski, L. pp. 223-242. Cambridge University Press.

Górecki, A. (1968) Metabolic rate and energy budget in the bank vole. *Acta Theriologica,* **13**, 341-365.

Górecki, A. (1971) Metabolism and energy budget in the harvest mouse. *Acta Theriologica,* **16**, 213-220.

Górecki, A. (1977) Energy flow through the common hamster population. *Acta Theriologica,* **22**, 25-66.

Grodziński, W. (1961) Metabolism rate and bioenergetics of small rodents from the deciduous forest. *Bulletin de l'Academie Polonaise des Sciences,* Cl. 11, **9**, 493-499.

Grodziński,W. (1966) Bioenergetics of small mammals from the Alaskan taiga forest. *Lynx,* **6**, 51-55.

Grodziński, W. (1968) Energy flow through a vertebrate population. In *Methods of Ecological Bioenergetics.* eds. Grodziński, W. & Klekowski, R.Z. pp. 239-252. Warsaw: Polish Academy of Sciences.

Grodziński, W. (1971a) Energy flow through populations of small mammals in the Alaskan taiga forest. *Acta Theriologica,* **16**, 231-275.

Grodziński, W. (1971b) Food consumption of small mammals in the Alaskan taiga forest. *Acta Theriologica,* **16**, 231-275.

Grodziński, W. (1975) Energy flow through a vertebrate population. In *Methods for Ecological Bioenergetics.* eds. Grodziński, W., Klekowski, R.Z. & Duncan, A. pp. 65-96. IBP Handbook No. 24, Oxford: Blackwell Scientific Publications.

Grodziński, W., Bobek, B., Drożdż, A. & Górecki, A. (1970) Energy flow through small rodent populations in beech forest. In *Energy Flow in Small Mammal Populations.* eds. Petrusewicz, K. & Ryszkowski, L. pp. 291-298. Warsaw: Polish Scientific Publishers.

Grodziński, W. & Górecki, A. (1967) Daily energy budgets of small rodents. In *Secondary Productivity of Terrestrial Ecosystems.* ed. Petrusewicz, K. Vol. 1, pp. 295-314. Warsaw: Polish Academy of Sciences.

Grodziński, W., Górecki, A., Janas, K. & Migula, P. (1966) Effect of rodents on the primary productivity of alpine meadows in Bieszczad Mountains. *Acta Theriologica,* **11**, 419-431.

Grodziński, W., Klekowski, R.Z. & Duncan, A. (1975) Methods for Ecological Bioenergetics. *IBP Handbook,* **24**. Oxford: Blackwell Scientific Press.

Grodziński, W., Makomaska, M., Tertil, R. & Wiener, J. (1977) Bioenergetics and total impact of vole populations. *Oikos,* **29**, 494-510.

Grodziński, W., Pucek, Z. & Ryskowski, L. (1966) Estimation of rodents numbers by means of prebaiting and intensive removal. *Acta Theriologica,* **11**, 297-314.

Grodziński, W. & Wunder, B.A. (1975) Ecological energetics of small mammals. In *Small Mammals: their productivity and population dynamics.* eds. Golley, F.B., Petrusewicz, K. & Ryszkowski, L. Cambridge University Press.

Handley, C.O. Jr. (1967) Bats of the canopy of an Amazonian forest. *Atlas do Simposia Biota Amazonia,* **5**, 211-215.

Hansson, L. (1971a) Estimates of the productivity of small mammals in a South Swedish spruce plantation. *Annales Zoologici Fennici,* **8**, 118-126.

Hansson, L. (1971b) Small rodent food, feeding and population dynamics. A comparison between granivorous and herbivorous species in Scandinavia. *Oikos,* **22**, 183-198.

Hansson, L. & Grodziński, W. (1970) Bioenergetic parameters of the field vole, *Microtus agrestis* L. *Oikos,* **21**, 76-82.

Hanney, P. (1965) The Muridae of Malawi (Africa: Nyasaland). *Proceedings of the Zoological Society of London,* **146**, 577-633.

Hart, J.S. (1971) Rodents. In *Comparative Physiology of Thermoregulation.* Vol. III Mammals. ed. Whitton, G.C. pp. 1-149. New York: Academic Press.

Heady, H.F. (1976) *Range Management.* New York: McGraw Hill.

Hodashova, K.S. (1950) Prirodnaya sreda i zhivotnyi mir glinistykh polupustyn Zavolzhya. *Izvestiya Akademii Nauk S.S.S.R.* p. 130.

Hodashova, K.S. (1970) Vozdeistvie pozvonochnykh fitogagov na biologi cheskuyu producktivnost' i krugovorot veshchestv v lesostypnykh landshftakh. In *Sredoabrazuyushchaya deyatelnost' zhivotnykh*. ed. Isakov, J.A. pp. 48-52. Izdaniya Moskvoskogo University Moscow (In Russian).

Holling, C.S. (1959) The components of predation as revealed by a study of small mammal predation of the European sawfly. *Canadian Entomologist,* **91**, 293-320.

Hood, G.A., Nass, R.D. & Lindsey, G.D. (1970) The rat in Hawaiian sugarcane. *Proceedings of the IVth Vertebrate Pest Conference* pp. 34-37.

Horn, E.E. & Fitch, H.C. (1942) Inter-relations of rodents and other wildlife of the range. In *The San Joaquin Experimental Range*. eds. Hutchinson, C.B. & Otok, K. pp. 96-129. *University of California College of Agriculture Bulletin,* **663**, 1-145.

Janzen, D. (1970) Herbivores and the number of tree species in tropical forests. *American Naturalist,* **104**, 501-568.

Janzen, D.H. (1971a) Seed predation by animals. *Annual Review of Ecology and Systematics,* **2**, 465-492.

Janzen, D.H. (1971b) Escape of juvenile *Dioclea megacarpa* (Leguminosae) vines from predators in a deciduous tropical forest. *American Naturalist,* **105**, 97-112.

Janzen, D.H. (1972) Escape in space by *Sterculia apetala* seeds from the bug *Dysdercus fasciatus* in a Costa Rican deciduous forest. *Ecology,* **53**, 350-361.

Kaczmarski, F. (1966) Bioenergetics of pregnancy and lactation in the bank vole. *Acta Theriologica,* **11**, 409-417.

Kalabukhov, N.I. (1969) Periodical (seasonal and annual) changes in the organism of rodents, their reasons and consequences. *Izdaniya Nauka Leningrad*. (In Russian).

Kalela, O., Kilpeläienn, L., Koponen, T. & Taast, J. (1971) Seasonal differences in habitats of the Norwegian lemming *Lemmus lemmus* (L.) Suomalainan Tied. Toim. (Annales Academicae scientarum Fennicae), Ser. A, IV Biology, **55**, 1-72.

Kalela, O. & Koponen, T. (1971) Food consumption and movement of the Norwegian lemming in areas characterized by isolated fells. *Annales Zoologici Fennici,* **8**, 80-84.

Kanervo, V. & Myllymäki, A. (1970) Problems caused by the field vole *Microtus agrestis* (L.) in Scandinavia. *EPPO Publications, Ser. A,* **58**, 11-26.

King, C.M. (1975) The home range of the weasel (*Mustela nivalis*) in an English woodland. *Journal of Animal Ecology,* **44**, 639-669.

King, J.R. & Farner, D.S. (1961) Energy metabolism, thermoregulation and body temperature. In *Biology and Comparative Physiology of Birds. II.* ed. Marshall, A.J. pp. 215-288. New York: Academic Press.

Kleiber, M. (1961) *The fire of life: an introduction to animal energetics.* New York: Wiley.

Koford, C.B. (1958) Prairie dogs, whitefaces and blue grama. *Wildlife Monographs* No. 3.

Koford, C.B. (1960) The prairie dog of the North American plains and its relations with plants, soil and land use. In *Ecology and Management of Grazing Animals in Temperate Zones.* ed. Bourlière, F. Symp. of IUCN 8th Tech. Meeting, Warszawa, July 15-24, Morges, Switzerland.

Komarek, E.V. & Komarek, R. (1938) Mammals of the Great Smoky Mountains. *Bulletin of the Chicago Academy of Science,* **5**, 137-162.

Koshkina, T.V. (1962) Migrations of *Lemmus lemmus. Zoologicheskii Zhurnal,* **41**, 1859-1874. (In Russian).

Krebs, C.J. (1966) Demographic changes in fluctuating populations of *Microtus californicus*. *Ecological Monographs,* **36**, 23-273.
Kuckeruk, V.V. (1963) Vozdeist vie travoyadnykh mlekopitayushchikh na produktivnost' travostoyo stepi i ikh znacherie v obrazovanii organicheskoi chasti stepnyk pochv. *Transactions of the Moscow Society of Naturalists,* **10**, 157-193. (In Russian).
Kuznetzov, G.B. (1970) O voynschey detatielnosti kav kaz kogo krota (*Talpa caucasica*). *Zoologicheskii Zhurnal,* **49**, 1245-1247. (In Russian).
Liat, L.B. (1966) Abundance and distribution of Malaysian bats in different ecological habits. *Journal of the Federal Museum,* **11**, 61-76.
Lidicker, W.Z. Jr. (1975) The role of dispersal in the demography of small mammals. In *Small Mammals: their productivity and population dynamics.* eds. Golley, F.B., Petrusewicz, K. & Ryszkowski, L. Cambridge University Press.
Mann, K.H. (1969) The dynamics of aquatic ecosystems. *Advances in Ecological Research,* **6**, 1-81.
Maclean, S.F., Fitzgerald, B.M. & Pitelka, F.A. (1974) Population cycles in arctic lemmings: winter reproduction and predation by weasels. *Arctic and Alpine Research,* **6**, 1-12.
McNab, B.K. (1963) A model of the energy budget of a wild mouse. *Ecology,* **44**, 521-532.
McNeil, S. & Lawton, J.H. (1970) Animal production and respiration in animal populations. *Nature,* **225** (5231), 472-474.
Marshall, D.R. & Jain, S.K. (1970) Seed predation and dormancy in the population dynamics of *Avena fatua* and *Avena barbata*. *Ecology,* **51**, 886-891.
Mathies, J.B., Dunaway, P.B., Schneider, G. & Auerbach, S.I. (1972) *Annual consumption of cesium-137 and cobalt-60 labeled pine seeds by small mammals in an oak-hickory forest*. TM-3912. Oak Ridge National Laboratory, Tennessee.
Medwecka-Kornaś, A., Łomnicki, A. & Bandoła-Gołczyk, E. (1973) Energy flow in the deciduous woodland ecosystem, Ispina project, Poland. In *Modeling Forest Ecosystems,* pp. 144-150. Report of International Workshop, IBP/PT Oak Ridge, Tennessee.
Migula, P. (1969) Bioenergetics of pregnancy and lactation in European common vole. *Acta Theriologica,* **13**, 167-179.
Moors, P. (1974) The annual energy budget of a weasel population in farmland. Ph.D. Thesis, University of Aberdeen.
Moors, P. (1977) Studies on the metabolism, food consumption and assimilation efficiency of a small carnivore the weasel (*Mustela nivalis* L.) *Oecologia (Berlin),* **27**, 185-202.
Morhardt, J.E. & Morhardt, S.S. (1971) Correlations between heart rate and oxygen consumption in rodents. *American Journal of Physiology,* **221**, 1580-1586.
Morrison, P.R., Ryser, F. & Dawe, A. (1959) Studies on the physiology of the masked shrew, *Sorex cinereus*. *Physiological Zoology,* **32**, 256-271.
Mullen, R.K. (1973) The $D_2{}^{18}O$ method of measuring the energy metabolism of free living animals. In *Ecological Energetics of Homeotherms – a view compatible with ecological modelling*. ed. Gessaman, J.A. pp. 32-43. Logan, Utah: Utah State University Press.
Muul, I. (1968) Behavioural and physiological influences on the distribution of the flying squirrel, *Glaucomys volans*. *Miscellaneous Publications of the Museum of Zoology of University of Michigan,* **134**, 1-66.

Myrcha, A., Ryszkowski, L. & Walkowa, W. (1969) Bioenergetics of pregnancy and lactation in the white mouse. *Acta Theriologica,* **13**, 391-400.

Naumov, N.P. (1975) The role of rodents in ecosystems of the northern deserts of Eurasia. In *Small Mammals: their productivity and population dynamics.* eds. Golley, F.A., Petrusewicz, K. & Ryszkowski, L. Cambridge University Press.

Novikov, G.A. & Petrov, O.V. (1953) K ecologii podzemnoi potevki v lesostepnyk dubravakh. *Zoologischeskii Zhurnal,* **32**, 130-139.

Obtrel, R., Zejda, J. & Holisová, V. (1978) Impact of small rodent predation on an overcrowded population of *Diprion pini* during winter. *Folia Zoologica,* **27**, 97-110.

Odum, E.P., Connell, C.E. & Davenport, L.B. (1962) Population energy flow of three primary consumer components of old-field ecosystems. *Ecology,* **43**, 88-95.

Osborn, B. (1942) Prairie dogs in shinnery (oak scrub) savannah. *Ecology,* **23**, 110-115.

Packard, R.L. (1968) An ecological study of the fulvous harvest mouse in eastern Texas. *American Midland Naturalist,* **79**, 68-88.

Pearson, O.P. (1960) The oxygen consumption and bioenergetics of harvest mice. *Physiological Zoology,* **33**, 152-160.

Pearson, O.P. (1963) History of two local outbreaks of feral house mice. *Ecology,* **44**, 540-549.

Pearson, O.P. (1964) Carnivore-mouse predation: an example of its intensity and bioenergetics. *Journal of Mammalogy,* **45**, 177-188.

Pearson, O.P. (1966) The prey of carnivores during one cycle of mouse abundance. *Journal of Animal Ecology,* **35**, 217-233.

Pearson, O.P. (1971) Additional measurements of the impact of carnivores on California vole (*Microtus californicus*). *Journal of Mammalogy,* **52**, 41-49.

Peiper, R. (1963) *Production and chemical composition of arctic tundra vegetation and their relation to the lemming cycle.* Ph.D. Thesis, University of California, Berkeley.

Pernetta, J.C. (1973) *Field and laboratory experiments to determine the feeding ecology and behaviour of* Sorex araneus *L. and* Sorex minutus *L.* D.Phil. Thesis, University of Oxford.

Pernetta, J.C. (1976a) Bioenergetics of British shrews in grassland. *Acta Theriologica,* **21**, 481-497.

Pernetta, J.C. (1976b) Diets of the shrews *Sorex araneus* L. and *Sorex minutus* L. in Wytham grassland. *Journal of Animal Ecology,* **45**, 899-912.

Petrusewicz, K., Andrezejewski, R., Bujalska, G. & Gliwicz, J. (1968) Productivity investigation of an island population of *Clethrionomys glareolus* (Schreber, 1780). IV. Production. *Acta Theriologica,* **13**, 435-445.

Petrusewicz, K., Bujalska, G., Andrezejewski, R. & Gliwicz, J. (1971) Productivity processes in an island population of *Clethrionomys glareolus. Annales Zoologici Fennici,* **8**, 127-132.

Petrusewicz, K. & Grodziński, W. (1975) The role of herbivore consumers in various ecosystems. In *Productivity of World Ecosystems.* eds. Reichle, D.E. & Goodall, D.M., pp. 64-70. National Academy of Sciences, Washington, 166pp.

Petrusewicz, K. & Hansson, L. (1975) Biological production in small mammal populations. In *Small Mammals: their productivity and population dynamics.* eds. Golley, F.B., Petrusewicz, K. & Ryszkowski, L. Cambridge University Press.

Petrusewicz, K. & Macfadyen, A. (1970) Productivity of Terrestrial Animals:

Principles and Methods. *IBP Handbook,* **13**. Oxford: Blackwell Scientific Publications.

Petrusewicz, K., Markov, G., Gliwicz, J. & Christov, L. (1972) A population of *C. glareolus pirineus* on the Vithosa Mountains, Bulgaria. IV. Production, *Acta Theriologica,* **17**, 437-453.

Phillipson, J. (1973) The biological efficiency of protein production by grazing and other land based systems. In *The Biological Efficiency of Protein Production*. ed. Jones, J.G.W. pp. 217-235. Cambridge University Press.

Phillipson, J. (1975) Rainfall, primary production and 'carrying capacity' of Tsavo National Park (East), Kenya. *East African Wildlife Journal,* **13**, 171-201.

Phillipson, J., Putman, R.J., Steel, J. & Woodell, S.R.J. (1975) Litter input, litter decomposition and the evolution of carbon dioxide in a beech woodland – Wytham Woods, Oxford. *Oecologia,* **20**, 203-217.

Pitelka, F.A. (1957) Some characteristics of microtine cycles in the Arctic. In *Arctic Biology*. ed. Hansen, H.P. pp. 153-184. Oregon State University.

Pitelka, F.A. & Schultz, A.M. (1964) The nutrient-recovery hypothesis for arctic microtine rodents. In *Grazing in Terrestrial and Marine Environments*. ed. Crisp, D. pp. 55-68. Oxford: Blackwell Scientific Publications.

Pjastalova, O.A. (1972) The role of rodents in the energetics of biogeocoenoses of forest tundra and southern tundra. *Proceedings of the IV International Meeting on the Biological Productivity of Tundra*. eds. Wielgolasky, F.W. & Rosswall, T., Leningrad.

Ponugaeva, A.G. (1961) *Fizjologiceskie isseldovanija instinktov u mlekopitajusick*. Leningrad: Moskova.

Quay, W.B. (1951) Observations on mammals of the Seward Peninsula, Alaska. *Journal of Mammalogy,* **32**, 88-99.

Radvanyi, A. (1966) Destruction of radio-tagged seeds of white spruce by small mammals during summer months. *Forest Science,* **12**, 307-315.

Randolph, J.C. (1973) The ecological energetics of a homeothermic predator, the short-tailed shrew. *Ecology,* **54**, 1166-1187.

Ryszkowski, L. (1969) Estimates of consumption of rodent populations in different pine forest ecosystems. In *Energy Flow Through Small Mammal Populations*. ed. Petrusewicz, K. & Ryszkowski, L. pp. 281-289. Warsaw: Polish Scientific Publishers.

Ryszkowski, L., Goszczynski, J. & Truszkowski, J. (1973) Trophic relationships of the common vole in cultivated fields. *Acta Theriologica,* **18**, 125-165.

Schmidt-Nielsen, K. (1972) Locomotion: energy cost of swimming, flying and running. *Science,* **177**, 222-228.

Schultz, A.M. (1969) A study of an ecosystem: the arctic tundra. In *The Ecosystem Concept in Natural Resource Management*. ed. Van Dyne, G. pp. 77-93. New York: Academic Press.

Shaw, M.W. (1968) Factors affecting the natural regeneration of sessile oak (*Quercus patraea*) in north Wales. II. Acorn losses and germination under field conditions. *Journal of Ecology,* **56**, 647-660.

Sheppe, W. (1972) The annual cycle of small mammal populations on a Zambian floodplain. *Journal of Mammalogy,* **53**, 445-460.

Shvarts, S.S. (1969) Évolyutsionnoya ékologiya zhivotnykh. Ural Branch of Academy of Science, Sverdlosk.

Shvarts, S.S. (1975) Morpho-physiological characteristics as indices of population processe. In *Small Mammals: their productivity and population dynamics*. eds. Golley, F.B., Petrusewicz, K. & Ryszkowski, L. Cambridge University Press.

Shvarts, S.S., Pokrowski, A.V., Istchenko, V.G., Olenjey, V.G.,

Outschinnikova, N.A. & Pjastalova, O.A. (1964) Biological peculiarities of seasonal generation of rodents with special reference to the problem of senescence in mammals. *Acta Theriologica,* **8**, 11-43.

Sinclair, A.R.E. (1975) The resource limitation of trophic levels in tropical grassland ecosystems. *Journal of Animal Ecology,* **44**, 497-520

Skvorcova, V.K. & Ulenin, V.D.V. (1969) Vliyanie voyushchei deyatelnosh slepusha (*Spalox micropthalamus*) na vidovoy sostav i produktovnost travoyanistykh fitocenozov tesostepi. *Biogeografia,* **3**, 7-10. (In Russian).

Smirnov, V.S. & Tokmakova, S.G. (1972) Influence of consumers on natural phytocenoses' production variation. In *Tundra Biome.* Proceedings of the IV International Meeting on the Biological Productivity of Tundra. eds. Wielgolaski, F.E. & Rosswall, T. pp. 122-127. Leningrad.

Smirnov, V.S. & Tokmakova, S.G. (1971) Preliminary data on the influence of different numbers of voles upon the forest tundra vegetation. *Annales Zoologici Fennici,* **8**, 154-156.

Smith, C.F. & Aldous, S.E. (1947) The influence of mammals and birds in retarding artificial and natural reseeding of coniferous forests in the United States. *Journal of Forestry,* **45**, 361-369.

Smith, M.H., Gardner, R.H., Gentry, J.B., Kaufman, D.W. & O'Farrell, M.H. (1975) Density estimations of small mammal populations. In *Small Mammals: their productivity and population dynamics.* eds. Golley, F.B., Petrusewicz, K. & Ryszkowski, L. Cambridge University Press.

Soholt, L.F. (1973) Consumption of primary production by a population of kangaroo rats (*Dipodoms merriami*) in the Mojave desert. *Ecological Monograph,* **43**, 357-376.

Spencer, P.R. (1955) The effects of rodents on reforestation. *Proceedings of the Society of American Foresters, 1955.* pp. 125-128.

Spitz, F. (1968) Interactions entre la végétation épigée d'une Luzernière et les populations enclose ou non enclose de *Microtus arvalis* (Pallas). *Terre et Vie,* **3**, 274-306.

Stark, N. (1973) *Nutrient cycling in a Jeffrey pine forest ecosystem.* Institute for Microbiology, University of Montana, Missoula.

Stenmark, A. (1963) Guventering an sorkskadoruas ebonomiska bety delse for tradgardsodlingen under 1962. *Vaxtskyddsnotiser,* **27**, 24-29. (In Swedish).

Sviridenko, P.A. (1940) Pitanie myshevidnykh gryzunov i znachenie ikh v problemlevozobnovleniya lesa. *Zoologischeskii Zhurnal,* **19**, 680-703. (In Russian).

Sviridenko, P.A. (1957) Zapasanie korma shivotnymi. *Izvestiya Akademii U.S.S.R., Kiev.* (In Russian).

Tahon, J. (1969) Non-related criteria between food consumption by small mammals and waste caused to vegetation. In *Energy Flow Through Small Mammal Populations.* eds. Petrusewicz, K. & Ryszkowski, L. pp. 159-165. Warsaw: Polish Scientific Publishers.

Talbot, M.W., Biswell, H.H. & Hormay, A.L. (1939) Fluctuation in the annual vegetation of California. *Ecology,* **20**, 394-402.

Tanton, M.T. (1965) Acorn destruction potential of small mammals and birds in British woodland. *Quarterly Journal of Forestry,* **59**, 230-234.

Tast, J. & Kalela, O. (1971) Comparisons between rodent cycles and plant production in Finnish lapland. *Annales Academiae Scientarum, Fennicae IV. Biologica,* **186**, 1-9.

Teshima, H. (1970) Rodent control in the Hawaiian sugar industry. *Proceedings of the IVth Vertebrate Pest Conference,* pp. 38-40.

Tevis, L. Jr. (1953) Interrelations between the harvester and *Veromessa pergandei* (Mayr) and some desert ephemerals. *Ecology,* **39**, 695-704.

Tikhomirov, B.A. (1959) *Relationship of the animal world and plant cover of the tundra.* Botanical Institute, Academy of Sciences, U.S.S.R. (Trans. E. Issakoff & T.W. Barry, Boreal Institute, University of Alberta).

Trojan, P.J. (1965) Intrapopulation relations and regulation of numbers in small forest rodents. *Ekologia Polska, Ser. A.* **13**, 143-168.

Trojan, P. (1969) Energy flow through a population of *Microtus arvalis* (Pall.) in an agrocoenosis during a period of mass occurrence. In *Energy Flow Through Small Mammal Populations.* eds. Petrusewicz, K. & Ryszkowski, L. pp. 267-279. Warsaw: Polish Scientific Publishers.

Trojan, P. & Wojciechowska, B. (1967) Resting metabolism rate during pregnancy and lactation in the European common vole, *Microtus arvalis* (Pall.) *Ekologia Polska, Ser. A.* **15**, 811-817.

Ulehla, J., Pelikan, J. & Zichoza, L. (1974) Rodent burrowing activity and heterogeneity of a lucerne stand. *Zoologické Listy,* **23**, 113-121.

Van Soest, P.J. (1966) Non-nutritive residues: a system of analysis for replacement of crude fibre. *Journal of the Association of Official Agricultural Chemists,* **49**, 546.

Varley, G.L. & Gradwell, G.P. (1968) Population models for the winter moth. In *Insect Abundance.* ed. Southwood, T.R.E. *Symposium of the Royal Entomological Society of London* No. 4. September 1967, pp. 132-142. Oxford, 160pp.

Vorhies, C.T. & Taylor, W.P. (1940) Life history and ecology of the white throated woodrat, *Neotoma albigula albigula* Harley, in relation to grazing in Arizona. *Bulletin of the Arizona Agricultural Experimental Station,* **86**, 453-529.

Voronov, N.P. (1953) Iz nablyudenii nad royushchei deyatelnostyu grysunov v lesu. *Pochvovedenie,* **10**, 61-74. (In Russian).

Watts, C.H.S. (1968) The foods eaten by wood mice (*Apodemus sylvaticus*) and bank voles (*Clethrionomys glareolus*) in Wytham Woods, Berkshire. *Journal of Animal Ecology,* **37**, 25-41.

West, N.E. (1968) Rodent influenced establishment of Ponderosa pine and bitter-bush seedlings in central Oregon. *Ecology,* **49**, 1009-1011.

Whittaker, R.H. & Feeny, P.D. (1971) Allelochemics: chemical interactions between species. *Science,* **171**, 757-770.

Wiegert, R.G. (1961) Respiratory energy loss and activity patterns in the meadow vole, *Microtus pennsylvanicus pennsylvanicus. Ecology* **42**, 245-253.

Wilson, D.E. (1973) A trophic comparison of bat faunas. *Systematic Zoology,* **22**, 14-29.

Wilson, D.E. & Janzen, D.H. (1972) Predation on *Scheelea* palm seeds by bruchid beetles: seed density and distance from parent palm. *Ecology,* **53**, 954-959.

Winberg, G.G. (1962) Energeticeskij princip izucenija troficeskih svajezej i produktivnosti ekologiceskih sistem. Energetic principle in investiation of trophic relations and productivity of ecological systems. *Zoologicheskii Zhurnal* **41**, 1618-1630. English summary.

Woodmansee, R.G., Dodd, J.L., Bowman, R.A., Clark, F.E. & Dickinson, C.E. (1978) Nitrogen budget of a shortgrass prairie ecosystem. *Oecologia,* **34**, 363-376.

Wunder, B.A. (1975) A model for estimating metabolic rate of active or resting mammals. *Journal of Theoretical Biology,* **49**, 345-354.

Zhiryakov, V.A. (1977) Role of mammals as consumers of Schrenk's spruce. *Soviet Journal of Ecology,* **8**, 165-168.

Zhukov, A.B. (1949) Dubravy USSR i sposoby ikh vosstano vleniya. In *Dubravny SSSR, I Ves. nauchno issled.* pp. 30-352. Institut lesn. Khoz. 28.
Zimina, R.P., Pogodina, G.S. & Urushadze, T.F. (1970) Landshaftoobrazuyush-chaya rol surkov v aridnykh vysokogoryakh Tyan-Shanya i Pamira. *Materialy k poznanyu fauny i flory. SSSR (MOIP)*, **45**, 177-191. (In Russian).
Zlotin, R.J. (1975) Ocenka voz destvia źivotnych fitofagov na pervićnuju produkcju lugovo-slepnogo pastiśćá (In Russian). In *Rol' źivotnych v funkcjonirovanii ekosistem*. eds. Zlobin, R.J., Isakov, J.A. & Hodasova, K.S. Izd. Nauka, Moskav, pp. 18-22.

Ecological importance of small mammals as reservoirs of disease 5

F. E. G. COX

5.1 Introduction

Disease is a state in which an individual's activities are reduced to levels below the normal day to day fluctuations and is a concept easier to understand than to define. Diseases are divided, for convenience, into two major categories, those that cannot be transmitted from one individual to another and those that can. The former include such diseases as those of physiological or psychological disfunction while the latter embrace most of the better known diseases of man and his domesticated animals. Diseases which can be passed from one individual to another are known as infectious, transmissible or communicable diseases and are caused by infectious agents. Transmission can be brought about by direct contact or through the agency of some intermediary, usually an insect.

The connection between certain animals and human disease has been realized since the earliest times. Aristotle wrote in the fourth century B.C. of madness (rabies) being transmitted through the bites of dogs and it is now known that over 200 diseases can pass from animals to man. Such diseases, known as zoonoses, are rightly regarded as being among the major health hazards faced by man and the majority involve small mammals, particularly rodents.

Zoonoses are defined as 'those diseases and infections which are naturally transmitted between vertebrate animals and man' (WHO, 1967a). A number of attempts have been made to extend this definition and to introduce new subdivisions of zoonoses but these all tend to confuse the definition rather than clarify it.

In a classical zoonosis, the infection cycles in its wild hosts and man is incidentally involved. Man's involvement may be slight, the occasional person being infected often well away from human habitation as in rabies, or the disease, having reached man, may continue to cycle independently in the human population as in plague. There is often an intermediate link between the infection in the wild and that in man, the bridge usually being synanthropic wild mammals or pets that come and go between the house and the wild. In all these cases wild mammals are the ultimate source of human infection and they, and the intermediate animals, act as reservoirs of human infection.

Transmission between animals and between animals and man can be direct, usually by contact or by ingestion of infective stages, or indirect, the contact being by way of a vector, usually a biting arthropod, in which the infectious agent may undergo part of its development (cyclical transmission) or which carries the infectious agent passively (mechanical transmission). The importance of the vector is that it can bring an infectious agent from a wild animal to man without direct contact between the two being necessary. In such cases, it is often the ecology of the vector that governs the spread of the disease rather than the ecology of the mammal host.

5.2 Zoonoses involving small mammals

Excellent accounts of all the zoonoses are given in van der Hoeden (1964) and Hubbert, McCulloch and Schnurrenberger (1975) and an elementary account of bacterial and viral zoonoses is given by Andrewes and Walton (1977). The World Health Organisation has published a wealth of summarized information (WHO, 1967a) and there are a number of more specialized works on zoonoses including an annotated bibliography (Gluckstein, 1974) and zoonoses for veterinarians (Handler, 1975). These should be read in conjunction with relevant standard text books on microbiology (Buchanan and Gibbons, 1974; Dubos and Hirch, 1965; Evans, 1977; Fenner and White, 1976; Horsfall and Tamm, 1965), mycology (Al-Doory, 1976), parasitology (Nobel and Noble, 1977; Olsen, 1974) and epidemiology (MacMahon and Pugh, 1970; Le Riche and Milner, 1971).

Not all the zoonoses involve small mammals and the most important of those that do are listed in Tables 5.1 and 5.2. This list is not complete because new zoonoses are continually being identified and some that have been described in the past are of such little importance to man that they are best ignored. In addition, it is becoming clear that most of the infectious agents implicated in zoonoses exist in a variety of forms (*see* 5.4.2) and the isolation of seemingly similar agents from a wild animal and from man does not necessarily mean that transmission from animals to man does or could occur. Nevertheless, the list indicates the range of zoonoses involving small mammals but before they can be discussed further it is necessary to say something about the diseases themselves.

5.2.1 Bacterial zoonoses

It is convenient to consider bacterial zoonoses first because bacteria seem to be less host-specific than other microorganisms and can, unlike viruses, exist for long periods outside the host. These characteristics allow one to examine the whole spectrum of zoonoses from accidental and occasional infections of man to those in which there is a constant overflow from the wildlife cycle to man.

Salmonellosis

Of over 1000 serotypes of *Salmonella*, one, *Salmonella typhimurium,* causes the majority of human cases of food poisoning. The bacteria are found in numerous species of mammals, birds, reptiles and some insects. Man usually becomes infected from other humans, pets, domesticated animals and animal products such as eggs and meat. Small mammals, particularly rodents, are commonly infected and can contaminate food or infect people who handle them. Small mammals therefore act as reservoirs of infection but their role is minimal in a disease which is so widespread (*see* also Taylor, 1969).

Anthrax and brucellosis

These are diseases of cattle, sheep and goats but virtually all mammals can be infected. Man is infected by handling animal products or, in the case of brucellosis, drinking milk. Small mammals, particularly rodents, can carry these diseases but their role in the infection of man is minimal. (*see* also McCaughey, 1969).

Glanders and melioidosis

Glanders is a classical zoonoses with the disease passing directly from horse to horse; man and other animals being infected incidentally. Melioidosis is rare in man but as well as horses a number of species of mammals and birds are susceptible to infection. Rodents, in tropical and subtropical areas, can occasionally pass both these diseases on to man.

Listeriosis

Listeria monocytogenes infects many species of mammals and birds, and is also found in fish, soil and sewage. Infected rodents very occasionally pass the infection on to man.

Pseudotuberculosis

This is an important natural disease of many wild animals mainly in Western Europe. Hares, in which it causes serious epizootics, hedgehogs, rodents and birds constitute the main reservoir hosts of both human and farm animal infections which are transmitted by direct contact or in the faeces or urine. Man becomes infected from his farm animals, pets or contaminated foodstuffs and wild small mammals constitute a serious health threat (*see* also Mair, 1969). Another bacterium *Yersinia enterocolitica* has been recognized as a pathogen of hares and other small mammals in Europe, Africa and America since 1953 but its significance for human health is not known.

Leptospirosis

There are a number of serotypes of *Leptospira* and it is usual to refer to some of these as species.

L. icterohaemorrhagicae occurs in fish, amphibia, birds and many small mammals, particularly rats. Infection is usually through the urine which may contaminate water. So many animals can be infected that it is difficult to say which constitute important reservoirs of the human disease but small mammals certainly do play a role in passing the infection on to man in particular situations such as paddy fields, sewers and other damp places and as a result of urine entering skin wounds when animals are being handled. (*see* WHO 1976b).

Pasteurellosis

Human pasteurellosis is acquired from birds or mammals, usually through cat or dog bites or scratches but rarely directly from rodents although they play a peripheral role in maintaining the infection in the wild.

Erysipelas

Many species of mammals and birds can be infected but the bacteria can also survive in manure or in the slime on the outside of fish, which is the main source of human infection, the organisms entering the body through skin wounds. Rodents can harbour the infection and are sometimes severely affected themselves but their role as reservoirs of the human disease is minimal.

Rat bite fevers

Rat bite fevers are caused by two different bacteria transmitted to man through the bites of rats which are, in both cases, the sole reservoirs of the human infection.

Tularemia

Tularemia is a vector-borne infection which affects a wide range of animals including over 100 species of wild mammals as well as domesticated animals, birds and molluscs. It is transmitted by the bites of ticks, deerfly, mosquitoes and other blood-sucking arthropods, by the ingestion of the faeces or urine of infected animals or by drinking water contaminated by them. The distribution of tularemia is essentially northern Europe, Eurasia and America and this coincides with the distribution of the principal hosts, lagomorphs. *Sylvilagus* spp. and *Lepus californicus* are the most frequently infected hosts and many human cases result from handling infected animals. In some parts of Eastern Europe water contaminated by water voles (*Arvicola terrestris*) causes local epidemics.

Plague

Plague is essentially a sylvatic disease transmitted among squirrels, mice and voles by fleas but over 200 species of small mammals are infected naturally in various parts of the world and the disease can cause considerable mortality in these populations. It is when rats enter the cycle and carry plague to human settlements that it becomes a health hazard. The domestic cycle involves rats and the rat flea *Xenopsylla cheopis* but this flea also bites man and the disease is transmitted to him. Thereafter plague can be transmitted directly from man to man by droplet infection and, under these conditions, can spread rapidly (*see* also Pollitzer, 1954).

Tick-borne relapsing fever

In North Africa, the Mediterranean region, Central Asia and America, relapsing fever is essentially a disease of rodents transmitted by soft ticks. The rodents seldom come into contact with man who becomes infected when he enters an area where infected ticks are present. In Central, East and South Africa, the disease is transmitted directly from man to man and there is no animal reservoir.

5.2.2 Rickettsial zoonoses

Tick-borne rickettsial diseases

Rocky Mountain spotted fever, is a classical zoonosis involving lagomorphs and rodents and is transmitted by ixodid ticks that occasionally bite man and pass on the infection. *Boutonneuse fever*, in the Mediterranean region, involves rabbits and dogs and is transmitted by the tick *Rhipicephalus sanguineus*. In Kenya, a similar disease, *Kenya tick typhus,* involves a whole range of rodents and is transmitted by the larvae of the ticks *Haemaphysalis leachii* and *Rhipicephalus simus*. In Malaysia, both sylvatic and synanthropic rodents are involved. The North Asian form of tick typhus is a natural infection of rodents and is transmitted by ixodid ticks. In all these cases, man becomes infected when he intrudes into the natural cycle and is bitten by an infected tick. *Q. fever* is primarily a rodent-tick infection. However it may be spread to a wide range of animals including rodents, domesticated ungulates, birds and tortoises. It is now primarily a disease of cattle, sheep and goats and it is from these, or their products, that man becomes infected. Occasionally man may become infected through the bite of a tick, but this is exceptional.

Murine (flea-borne) typhus

The natural hosts of murine typhus are *Rattus rattus* and *R. norvegicus* and the vector is the flea *Xenopsylla cheopis*. In areas where *X. cheopis* is absent other insects can transmit the disease to other small mammals. Man becomes infected by being bitten by fleas that have strayed from their natural hosts.

Scrub typhus

Scrub typhus, in Asia and Australasia, involves rodents and mites and man becomes infected when he enters an infested area. This is a particularly serious disease in war time when troops and refugees enter undisturbed areas and the disease takes a large toll of those infected.

Rickettsial pox

This is a natural infection of house mice and mites *Liponyssoides sanguineus*, which abound in mouse-infested houses. Man becomes infected when bitten by mites. This disease is essentially an urban one in North America and Russia but the same, or similar, diseases occur in non-urban areas of South America and Korea where the reservoir hosts cannot be mice. In Korea they are voles.

5.2.3 Viral zoonoses

For practical purposes it is convenient to discuss the viral zoonoses under two headings, those in which infection is by direct contact and those that require vectors which are ticks, mosquitoes or sandflies. The importance of vectors in viral zoonoses cannot be over-emphasized because viruses are unable to survive for any period outside a host of some kind.

Rabies

Rabies exists in a number of distinct ecological forms each with a characteristic major host which is usually a carnivore, for example the polar fox, *Alopex lagopus*, in the Arctic, the red fox *Vulpes vulpes* in Central Europe, where badgers, martens and weasels are also infected, and the racoon dog, *Nyctereutes procyonoides*, in Eastern Europe. In the New World, endemic rabies is maintained in racoons and skunks and in the tropics, mongooses, jackals and wolves are involved. The danger to man comes when the infection overflows to domesticated animals such as dogs. Small mammals are not usually involved in the transmission of the disease to man and domesticated animals except in the New World where vampire and insectivorous bats are important. Vampire bats, particularly *Desmodus rotundus*, transmit rabies when they feed on cattle and there have been a number of human cases. Insectivorous bats can also transmit the disease, presumably through droplets, in the caves in which they live. No small mammals other than bats are known to act as reservoirs of rabies in the New World. In Europe, a number of small mammals, including *Microtus arvalis*, *Clethrionomys glareolus*, *Apodemus flavicollis* and *Mus musculus* have been found to harbour a virus similar to that of rabies but whether this is capable of infecting man is not known. (For a comprehensive account of rabies see Baer, 1975a and for a general account see Kaplan, 1977, and WHO, 1973).

Encephalomyocarditis (ECM)

The virus is found in rats in which it causes no harm but occasionally humans become infected presumably through direct contact with the faeces of infected rats.

Lymphocytic choriomeningitis (LCM)

The virus of LCM is harmless in its natural hosts, house mice, but may cause serious diseases in laboratory mice and guinea pigs. It is acquired by the inhalation of infective particles passed in the faeces or urine and human infections, especially among laboratory workers, occasionally occur.

Haemorrhagic fevers

Four important haemorrhagic fevers have been recently recognized. *Bolivian haemorrhagic fever* is an urban disease transmitted to man from *Calomys callosus* which lives, like house mice, in dwellings in South America. *Argentinian haemorrhagic fever* is a rural disease mainly of those working in the Pampas and has as its reservoir *Calomys laucha*. In Korea, epidemiological evidence suggests a rodent reservoir and, in Africa, the main reservoir of *Lassa fever* is *Mastomys natalensis* but the disease can also be transmitted directly from man to man.

Marburg and Ebola viruses

Marburg virus is mentioned here because, although it is a virus of African monkeys transmitted to man and from man to man, there is epidemiological evidence that suggests that monkeys cannot be the main reservoir of human infections and that small mammals may be implicated. Ebola virus is a similar virus. (*See* Simpson and Zukerman, 1977).

Tick-borne viruses

Over 70 viruses are known to be transmitted by ticks but the actual number transmitted to man is not known for certain. Those that are include three distinct forms of tick-borne encephalitis (TBE), louping ill, Central European TBE and Far Eastern TBE (Russian Spring-Summer encephalitis). *Louping ill* occurs in a number of animals including shrews, rodents, deer and red grouse in Britain and is transmitted from these to sheep by the sheep tick *Ixodes ricinis* and shepherds occasionally become infected. *Central European TBE* virus is found in murine rodents, insectivores and birds and, like louping ill, is transmitted to sheep by *I. ricinis*. *Far Eastern TBE* virus occurs in many mammals and birds and is transmitted to man by *I. persulcatus*. This is a serious disease of man and mainly affects foresters and others working in forest clearings.

Powassan virus is found in North American rodents, especially woodchucks, and is transmitted between them and to man by *Ixodes* spp. and *Dermacentor* spp. *Omsk haemorrhagic fever* occurs in Russia and the virus infects indigenous small mammals such as *Microtus* spp. but can also be transmitted to non-native musk rats by *Dermacentor* spp. and from these to man, the disease being essentially one of those working in the fur trade. *Kyasanur forest disease* has a scattered distribution in Southern India and small mammals, including rats of the genus *Rattus*, shrews of the genus *Suncus* and bats of the genus *Rhinolophus*, form the reservoir of the disease which is transmitted to man by various species of ticks.

Colorado tick virus is a natural virus of rodents, especially chipmunks and ground squirrels, in North America. It is transmitted by ticks which also infect foresters and cattlemen who enter the infested area. *Nairobi sheep disease* is a disease of sheep and goats. The virus is also found in *Arvicanthus* spp. and is transmitted by *Rhipicephalus appendiculatus*. Man is sometimes affected but whether the infection passes from the sheep or the rodents is not clear.

Mosquito-borne viruses

Mosquitoes tend to be less selective in their choice of hosts than ticks, therefore mosquito-borne viruses have wider host ranges than tick-borne viruses. *Western equine encephalitis* (WEE) is an important disease of horses in North and South America. It is basically an infection of birds transmitted by *Culex* spp. but man, horses, rodents and squirrels can also be infected. *Eastern equine encephalitis* (EEE) is rarer than the western form but causes a more serious disease in horses. The disease occurs in Northern and Eastern U.S.A. and S. America and is naturally transmitted between birds by various mosquitoes. Other animals including man, horses, small mammals and reptiles can be infected. *Venezuelan equine encephalitis* is a serious disease of horses now spreading northwards from South America. The virus appears to be a natural one of birds, rodents, bats and oppossums between which it is transmitted by mosquitoes of several genera. Pigs and cattle are infected but are unaffected by the disease whereas equines are seriously affected and there have been a number of epidemics in man. Man may be infected with any of these viruses either from horses or from the wild.

California virus disease is caused by a natural virus of rabbits and hares and is transmitted by several species of mosquitoes found in North America. Man occasionally becomes infected. *Rift Valley fever* is caused by a virus which is naturally transmitted between forest rodents by mosquitoes in South and East Africa and causes serious disease in lambs. Man is occasionally infected.

As well as these viruses there are a number of less important ones that are naturally transmitted among wildlife by mosquitoes in many parts of the world and sometimes infect man.

Sandfly-borne viruses

Sandfly fever occurs in several forms in India, East Africa and the Mediterranean regions. The virus seems to be a natural one of sandflies which they can transmit to gerbils, man and other animals. A number of related viruses have been recorded in various parts of the world where sandflies live.

Vesicular stomatitis viruses

The vesicular stomatitis viruses constitute a group of viruses transmitted between a number of mammals by a range of blood-sucking arthropods including mosquitoes, midges, sandflies and mites in the New World. Vesicular stomatitis is an important disease of cattle, sheep, pigs, horses, deer and man and the virus has been isolated from rodents. The significance of rodents as reservoirs of human and cattle disease is not at all clear but most human infections are acquired from cattle. A virus of the vesicular stomatitis type occurs in sandflies, gerbils and humans in Iran and is known as Isfahan virus.

5.2.4 Fungal zoonoses

Fungi are important pathogens of man and domesticated animals and many can be acquired from the soil or from animal products. The fungal zoonoses are usually considered under two headings, those that infect the skin (dermatophytes) and those that invade the deeper tissues (systemic) (*see* Austwick, 1969).

Dermatophytoses

The dermatophytes invade the skin causing diseases known as ringworm. *Trichophyton mentagrophytes*, of small mammals occurs in several forms and man is frequently infected, in the wild.

Systemic mycoses

Most of the systemic mycoses are caused by fungi that spend part of their lives in the soil from which man and other animals acquire the infections independently.

Histoplasmosis has been recorded from man and many species of wild and domesticated animals including shrews, rats, spiny rats and woodchucks. *Coccidioidomycosis* occurs in the New World and it was at one time thought that rodents were the reservoirs of human infections. *Actinomycosis* is a chronic disease of man and occurs in many animals. *Cryptococcosis* affects man and a number of animals including small mammals. The fungus thrives in bird droppings and this may contribute to its spread. *Sporotrichosis* is a rare disease of man and occurs in rats. *Adiaspiromycosis* affects man and over 50 species of small mammals most of which live in burrows.

None of these infections can be considered as true zoonoses in the sense that small mammals act as reservoirs of human disease because the soil is the source of both the human and animal infections. Nevertheless, rodents play a role in distributing the fungi and may indirectly contribute to the infection in humans.

5.2.5 Protozoal zoonoses

In general, protozoa tend to be more host specific than viruses, bacteria or fungi and only four important protozoal diseases are acquired by man from small mammals, one with a direct and three with indirect life cycles.

Toxoplasmosis

Toxoplasma gondii has a wider host range than any other protozoan. It is a harmless parasite of cats passed from cat to cat in the form of a resistant oocyst in the faeces. In virtually all mammals other than the cat, many birds and some reptiles, the oocyst gives rise to proliferative stages that are only passed on when the host is eaten. Man becomes infected from cats or by eating infected meat and can therefore only acquire an infection directly from a small mammal by eating it in an undercooked state. However, small mammals are important in maintaining the infection in the cat and thus indirectly serve as a reservoir of human infections (*see* also Beverley, 1974).

Leishmaniasis

Human leishmaniasis is caused by some 15 species or subspecies of *Leishmania* transmitted by sandflies mainly in the tropics and subtropics. Several small mammals act as reservoirs. In the Old World, gerbils, *Rhombomys opimus*, are the main rodent reservoirs of *L. tropica* which causes cutaneous leishmaniasis. A variety of rodents act as reservoirs for visceral leishmaniasis, caused by *L. donovani*, in East Africa but in India and the Mediterranean regions rodents are not involved, dogs being the reservoirs in the Mediterranean and man himself in India. In the New World, *Otolomys*, *Oryzomys*, *Proechimys* and other genera of rodents are the reservoir hosts of *L. mexicana* and *L. braziliensis* which cause cutaneous leishmaniasis but the reservoir of *L. donovani* is a fox (*see* also Bray, 1974).

Chagas' disease

Over 150 species of mammals have been recorded as hosts of this trypanosome which is transmitted by a number of species of triatomid bugs. Both the infection and the bugs are sylvatic in origin but various degrees of association between particular mammals, bugs and man exist bringing the infection into villages and urban areas and most people become infected through the bite of a bug in their houses. *Didelphis* spp. are the most important reservoirs in the areas where they are found (*see* also Zeledon, 1974).

Piroplasmosis

Human piroplasmosis, caused by *Babesia microti,* is a recently recognized and rare disease. *B. microti* is a natural parasite of small mammals in Europe and America and is transmitted by ticks. Man presumably becomes infected when bitten by a tick. (*see* Spielman, 1976).

5.2.6 Helminth zoonoses

Trichinosis

Trichinosis is mainly North-Temperate in its distribution and has the widest host range of any helminth parasite, infecting over 100 species of mammals including about 30 rodent species. Pigs, which become infected by being fed pork scraps, are the most important hosts and man becomes infected by eating uncooked pork. Various mammals become infected by feeding on pork scraps and may spread the disease among themselves by cannibalism and this is often the case with rats. Small mammals play little part in the direct transmission of trichinosis to man unless they are eaten.

Angiostrongyliasis

The rat lungworm infects rats of several genera in the Pacific area. Man becomes infected by ingesting the infective larvae which may occur in a variety of hosts including molluscs.

Capillariasis

This natural infection of rodents can only be passed on when the host dies and eggs released from the liver are ingested. Rats, voles and lemmings are all infected but human cases are very rare.

Hymenolepiasis

Two species, *Hymenolepis nana* and *H. diminuta*, are found as adults in man. *H. nana* is a natural parasite of man and rats and mice are alternative hosts. *H. diminuta* is a rat tapeworm and man is occasionally infected when insects harbouring the larval stages of the parasite are eaten.

Echinococcosis

The hosts of the adult stages of *Echinococcus multilocularis* are foxes and dogs in Central Europe and Asia, and in Japan. The larval stage normally develops in microtine rodents and this stage, which can also infect man, can only be acquired by ingesting the eggs in the faeces of the final host. Man is therefore a parallel host with rodents and cannot be infected directly from them although when the infection passes from mice to cats it creates a health risk (*see* also Euzeby, 1974).

Inermicapsifer infection

Inermicapsifer spp. are tapeworms of African rodents that occasionally occur in children.

Sparganosis

Various tapeworms belonging to the genus *Spirometra* occur as adults in carnivores and use arthropods, especially *Cyclops* spp., as intermediate hosts. These intermediate hosts are eaten by a number of animals including rodents and, within these, the larval stage remains until it is eaten by the final host. Man is infected by swallowing *Cyclops* spp. and is therefore a parallel host with rodents and a dead end for the infection (*see* also Euzeby, 1974).

Schistosomiasis

The intermediate hosts for *Schistosoma* spp. are aquatic snails. *S. japonicum* is a well-established zoonosis in the Far East and many species of mammals, including rats and mice, may be infected. Both man himself and these mammals act as maintenance hosts for the infection. *S. mansoni* in the Middle East, Africa and South America, has also been recorded from a variety of small mammals but it is unlikely that these play any major role in the transmission of schistosomiasis to man.

5.3 Epidemiology of human diseases involving small mammals

5.3.1 Foci of infection

The diseases of small mammals exist in well-defined foci of infections in which the infectious agent, its hosts and its vectors exist as part of a biotic community or biocenosis (Pavlovsky, 1966). These are characterized by their geographical situations, soil type, flora and fauna and are relatively stable changing only with the normal progression of the seasons. Diseases constitute integral parts of these biotic communities and, like the communities themselves, are relatively stable.

Within any focus of infection there is also an area where the conditions are most suitable for the maintenance of the infection. This is called an elementary focus and is particularly stable. It is now generally accepted that there are two distinct kinds of elementary foci, the restricted type, which is precisely limited by geographical or biotic factors, and the diffuse type, which has less clearly defined boundaries (Rosicky, 1967). It is these elementary foci that provide the conditions most suitable for the maintenance of disease and it is not unusual to find several infections co-existing in the same focus. For example, in Iran sandfly fever, Isfahan virus and leishmaniasis all exist in the same area, the focus being rodent burrows inhabited by sandflies (Tesh *et al.*, 1977).

Because they offer particularly suitable conditions, it is to the elementary foci that other animals move when the original occupants die or are killed by a disease, thus ensuring the continued existence of the focus.

Most of the diseases studied in the context of foci of infections are those transmitted between rodents and from rodents to man by biting arthropods, particularly tick-borne and mosquito-borne encephalitis, rickettsial diseases, plague, Chagas' disease and leishmaniasis but it has also been found that diseases such as leptospirosis and fungal diseases can be studied in the same way.

5.3.2 Spread of infections from foci

Evolution has ensured the stability of foci of infections and spread from these should not normally occur because, as one moves from the elementary focus to the focus proper and then outside, the conditions that facilitate transmission becomes increasingly diminished. Various characteristics of the biotic community can influence spread from a focus, for example the fewer animals and vectors involved the less likely it is that a disease will spread, while with increasing numbers of potential hosts and vectors the possibilities of spread become greater. Similarly, the more specialized or isolated the habitat the more likely it is that a disease will be contained within it.

The usual way in which humans become infected is by intrusion into foci of infections. Such intrusions may be occasional ones or on a large scale. Occasional intrusions are well documented for example the chicle collectors and wild-rubber tappers who enter forests and become infected with cutaneous leishmaniasis in South America (Lainson and Strangeways-Dixon, 1963) and children who graze their goats on cliff faces in Ethiopia and there become infected with leishmaniasis (Bray, 1974). Many tourists, campers, explorers and official and unofficial soldiers come into this category and tick-borne rickettsial diseases present particular hazards.

The building of roads and the extension of farmland are examples of large scale intrusions. These not only disturb or destroy old foci but also create new ones. Forest clearing is always hazardous and many epidemics have resulted from such activity. The creation of farmland brings humans closer to natural foci of infections and often creates ideal conditions for the spread of disease and the establishment of new foci. The replacement of deciduous trees with spruce in Eastern Europe has created a much more uniform habitat and thus particular foci have been able to extend. In farmland, infections are spread from the natural foci by the intrusion of domestic animals, dogs or synanthropic mammals, often by way of intermediate foci such as farm buildings. This is well exemplified by the spread of ringworm from *Clethrionomys glareolus* and *Apodemus sylvaticus* in woods and forests to *Microtus arvalis* and then to *Mus musculus* and finally man in Eastern Europe (Rosicky, 1972).

The spread of disease from sylvatic foci to urban foci usually involves house mice or rats, the latter being particularly important because they tend to follow trade. In urban areas, foci involving synanthropic mammals are common and in some cases the corresponding sylvatic disease has not been identified.

5.3.3 Models of the spread of disease

In most mathematical models of infectious diseases the major components considered are the numbers of infective (I) and susceptible (S) individuals in the population and the single most important estimate is the rate at which there is a transition from susceptible to infected (and then infective) individuals (Bailey, 1975). In zoonoses, there is the added complication that the infective population is not only the human one but also an animal one and therefore the number of infectives may be impossible to establish. Several attempts have been made to quantify this problem but most observers agree that the available knowledge about small mammal reservoirs and the vectors is inadequate (*see*, for example, Hairston, 1972). Even in the case of plague, the role of the rat is seldom considered (Noble, 1974). Rabies in which small mammals are seldom involved in infecting humans has, however, received considerable attention (Berger, 1976).

5.4 The importance of small mammals as reservoirs of disease

5.4.1 Overall assessment

There is no doubt that small mammals are very important as reservoirs of

Table 5.1 *A list of the more important diseases of man that can be acquired from small mammals. Diseases indicated as (c) are cosmopolitan in their distribution. The distribution of the other diseases is indicated by their names or in the text.*

Disease		Causative organism
VIRUSES		
Encephalomyocarditis		Picornavirus
Western equine encephalomyelitis (WEE)		Alphavirus
Eastern equine encephalomyelitis (EEE)		Alphavirus
Venezuelan equine encephalomyelitis		Alphavirus
California virus disease		Bunyavirus
Rift Valley fever		?
Sandfly fever		Bunyavirus–type
Louping ill		Flavivirus
Central European tick-borne encephalitis		Flavivirus
Far Eastern tick-borne encephalitis		Flavivirus
Omsk haemorrhagic fever		Flavivirus
Powassan virus disease		Flavivirus
Kyasanur forest disease		Flavivirus
Colorado tick fever		Orbivirus
Nairobi sheep disease		?
Lymphocytic choriomeningitis (LCM)		Arenavirus
Argentinian haemorrhagic fever		Arenavirus
Bolivian haemorrhagic fever		Arenavirus
Korean haemorrhagic fever		Arenavirus
Marburg disease		?
Ebola virus disease		?
Rabies	(c)	Rhabdovirus
Vesicular stomatitis		Rhabdovirus
BACTERIA		
Anthrax	(c)	*Bacillus anthracis*
Brucellosis	(c)	*Brucella abortus*
Salmonellosis	(c)	*Salmonella* spp.
Erysipelas	(c)	*Erysipelothrix rhusiopathiae*
Melioidosis		*Pseudomonas pseudomallei*
Glanders		*Pseudomonas mallei*
Listeriosis	(c)	*Listeria monocytogenes*
Pasteurellosis	(c)	*Pasteurella multocida*
Plague	(c)	*Yersinia pestis*
Tularemia		*Francisella tularensis*
Pseudotuberculosis		*Yersinia pseudotuberculosis*
Rat bite fever		*Spirillum minus Streptobacillus moniliformis*

Table 5.1 (*cont.*)

Disease		Causative organism
BACTERIA (*cont.*)		
Tick-borne relapsing fever		*Borrelia duttoni*
Leptospirosis	(c)	*Leptospira* spp.
Rocky Mountain spotted fever		*Rickettsia rickettsiae*
Boutonneuse fever		*Rickettsia conorii*
Kenya typhus		*Rickettsia conorii*
North Asian typhus		*Rickettsia sibirica*
Rickettsial pox		*Rickettsia akari*
Scrub typhus		*Rickettsia tsutsugamushi*
Murine typhus	(c)	*Rickettsia mooseri*
Q fever	(c)	*Coxiella burnetti*
FUNGI		
Dermatophytoses		*Trichophyton mentagrophytes*
Coccidiodomycoses		*Coccidium immitis*
Histoplasmosis		*Histoplasma capsulatum*
Sporotrichosis	(c)	*Sporothrix schenckii*
Actinomycoses		*Actinomyces israeli*
Adiaspiromycosis		*Emmonsia parva E. crescens*
Cryptococcosis	(c)	*Cryptococcus neoformans*
PROTOZOA		
Leishmaniasis		*Leishmania* spp.
American trypanosomiasis		*Trypanosoma cruzi*
Toxoplasmosis	(c)	*Toxoplasma gondii*
Piroplasmosis		*Babesia microti*
HELMINTHS		
Trichinosis		*Trichinella spiralis*
Capillariasis		*Capillaria hepatica*
Angiostrongyliasis		*Angiostrongylus cantonensis*
Echinococcosis		*Echinococcus multilocularis*
Hymenolepiasis		*Hymenolepis nana H. diminuta*
Inermicapsifer infection		*Inermicapsifer madagascariensis*
Sparganosis		*Spirometra* spp.
Schistosomiasis		*Schistosoma japonicum S. mansoni*

diseases of man and his domesticated animals, (*see* Table 5.2) but the overall cost in terms of mortality, morbidity and economic loss is impossible to assess. It is not even possible to say which of the 100 or so zoonoses are the most important. Rabies is greatly feared but the number of human cases is relatively small. Vampire bats are probably the most important reservoirs and 150 (Baer, 1975c) or 170 (WHO, 1975) human deaths have been directly attributed to this source and these have been mainly in Trinidad. Human rabies attributed to non-haematophagous bats has been confirmed in about ten cases (Baer, 1975b). On the other hand bovine paralytic rabies acquired from vampire bats accounts for 'hundreds of thousands' of deaths in North America and Argentina (Baer, 1975c). Far Eastern tick-borne encephalitis is a much more serious disease than rabies causing 20–30 per cent mortality and Kyasanur forest disease causes serious epidemics that attract far less attention than rabies.

The importance of infections with small mammal reservoirs is always relative, rare diseases in some countries receiving considerably more attention than common diseases in others. In general, such diseases create local problems which will only be solved when the particular characteristics of the infection and its transmission have been fully elucidated.

One of the most disturbing features of zoonoses is that they are continually emerging as new diseases (WHO, 1967a). Kyasanur forest disease, 1957, Bolivian haemorrhagic fever, 1959, Marburg virus, 1967, and Lassa fever, 1970, are all recently identified important diseases and less important ones are being identified all the time as techniques for their detection improve.

Despite increasing knowledge of zoonoses they are not being contained; the spread of rabies across Europe is well known (Kaplan, 1977) and equally important diseases are still spreading elsewhere. For example, Venezuelan equine encephalitis is gradually moving into North America causing tens of thousands of cattle deaths annually and infecting an increasing proportion of the human population.

Many of the diseases listed in Table 5.1 can be classified as occupational diseases and the occupations most likely to be at risk are farming, road building, forest clearing, animal handling, animal-product handling, veterinary work and laboratory work and tourists, explorers, soldiers and guerrillas may also be involved. As well as these occupations, many other people are at risk because of modern forms of road and air transport which can bring an infected person from the field to a crowded town or city before he knows he is ill.

5.4.2 Problems of assessing the role of small mammals as reservoirs of disease

There are two basic problems that make the assessment of the role of small mammal reservoirs of human disease difficult; not enough is known about the ecology and behaviour of the mammals that can and do act as reservoirs (WHO, 1974) and in many cases the organisms that cause the diseases cannot be adequately identified. Many of the infectious agents are now classified below the species level as serotypes or even biochemically distinct strains. There are, for example, over 60 serotypes of *Leptospira* (WHO, 1976b) and more are continually appearing. In Europe, rabies-like viruses have been isolated from rodents (Sodja *et al.*, 1971) but it is unlikely that these can actually spread to carnivores and man (Lloyd, 1976) and this is also the case in other parts of the World (Turner, 1976). In Brazil, biochemically distinct though morphologically identical strains of *Trypanosoma cruzi* co-exist (Miles *et al.*, 1977) and the problem of distinguishing between morphologically similar trypanosomes has been discussed by Godfrey (1977). Many of the arthropod transmitted viruses isolated from man differ serologically from those of small mammals thought to be reservoirs of the human infection (*see* Berge, 1975) and it may well be that the human and small mammal infections cycle independently of one another. A final problem is that many viruses cross-react with one another making them impossible to identify with certainty. The virus of California encephalitis is one example (Berge, 1975).

5.5 The control of diseases with small mammal reservoirs

Measures to control diseases with small mammal reservoirs include immunization against and treating the disease itself and eradicating the vector or the reservoir species. A number of vaccines effective against the more important zoonotic diseases are now available and most of the bacterial diseases respond well to drugs. Viral, mycotic, protozoal and helminth diseases are less easy to treat. The control of the vectors is generally agreed to be the most effective way of dealing with arthropod transmitted diseases and, despite the problems of resistance and ecological objections, insecticides are, and will continue to be, the main defence against such diseases.

The control of small mammals creates few problems specific to such animals as reservoirs of disease. It is not possible, desirable or feasible to eradicate small mammals in sylvatic or other uninhabited areas. In urban areas, the control of synanthropic mammals is desirable for many reasons other than the prevention of disease and effective measures include good building design, rodent-proofing buildings, the disposal of refuse and the destruction of all man-made habitats that might be used by rodents. The use of repellent substances is also effective.

The main method of controlling small mammals is the use of rodenticides. Rodenticides can be classified under two headings, quick acting or single-dose and slow acting or multidose; the anticoagulants belonging to

Table 5.2 *The Families, Genera and common names of the rodent reservoirs of the more important human diseases together with their distribution and the diseases they carry. Reprinted with permission from the World Health Organization Technical Report No. 553 (1974)*

Classification	Number of genera	Number of species	Common names	General distribution	Representative genera (genera for which control data are required are marked with an asterisk (*))	Associated human diseases
Order Rodentia	354	1687		Cosmopolitan		
Sciuridae	41	261	Squirrels, marmots, and susliks	Cosmopolitan except for Australia and Oceania, polar regions, and parts of other continents	*Citellus* * (= *Spermophilus*), *Marmota*, *Sciurus*, *Funambulus*	Plague, tularaemia, listeriosis, erysipelas, pseudotuberculosis, leptospirosis, Chagas' disease, Q fever, scrub typhus and other rickettsial infections, western equine encephalitis, central European encephalitis, and other viral infections, histoplasmosis, actinomycosis, and dermatomycoses
Heteromyidae	5	75	Kangaroo rats and pocket mice	North America to northern South America	*Dipodomys*, *Perognathus*	Plague, tularaemia, coccidioidomycosis, Q fever
Geomyidae	8	40	Pocket gophers	North and Central America	*Geomys*, *Thomomys*	Q fever
Castoridae	1	2	Beavers	North America, Europe, and North Asia	*Castor*	Tularaemia, leptospirosis, pseudotuberculosis, adiaspiromycosis
Anomaluridae	4	12	Scaly-tailed squirrels	West and Central Africa	*Anomalurus*	Information not available
Pedetidae	1	2	Springhaas	Central and South Africa	*Pedetes*	Information not available

			and mice	America, Europe and North Asia	related genera,* *Peromyscus, Neotoma, Cricetus, Cricetulus, Sigmodon, Oryzomys**	Plague, tularaemia, leptospirosis, salmonellosis, brucellosis, paragonimiasis, leishmaniasis, Chagas' disease, Q fever, murine typhus, spotted fevers and related rickettsioses, Bolivian and Argentinian haemorrhagic fevers, lymphocytic choriomeningitis and other viral infections, coccidioidomycosis and other infections
Nesomyinae	(7)	(12)	Malagasy rats	Madagascar	*Nesomys*	Plague
Lophiomyinae	(1)	(1)	Maned rats	Africa	*Lophiomys*	Information not available
Microtinae	(18)	(105)	Voles and lemmings	North and Central America, Europe and North Asia	*Microtus,* Arvicola,* Lemmus, Ondatra*	Plague, tularaemia, pseudotuberculosis, listeriosis, leptospirosis, rickettsial pox and other richettsioses, tick-born encephalitis, rabies, and other viral infections, ringworm, sportrichosis and dermatomycoses, and other infections
Gerbillinae	(12)	(93)	Gerbils and jirds	Arid zones of Africa, Central and Middle Asia	*Meriones,* Tatera,*, Rhombomys, Nesokia*	Plague, cutaneous leishmaniasis, listeriosis, lymphocytic choriomeningitis and other viral diseases, and other infections
Spalacidea	1	3	Mole rats	Eastern Europe, Middle East to Iran	*Spalax*	Information not available
Rhizomyidae	3	18	Bamboo rats	East and South Asia to Sumatra and East Africa	*Rhizomys, Cannomys*	Penicilliosis
Gliridae	7	23	Dormice	Europe, parts of North Africa, Middle East and Central Asia and Africa, south of Sahara	*Glis, Dryomys*	Information not available
Seleviniidae	1	1	Dzhalmans	Central Asia near Lake Balkhash	*Selevinia*	Information not available
Platacanthomyidae	2	2	Spring dormice	India and southern China	*Platacanthomys*	Information not available

Table 5.2 *(cont.)*

Classification (families, subfamilies)	Number of genera	Number of species	Common names	General distribution	Representative genera (genera for which control data are required are marked with an asterisk (*))	Associated human diseases
Zapodidae	4	11	Jumping mice and birch mice	North America, eastern Europe and northern Asia	*Zapus* and *Sicista*	Tularaemia
Dipodidae	10	27	Jerboas	Deserts and plains of North Africa, Middle East, and Central Asia	*Allactaga, Jaculus*	Plague, tularaemia
Muridae	98	457	Old World rats and mice	Primarily South-East Asia, Africa, Europe, Central Asia, Malayan-Indonesian Archipelago, Australia and Pacific regions, and, by introduction, cosmopolitan		Plague, tularaemia, listeriosis, erysipelas, leptospirosis, pseudotuberculosis, salmonellosis, brucellosis, rickettsial pox, murine typhus, Q fever, scrub typhus and other rickettsioses, histoplasmosis, lymphocytic choriomeningitis, Lassa fever, rabies and other viral infections, Asian schistosomiasis, Chagas' disease, rat-bite fever and other infections
Murinae	(67)	(377)	True rats and mice	Essentially, distribution of Family Muridae	*Rattus** (including *R. exulens** and other Asian species of *Rattus**) *Mus,* * *Bandicota,* * *Arvicanthis,* * *Mastomys**	
Dendromurinae	(8)	(30)		Africa south of Sahara	*Dendromus*	Information not available
Otomyinae	(2)	(10)		Africa south of Sahara	*Otomys*	Plague
Phloeomyinae	(7)	(22)		South-East Asia to Australia	*Chiropedomys*	Information not available
Rhynchomyinae	(1)	(1)		Philippines	*Rhynchomys*	Information not available

				Philippines		
Erethizontidae	4	8	New World porcupines	North and South America	*Erethizon* and *Coendu*	Tularaemia, Q fever and other rickettsial infections
Caviidae	5	12	Guinea-pigs	South America	*Cavia*	Plague (and many infections in domestic forms)
Hydrochoeridae	1	2	Capybaras	South America	*Hydrochoerus*	Actinomycosis
Dinomyidae	1	1	Pacarana	South America	*Dinomys*	Information not available
Heptaxodontidae	2	2	None	Puerto Rico and Hispaniola	*Elasmodontomys*	Information not available
Dasyproctidae	4	11	Agoutis and pacas	Central and South America	*Dasiprocta*, *Cuniculus*	Pseudotuberculosis, leptospirosis
Chinchillidae	3	6	Chinchillas	South America	*Chinchilla*	Pseudotuberculosis, listeriosis, coccidioidomycosis
Capromyidae	3	11	Hutias	West Indies	*Capromys*	Information not available
Myocastoridae	1	1	Nutria or coypu	South-eastern South America and widely introduced	*Myocastor*	Pseudotuberculosis, anthrax, cryptococcosis
Octodontidae	5	8	Hedge rats	South America	*Octodon*	Information not available
Ctenomyidae	1	26	Tuco-tucos	Southern South America	*Ctenomys*	Information not available
Abrocomidae	1	2	Chinchilla rats	South America	*Abrocoma*	Information not available
Echimyidae	14	43	Spiny rats	Central and northern South America and parts of Antilles	*Echimys*, *Proechimys*	Leptospirosis, trypanosomiasis
Thryomyidae	1	6	Cane rats	Southern and central Africa	*Thryonomys*	Information not available
Petromyidae	1	1	Dassie rats	South-West Africa	*Petromus*	Information not available
Bathyergidae	5	22	Mole rats	Southern and central Africa	*Bathyergus* and *Cryptomys*	Information not available
Ctenodactylidae	4	8	Gundis	Northern Africa	*Ctenodactylus*	Information not available

the latter category. Despite the fact that some rodents can develop resistance to rodenticides there are sufficient available to protect humans in most urban areas. One of the problems inherent in the use of rodenticides is that if the rodent dies its ectoparasites will usually leave it and seek blood meals elsewhere as will other blood-sucking arthropods deprived of their normal source of food. In plague, for example, if the rat dies, the flea will leave it and feed on man thus increasing the chances of human infection and precipitating man to man transmission. In such cases anti-arthropod and anti-rodent measures should be carried out together. This example shows how essential it is to understand fully the ecology of the reservoir host the infectious agent and its vectors before beginning a control scheme.

The vampire bats that transmit rabies are difficult to control but can be markedly reduced in numbers by manipulating the bats' behaviour patterns. Bats are caught and smeared with an anticoagulant which is spread through a colony by mutual grooming (WHO, 1973; Linhart, 1975). An alternative method is to rub strychnine onto vampire bites on cattle. Vampires return to the same sites to feed and there pick up the poison (Linhart, 1975).

5.6 Conclusions

The study of small mammals as reservoirs of disease is expanding as more and more efficient methods of diagnosing human and animal diseases become available. The problem no longer lies in the diagnosing of the disease but in trying to prevent man and his domesticated animals from becoming infected. The extent of the problem can be seen in Europe where rabies is spreading across the continent despite the amount of knowledge and money available for its control. Rabies is a disease of carnivores which are large and relatively few in number. How much greater therefore is the problem of disease associated with small mammals, particularly rodents. The only way to solve the problem is to find out as much as possible about the ecology and behaviour of small mammals before they become important as reservoirs of disease and not afterwards.

5.7 References

Al-Doory, Y. ed. (1976) *The Epidemiology of Human Mycotic Diseases.* Thomas, Springfield, Illinois.

Andrewes, C.H. & Walton, J.R. (1977) *Viral and Bacterial Zoonoses.* Balliere Tindall, London.

Austwick, P.K.C. (1969) Mycotic infections. In: *Diseases in Free-Living Wild Animals.* Symposia of the Zoological Society of London No. 24, McDiarmid, A. (ed.) pp. 249-268. Academic Press, London.

Baer, G.M. ed. (1975a) *The Natural History of Rabies* Vol. 2. Academic Press, New York.

Baer, G.M. (1975b) Rabies in nonhematophagous bats. In: *The Natural History of Rabies,* Baer, G.M. (ed.) Vol. 2, pp. 77-97. Academic Press, New York.

Baer, G.M. (1975c) Bovine paralytic rabies and rabies in the vampire bat. In: *The Natural History of Rabies,* Baer, G.M. (ed.) Vol. 2, pp. 155-175. Academic Press, New York.

Bailey, N.T.J. (1975) *The Mathematical Theory of Infectious Diseases.* 2nd Edn. Charles Griffin, London.

Berge, T.O. ed. (1975) *International Catalogue of Arborviruses.* U.S. Department of Health Education and Welfare.

Berger, J. (1976) Model of rabies control. In: *Mathematical Models in Medicine,* Berger, J., Bühler, W., Repges, R. & Tautu, P. (eds.) pp. 75-88. Springer-Verlag, Berlin.

Beverley, J.K.A. (1974) Some aspects of toxoplasmosis, a world wide zoonosis. In: *Parasitic Zoonoses Clinical and Experimental Studies,* Soulsby, E.J.L. (ed.) pp. 1-25. Academic Press, New York.

Bray, R.S. (1974) Epidemiology of leishmaniasis: some reflections on causation. *Ciba Foundation Symposium,* **20**, 87-100.

Buchanan, R.E. & Gibbons, N.E. (1974) *Bergey's Manual of Determinative Bacteriology.* 8th Edn. Williams and Williams, Baltimore.

Dubos, R.J & Hirsch, J.G. (1965) *Bacterial and Mycotic Infections of Man.* 4th Edn. J.B. Lippincott, Philadelphia.

Euzeby, J.A. (1974) Zoonotic cestodes. In: *Parasitic Zoonoses Clinical and Experimental Studies,* Soulsby, E.J.L. (ed.) pp. 151-178. Academic Press, New York.

Evans, A.S. ed. (1977) *Viral Infections of Humans. Epidemiology and Control.* Wiley, London.

Fenner, F.J. & White, D.O. (1976) *Medical Virology.* 2nd Edn. Academic Press, New York.

Gluckstein, F.P. ed. (1974) *Zoonoses: A Selected Annotated Bibliography.* Reference Service Division: Reference Section, U.S. Library of Medicine.

Godfrey, D.G. (1977) Problems in distinguishing between the morphologically similar trypanosomes of mammals. *Protozoology,* 3, 33-49.

Hairston, N.G. (1972) Population Ecology and epidemiological problems. In: *Ciba Foundation Symposium on Bilharziasis,* Wolstenholme, G.E.W. & O'Connor, M. (eds.) pp. 36-62. Churchill Livingstone, London.

Handler, J. (1975) The worldwide fight against zoonoses. *Blue Book for the Veterinary Profession.* No. 26, 185-192.

van der Hoeden, J. ed. (1964) *Zoonoses.* Elsevier, Amsterdam.

Horsfall, F.L. & Tamm, I. (1965) *Viral and Rickettsial Infections of Man.* 4th Edn. J.B. Lippincott, Philadelphia.

Hubbert, W.T., McCulloch, W.F. & Schnurrenberger, P.R. eds. (1975) *Diseases Transmitted from Animals to Man.* 6th Edn. Thomas, Springfield Illinois.

Kaplan, C. ed. (1977) *Rabies: The Facts.* Oxford University Press, Oxford.

Lainson, R. & Strangeways-Dixon, J. (1963) *Leishmania mexicana*: the epidemiology of dermal leishmaniasis in British Honduras. *Transactions of the Royal Society of Tropical Medicine and Hygiene,* **57**, 242-265.

Le Riche, W.H. & Milner, J. (1971) *Epidemiology as Medical Ecology.* Churchill Livingstone, Edinburgh.

Linhart, S.B. (1975) The biology and control of vampire bats. In: *The Natural History of Rabies.* Baer, G.M. (ed.) Vol. 2. pp. 221-241. Academic Press, New York.

Lloyd, H.G. (1976) Wildlife Rabies in Europe and the British situation. *Transactions of the Royal Society of Tropical Medicine and Hygiene,* **70**, 179-187.

McCaughey, W.J. (1969) Brucellosis in wildlife. In: *Diseases in Free-Living Wild*

Animals. Symposia of the Zoological Society of London No. 24, McDiarmid, A. (ed.) pp. 99-105.

MacMahon, B. & Pugh, T.F. (1970) *Epidemiology: Principles and Methods*. Little Brown, Boston.

Mair, N.S. (1969) Pseudotuberculosis in free-living wild animals. In: *Diseases in Free-Living Wild Animals*. Symposia of the Zoological Society of London. No. 24, McDiarmid, A. (ed.) pp. 107-117. Academic Press, London.

Miles, M.A., Toye, P.J., Oswald, S.C.& Godfrey, D.G. (1977) The identification by isoenzyme patterns of two distinct strain-groups of *Trypanosoma cruzi*, circulating independently in a rural area of Brazil. *Transactions of the Royal Society of Tropical Medicine and Hygiene*, **71**, 217-225.

Nobel, E.R. & Nobel, G.A. (1977) *Parasitology: The Biology of Animal Parasites*. 4th Edn. Henry Kimpton, London.

Noble, J.V. (1974) Geographic and temporal development of plagues. *Nature, London*, **250**, 726-728.

Olsen, O.W. (1974) *Animal Parasites. Their Life-Cycles and Ecology*. University Park Press, Baltimore.

Pavlovsky, E.N. (1966) *Natural Nidality of Transmissible Diseases*. University Press, Urbana Illinois.

Pollitzer, R. (1954) *Plague*. World Health Organization (Monograph Series No. 22), Geneva.

Rosicky, B. (1962) Disease cycles in nature – natural foci of disease. In: *The Problems of Laboratory Animal Disease*, Harris, R.J.C. (ed.) pp. 27-37. Academic Press, London.

Rosicky, B. (1967) Natural foci of disease. In: *Infectious Diseases*, Cockburn, A. (ed.) pp. 108-126. Thomas, Springfield, Illinois.

Simpson, D.I.H. & Zuckerman, A.J. (1977) Marburg and Ebola: viruses in search of a relation. *Nature, London*, **226**, 217-218.

Sidja, I., Lim, D. & Matouch, O. (1971) Isolations of rabies-like virus from murine rodents. *Journal of Hygiene, Microbiology and Immunology*, **15**, 229-230.

Spielman, A. (1976) Human babesiosis on Nantucket Island: Transmission by nymphal *Ixodes* ticks. *American Journal of Tropical Medicine and Hygiene*, **25**, 784-787.

Taylor, J. (1969) Salmonella in wild animals. In: *Diseases in Free-Living Wild Animals*. Symposia of the Zoological Society of London. No. 24, McDiarmid, A. (ed.) pp. 53-73. Academic Press, London.

Tesh, R., Saidi, S., Jovadian, E., Loh, P. & Nadim, A. (1977) Isfahan virus a new vesiculovirus infecting humans, gerbils and sandflies in Iran. *American Journal of Tropical Medicine and Hygiene*, **26**, 299-306.

Turner, G.S. (1976) A review of the world epidemiology of rabies. *Transactions of the Royal Society of Tropical Medicine and Hygiene*, **70**, 175-178.

World Health Organization (1967a) *Joint FAO/WHO Export Committee on Zoonoses. Third Report*. World Health Organization Technical Report Series No. 378.

World Health Organization (1967b) *Current Problems in Leptospirosis Research*. World Health Organization Technical Report Series No. 380.

World Health Organization (1973) *Export Committee on Rabies. Sixth Report*. World Health Organization Technical Report Series No. 523.

World Health Organization (1974) *Ecology and Control of Rodents of Public Health Importance*. World Health Organization Technical Report Series No. 553.

Zeledon, R. (1974) Epidemiology, modes of transmission and reservoir hosts of Chagas' disease. *Ciba Foundation Symposium*, **20**, 51-77.

Importance of small mammals as pests in agriculture and stored products

6

ARVO MYLLYMÄKI

6.1 Introduction

Small mammals, especially rodents with their self-sharpening incisors and phenomenal capacity to increase in numbers, take a toll of man's provisions all over the globe. The history of rodent plagues can be traced back to biblical times, and the earliest historians. The first critical review of the phenomenon was compiled by Elton (1942). His ageless description of the course of events during a rodent outbreak:

> "The affair runs always along a similar course. Voles multiply. Destruction reigns. There is dismay, followed by outcry, and demands to Authority. Authority remembers its experts or appoints some: they ought to know. The experts advise a Cure. The Cure can be almost anything: golden mice, holy water from Mecca, a Government Commission, a culture of bacteria, poison, prayers denunciatory or tactful, a new god, a trap, a Pied Piper. The Cures have only one thing in common: with a little patience *they always work*. They have never been known entirely to fail. Likewise they have never been known to prevent the next outbreak. For the cycle of abundance and scarcity has a rhythm of its own, and the Cures are applied just when the plague of voles is going to abate through its own loss of momentum.

In the course of 37 years that has elapsed since the publication of *Voles, Mice and Lemmings*, huge piles of documents have accumulated on rodent ecology and rodent control. No one has even tried to synthesize this 'post-Eltonian' literature. In view of limitations of space and time, I shall not even attempt such a synthesis. Instead, I shall describe in short the ecological conditions in which major small mammal problems arise and call for countermeasures.

Regarding examples and references, the leading principle has been to illustrate the problem under discussion rather than to catalogue all individual references pertinent to the subject. To guide the reader to sources of further information, priority has been given to review articles and to any reports, where references are given to additional accounts of original research. Shortage of qualified scientific articles on special problems has sometimes necessitated references to preliminary reports and generalized descriptions.

6.2 Main types of small mammal problems

6.2.1 Damage by *Microtus arvalis* and subsidiary microtines to field crops of temperate regions.

Diffidently apologizing for not being a polyglot and having the great libraries at his command, Elton (1942) produced a unique panorama of the vole plagues in Europe and elsewhere. His penetration through the language barriers is still unsurpassed. Since his book, there has been no serious attempt at marshalling the information supplied in the numerous Russian contributions on *M. arvalis* and other irruptive microtines with the West European literature published since World War II (but *see* the brief review by Ryszkowski and Myllymäki, 1975).

The expression 'astronomical', repeatedly used by Elton (1942), is more than adequate to describe the numbers of the continental vole, the area covered by a plague, the amounts of crops destroyed by this pest or, ultimately, the sums of money lost. The list of crops damaged is a long one, and includes most cultivated field and garden plants, but the worst damage is done to cereals, particularly wheat, and to alfalfa and other fodder plants. The largest area affected in a particular outbreak was reportedly that of the 1932 outbreak of *M. arvalis* and several other rodent species in the USSR, when the plague was estimated to cover 10 million hectares, i.e., more than the total area of arable land in present-day Hungary. Hungary, in turn, has offered the most typical examples of the severe outbreaks in the past.

From time immemorial, it has been customary for Hungarian peasants to store cereals and fodder one year ahead as a precaution against famine during recurring vole plagues, when most of the harvest used to be consumed by voles. As recently as 1964–65, most of the cultivated fields of Hungary were overwhelmed by *M. arvalis*. One and a half million citizens, including most of the army, were sent to spread poison baits on the field. The area treated totalled 3.6 million ha (O. Nechay, *pers. comm.*), but the remedy came too late. Elton (1942, p. 50) was hardly exaggerating when he claimed that a severe and prolonged outbreak of the continental vole in 1917–18 influenced or accelerated the outcome of World War I!

The last serious outbreak of *M. arvalis* in Hungary coincided with the period of transition from the traditional small-farm system to the present co-operative industrialized enterprises. The patchy field mosaic was transformed into a vast, mechanically farmed monocultural landscape. There is a wide agreement among students of continental vole ecology (see references in Ryszkowski and Myllymäki, 1975) that permanent verges and other ruderal refuges play a decisive role as accumulators and starters of a local outbreak. An additional feature of a typical outbreak area is the proximity of sufficiently large – at least 1000 ha, according to van Wijngaarden (1957) – reception habitats (hayfields, extensively managed pastures, etc.), where the final phase of mass propagation occurs. In the olden days of private ownership the agricultural lands of Hungary undoubtedly provided all these conditions; now the reader is presumably curious to know what happened to *M. arvalis* in the course of the great reform.

Interestingly, the plagues in their traditional form have ceased, but *M. arvalis* has not lost its position as a key pest in Hungarian agriculture. The first prerequisite of a vole plague, abundance of permanent refuges, has been largely eliminated, but mass propagation continues in the 'secondary' habitats, alfalfa and wheat fields. Repeated disturbance by tilling and continuous control measures (some hundred thousand hectares are treated yearly against *M. arvalis*) keep the population below the threshold of a catastrophe, but, at the same time, create and maintain need for continued control. But it has been demonstrated that such measures reduce crop losses to a tolerable level.

In France, the situation is much the same as in Hungary. Agricultural practices are undergoing parallel changes, although at a different tempo. However, due to lack of a centralized organization, only one quarter of the infested fields are generally treated against the continental vole. Up to the mid-1960s, the total area of vole-infested fields during outbreaks was estimated at about 3 million hectares (Bouyx, 1967) and the toll then exacted from French agriculture at 1000 million francs. These figures are astronomical enough to show that *M. arvalis* is still a grave problem in France.

Similarly, the continental vole continues to cause problems in most of the Central European countries, in countries on both sides of the lower Danube, and in the USSR. Where outbreaks still occur, it is a matter of definition, whether they should still be termed plagues, but the menace of chaos and famine is nowadays consigned to history.

All the figures on continental vole damage quoted above were rounded empirical approximations. Recently, Ryszkowski *et al.* (1973) estimated more critically losses caused by *M. arvalis* in two important field crops, rye and alfalfa, basing their appraisal on the bioenergetics and the density of the voles. They concluded that damage to rye during the winter can be fully compensated by harvest time, while losses of alfalfa were proportional to the population density of the voles, ranging from 1.4 per cent at a low density (44 animals/ha) to 21.4 per cent at a high population density (774 animals/ha).

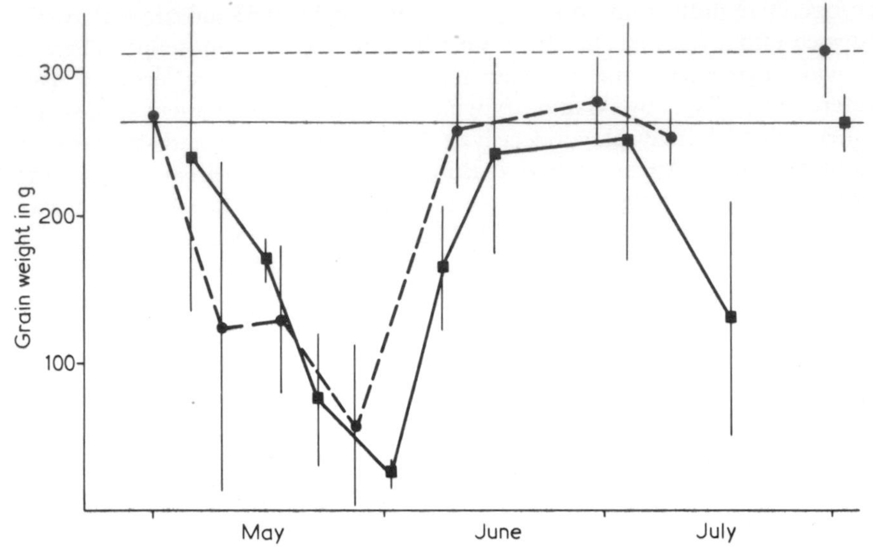

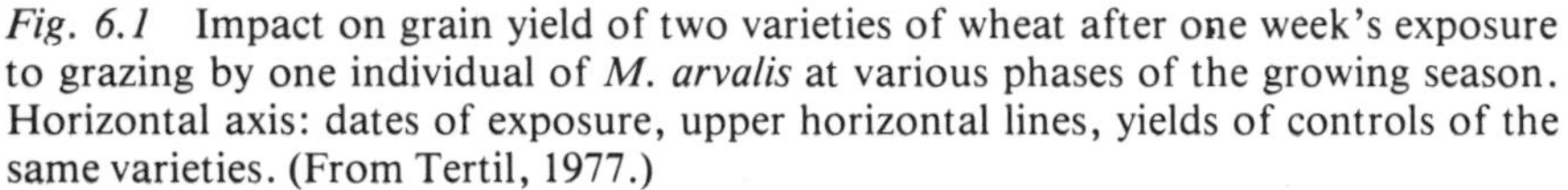

Fig. 6.1 Impact on grain yield of two varieties of wheat after one week's exposure to grazing by one individual of *M. arvalis* at various phases of the growing season. Horizontal axis: dates of exposure, upper horizontal lines, yields of controls of the same varieties. (From Tertil, 1977.)

That a purely bioenergetic approach neglects the proportion of plant biomass removed but not utilized, and so leads to underestimation of the damage, was pointed out by Myllymäki (1975a, Fig. 14.1). Consequently, Tertil (1977) tried to approach a general model of the total impact of *M. arvalis* on the yield of two crops, winter wheat and alfalfa. In the case of wheat, injury to the earing and flowering plants resulted in the greatest quantitative loss in both of the two varieties studied (Fig. 6.1), while the impact on the quality of the grain was largely dependent on the variety. In alfalfa, the impact increased from the first to the third yield, and the final losses were much higher than those observed by Ryszkowski *et al.* (1973). Earlier Spitz (1968) had shown that the total impact on alfalfa is aggravated by qualitative losses owing to a decrease in the leaf/stem ratio.

This series of experiments suggests new ways of objectively assessing damage to field crops. What is additionally needed is work relating the experimental situations to the actual field conditions, and a subsequent follow-up of the resulting pilot programmes for the determination of control thresholds.

M. arvalis has a wide range stretching southwards to the Po Valley, Croatia, Serbia and the Bulgarian plains. In the south of Italy, the Iberian peninsula and southernmost France, the continental vole is replaced by *Pitymys* spp.; of the 10 European species, Santini (1977) denominated three as agricultural pests. *P. savii* is an important pest of several crops in Italy.

Its harmfulness is accentuated by its pronounced predilection for valuable cash plants, such as artichoke, melon, fennel, onion, garlic, lettuce, etc. (Santini, 1977). In the southern parts of the Balkans, and especially in the Near East, the Levant vole, *Microtus guentheri*, replaces *M. arvalis* as the main rodent pest of field crops. The oldest descriptions of a vole plague, found in the Old Testament and in the texts of Herodotos and Aristotle, refer to the Levant vole. The serious pest character of the species is illustrated by the fact that two thirds of the cultivated area of Israel had to be treated against *M. guentheri* in 1949–50, when one of the most devastating outbreaks of this species took place and caused destruction of a wide variety of field crops, fruit trees and vegetables. The classical studies of Bodenheimer (1949) revealed that this species, like the more northern species of *Microtus,* showed earlier cyclic fluctuations. Therefore, the observations made later on in the Hula Valley region deserve special attention: In the mid-1950s, the swamp areas of the Hula Valley were drained, and large-scale cultivation of alfalfa was begun in the newly created fields. Soon the alfalfa was found to be heavily infested by *M. guentheri*, and this infestation has since persisted at a relatively high level without any marked annual, or even seasonal, fluctuation apart from short term and rapidly replaced depressions of the population after regularly conducted treatments with rodenticides. Despite these treatments, the annual losses of the alfalfa crop are 16–25 per cent; but omission of the treatments, or even their faulty timing, has led to losses of up to 50 per cent (Wolf, 1977). The Hula Valley story is thus an extreme example, on the Hungarian model, of new fluctuation patterns created by modern man for old rodent pests.

Towards the east, *M. arvalis* is replaced by microtine species like *M. socialis* in the southern Ukraine, the Crimea, the Caucasus, northern Iran and Transcaspia, *Lagurus lagurus* from Kazakhstan to the Altai, *M. gregalis* and *Arvicola terrestris* in southern Siberia, *M. brandti* in Mongolia and *M. fortis* in the far east. In Mongolia *M. brandti* and several other simultaneously irruptive microtines have recently caused complete devastation of pasture, as well as of the tiny patches of crops growing in between. North of the range of *M. arvalis* in Finland, the Scandinavian peninsula and Great Britain, the field vole, *M. agrestis*, partly occupies the niche of the continental vole. However, as *M. agrestis* is not primarily a pest of field crops, it will be discussed in detail in the following sub-section. The same applies to *A. terrestris*.

About half the vole species in the world are indigenous to North America (Elton, 1942). That continent has been the Land of Opportunity for microtines as well as for man, and when civilized man spread crops over the continent, many of the voles became pests in one place or another. None of these species has, however, shown a potential equal to that of the European *M. arvalis*. Local outbreaks by such species as *M. montanus, M. californicus* or

M. pennsylvanicus, have been recorded in the history, and some of these have been serious in character, like that which occurred in the Humboldt Valley, Nevada, in 1906–07 (Elton, 1942, pp. 107–108). Probably the most serious of recent outbreaks was due to *M. oregoni* in Oregon in 1957–58 (Anon. 1958). In addition to the above-mentioned species of *Microtus*, also *Sigmodon hispidus* is an important pest of several field crops in the southern parts of the USA.

Typically, the present outbreaks in the USA take place in irrigated areas with a dense network of canals and canal banks. These grassy verges serve as winter survival habitats, supplying the voles with green forage for most of the year and so providing conditions for starting the reproductive season well before the crops are ready to support the surplus population. In some instances, prophylactic measures, regular surveillance and poison treatments directed to the banks during the non-vegetative period, have prevented new outbreaks. But where, as in the Hula Valley, Israel, perennial fodder plants as alfalfa form a dominant part of the crop, such preventive measures are of limited value.

6.2.2 Major threats to cultivated trees: debarkers, root-cutters and seed depredators

The examples given in this subsection relate to species that are specialized pests of fruit trees, ornamentals and forest trees. Most of these species belong to microtines and forage on herbs and grasses during their breeding season, using bark or tree roots as a substitute feed during the non-breeding season. Their breeding habitats are also often distinct and sometimes far away from the areas where damage occurs, a feature that makes it hard to understand and cope with the problem.

In Europe, the most important debarker of trees is the field vole, *M. agrestis*. The history of field vole attacks on garden plants and forest trees can be traced back to the end of the nineteenth century (for the earliest references, *see* Myllymäki, 1977a), but it was not until the 1950s that the damage rose to an 'astronomical' level. In Central Europe, the peak of forest damage (*M. agrestis* is there a forest pest only) occurred in the 1950s, and nowadays the problem is crucial mainly in the Scandinavian countries. Detailed figures on damage and its economic importance are to be found in a recent review by Myllymäki (1977a), where the reader will also find a comprehensive list of original references.

In Scandinavia the field vole problem is closely connected with recent changes in agricultural and forestry practice and policy. The withdrawal of many farmers from traditional dairy farming and the abandonment of arable land and former pastures for other reasons have created new permanent grassland patches for *M. agrestis*. Corresponding changes in forestry practices led to the creation of vast clear-cut areas which, at least on fertile soil, soon turn grassy and suitable for the field vole. In Scandinavia as a whole in the early 1970s the total area of feral or afforested land that was earlier arable amounted to about 0.5 million ha, and that of clear-cut forest to at least 2 million hectares (Myllymäki, 1977a).

Parallel to these habitat changes, the numbers of vulnerable stands of trees have increased manyfold. For instance, all the saplings planted on afforested fields and pastures are new objects of damage; such stands were seldom planted in grassland until a couple of decades ago. Another trend worth noticing is that recently most forest regeneration in Scandinavia has been made by planting saplings, instead of by seeding. The saplings, in turn, are produced on a fertilized substrate under plastic, and such forced saplings are significantly more susceptible to attack by voles than seedlings in natural seeding areas. New and very expensive objects of damage are also the grafts in seed orchards of forest trees (cf. Myllymäki, 1977b); important innovations in forest breeding techniques.

A couple of examples may illustrate what power a tiny vole can exert even on national forest policy. In the 1950s, there was much enthusiasm in Finland for planting the hybrid aspen, a fast-growing hybrid between the Canadian and native aspen species. Largely owing to the destruction by *M. agrestis*, only a couple of the plantations then established have survived to harvestable age. Further, in the early 1970s, there was much propaganda for planting the abandoned fields with birch. However, wide-spread damage by voles to the first plantations has made farmers cautious, and the nurseries are now filled with over aged and unsold birch saplings optimistically produced to meet the predicted needs of the landowners!

Although there is no doubt that in Scandinavia, the principal pest both in horticulture and in forestry is *M. agrestis*, there is reason to believe (Hansson and Zejda, 1977) that some of the blame attaches to the bank vole, *Clethrionomys glareolus*. In Central Europe *C. glareolus* is known as a forest pest, and in the Far East, other species of *Clethrionomys*, as well as *Microtus*, are considered important pests both in horticulture and in forestry (Higuchi, 1976; Mizushima, 1976).

In the New World, gardening and forestry have their counterpart to *M. agrestis* in *M. pennsylvanicus* (Elton, 1942; Hayne and Thompson, 1965; Radvanyi, 1974). As the damage done and, presumably, the conditions in which it occurs are rather similar to those discussed above, there is no need to go into details. However, because most forest regeneration in the USA and Canada is done by direct seeding, the microtine problems in forestry are relatively less important than in Scandinavia. Damage to fruit trees by *M. pennsylvanicus* is concentrated in the area around the Great Lakes in Canada and the USA, while further south, along the Appalachians, the principal rodent pest of fruit trees is *Pitymus* (*Microtus*) *pinetorum* (Davis, 1977).

A pest specialized for cutting roots of cultivated trees is *Arvicola terrestris,* a vole widely distributed in Euro-Asia. There are two distinct ecological forms (cf. Meylán, 1977) of *Arvicola*, a fossorial one, usually called *A. t. sherman* and inhabiting a restricted range in the mountainous heart of Europe (Fig. 1 in Meylán, 1977), and the amphibious (*in sensu* Linnéus) water vole, *A. t. terrestris*. The former is strictly subterranean in its habits, whereas the latter adopts a subterranean mode of life during the non-breeding season only. Even in this case, it is primarily the fossorial phase of the animal's activity which brings it into conflict with man. During the breeding season, the amphibious form usually lives in shore or marsh vegetation, where it is quite harmless Panteleev, 1968; Maksimov, 1977).

In autumn, the water vole (both forms) hoards potatoes, root crops, rhizomes and bulbs of ornamental plants, and sometimes also cereals and legumes, in its underground galleries. The main problem is, however, that it attacks the roots of fruit trees and forest trees. The occurrence of this type of damage is often unpredictable, but may, at least locally, reach vast dimensions. Wieland (1973) reports destruction of 650 000 apple trees during the five-year period 1958–63 in DDR, the damage in monetary units exceeding 3 million U.S.$. In the USSR, especially in Siberia, *Arvicola* is also an important pest of field crops (Panteleev, 1968; Maksimov, 1977).

Although not comparable with microtine damage, the damage done by squirrels and other arboreal rodents should also be mentioned. A typical feature of squirrel damage is stripping bark off trees high above ground level. The reason for this behaviour is unclear but, almost certainly, it has very little to do with searching for food. Recently, detailed studies have been made on the damage done by an introduced species, *Sciurus carolinensis*, in Great Britain (Taylor, 1970).

A review of forestry problems in temperate regions, would be incomplete without discussing the problem of seed depredation. Owing to the fact that direct seeding predominates in North America, most of the literature on this problem is American (for a recent review and further references, *see* Radvanyi, 1973). Several rodent genera are involved in seed destruction in different parts of the USA and Canada, but the most typical depredators are *Peromyscus* spp., a convergent ecological type to the Old World *Apodemus*. Curiously enough, among the seed depredators are also several species of shrews (*Sorex, Blarina*). The recent confirmation of seed depredation by *Sorex araneus* in Finland (Myllymäki and Paasikallio, 1976), may prove important in northern Scandinavia, where murids are lacking and the dominant rodent species in clear-cut areas are not very prone to eat conifer seed. In the light of my own limited experience, I fully agree with Radvanyi (1973) in wondering how it is possible for conifer seed to germinate at all in view of the effectiveness with which small mammals examine the seeding spots, or the broadcast seed.

6.2.3 Pests emerging from the soil: hamsters, ground squirrels, gerbils, pocket gophers and moles

The problems discussed in this subsection are caused by small mammals, other than microtine rodents, with pronounced subterranean habits. Many of these animals emerge on the surface to search food, daily or seasonally, some of them hoarding food in caches and hibernating, or aestivating, over the unfavourable season, others being winter-active.

The European hamster (*Cricetus cricetus*) creates problems in the countries along the lower Danube and maybe in some parts of the southern USSR (for a general review, *see* Nechay, Hamar and Grulich, 1977). The hamster is unique among rodent pests in being favoured by the introduction of industrialized agriculture. The animal burrows deep in the soil and is well protected from any disturbance on the soil surface. It can survive periods of virtual absence of a vegetative cover, as during tillage and the seeding period of maize, or the harvesting of any crop, and inhabit wide monocultures, such as are typical of Hungarian agriculture today. The only condition which may prove fatal to this pest is a high ground water table, especially during hibernation. In this very respect, however, man has helped the hamster by regulating the water level, as in the programme for the Tisza River Valley in eastern Hungary.

The hamster has tenaciously demanded a share of the increased crop production, attacking almost literally every crop. At the culmination of the latest great hamster outbreak in 1975 in Hungary, up to 400 000 ha of cultivated fields were heavily infested by this pest and a corresponding outbreak in Slovakia in 1970–72 covered 200 000 ha (Nechay *et al.*, 1977). An example from a co-operative farm in the Tisza Valley may best illustrate the practical implications of a hamster outbreak: The farm has 3650 ha of cultivated field, the main crops being wheat, maize, sugarbeet, alfalfa and sunflower. In 1975, 9000 man-hours were spent on hamster control in these crops, the total area of the treatments (double and triple treatments of the same lot included) being 3850 ha. Despite these efforts losses were recorded in every crop, e.g. 500 kg/ha in maize and 1000 kg/ha in wheat. Some crops, e.g. an experimental plantation of soya bean, were completely destroyed.

As the hamster is a major pest and the subject of numerous publications (cf. the reference lists in Nechay *et al.*, 1977; and Gorecki, 1977), it is surprising how many key questions on hamster ecology are still open to speculation. For instance, basic figures on hamster density continue to be based on scholastic extrapolations of burrow counts, although no facts are available about the hamster's normal movement patterns, or the sociological structure of the population. So far no-one has tried systematic trapping as a research tool, a surprising omission in view of the fact that the hamster is generally trapped for its fur. As long as the biological fundamentals remain obscure, determination of control thresholds and evaluation of control success can hardly be more than guesswork.

East of the range of the European hamster several other hamster species extend to the Far East, and cause problems, but of much more practical importance are the ground-squirrels or sousliks (*Citellus* spp.). In the USSR, altogether 7 species of the genus *Citellus* are considered important as pests and subjected to regular surveillance and obligatory control (Poljakov, 1976) – the number of microtine species ranked similarly is six. Like the hamster, the sousliks are polyphagous pests of several crops. They also show hoarding behaviour and hibernate like the hamster. The range of the ground squirrel problems extends over the whole steppe region of the USSR.

The North American continent has a good collection of pest rodents comparable with those discussed above. For example, the Californian ground squirrel, *Citellus beecheyi*, is considered such a harmful pest in rangeland that control operations against it are carried out on up to 2 million hectares yearly.

Gerbillid rodents belonging to the genera *Meriones, Tatera, Rhombomys* etc. are dominant rodent pests of cultivated plants throughout the arid region from Morocco to India. The species generally occurring in plague numbers in the countries south of the Mediterranean Sea is *Meriones shawi;* Bernard (1977) lists 15 other harmful species in North Africa and the Near East. In India and Pakistan, the dominant pest is *M. hurrianea* (Prakash, 1976). The types of damage by gerbillids vary greatly, ranging from direct foraging on cereals, vegetable crops, olives, legumes, etc., damaging young olive saplings or forest trees, and hoarding of great quantities of crop products to indirect damage by burrowing which accelerates desertization. The gerbillids also compete for animal feed, which is indirectly manifested in animal production. No damage surveys have been made anywhere, but in the North African countries alone the examples of control treatments on 1–2 million ha (Bernard, 1977) clearly indicate that there are real problems, especially social ones due to periodic destruction of all the subsistence crops of the poorest part of the population.

A strictly North American group of rodents, the pocket gophers (Geomyidae), are highly specialized for fossorial life. Some of the 30 species are widely distributed pests of rangelands, field crops or forest regeneration areas. Howard and Childs (1959) regarded the pocket gophers as the most important agricultural rodent pests in California. Nuisance by one species, *Thomomys talpoides*, to range management led to the establishment of the Colorado Co-operative Pocket Gopher Project in the late 1950s. The main results of this project were summarized by Turner *et al.* (1973; includes a comprehensive list of references to gopher biology and control).

The pocket gophers, like *Arvicola*, are active all the year round. They live in subterranean galleries, but emerge on the surface for feeding, even under the snow in winter. In this respect they deviate from *Arvicola*, which normally stays in galleries even for feeding throughout the winter. Consequently damage to tree seedlings by the pocket gopher is mainly above ground, that by *Arvicola* chiefly to the roots. Quantitatively, the damage done by pocket gophers to conifer seedlings is far from negligible (Barnes, 1973).

Gopher damage to range vegetation has been summarized by Turner *et al.* (1973) who drew the conclusion that gophers can reduce the herbage yield available for livestock by 20 per cent. An interesting finding is that pocket gophers are highly dependent on dicotyledonous food and do not thrive well on grass alone; the practical implications of this finding are discussed in section 6.3.2.

By their burrowing activity and utilization of living plant material, pocket gophers expose the soil and have been claimed to be major agents in soil erosion. Turner *et al.* (1973) point out that this effect, as well as the gophers' influence on infiltration, runoff and soil moisture, varies with local conditions and consequently the ultimate balance may even be positive.

All burrowing rodents exert gopher-type effects, positive or negative, on soil conditions. This is also true of the European mole, *Talpa europea*, and its American counterparts. *Talpa* is strictly carnivorous and does not injure any crop directly. However, as a result of its burrowing activity, the mole is accused of at least the following types of economically significant damage (Lund, 1976): reduction of grazing area in pastures, mechanical damage to newly planted crops (cereals, beets, vegetables), damage to harvesting equipment, and aesthetic damage to lawns and golf links. For these reasons the mole is considered an important pest in several countries and actively controlled as a rule with poor success.

A semi-aquatic rodent, the muskrat, *Ondatra zibethica*, makes considerable damage by burrowing river and canal banks. The species was introduced in several European countries for its fur, it found an unoccupied niche and spread over wide areas. Especially in Western Europe much effort has to be devoted to muskrat control by means of trapping and poisoning. Continued extension of the species' range and increase in catch numbers (cf. EPPO, 1968) indicates that these measures have been far from effective. There are obvious parallels between the muskrat and hamster control operations, which may both turn beneficial to the pest populations.

6.2.4 Murid rodents: depredators of cereals and other subsistence crops

From Elton's (1942) descriptions of the rodent plagues in the steppe region of the USSR it is known that some species of mice, e.g., *Mus musculus, Apodemus agrarius* and *Micromys minutus*, often irrupt simultaneously with voles. Very high mouse densities, up to 50–70 per cubic metre, were found in haystacks. In Great Britain, the house mouse used earlier to be a common pest of corn ricks (Rowe, Taylor and Chudley, 1963). Crowcroft (1966, p. 114) wrote: 'the corn rick is really an outdoor enclosure provided by the farmer to enable the house-mouse to overwinter in sufficient numbers to be capable of efficiently re-infesting the fields in the following spring'. The combines deprived mice of this privilege.

A classical example of a plague due to feral *M. musculus* is that which occurred in Buena Vista Lake Basin, California, in 1926 (Hall, 1927). The mice bred in a dried-up lake bottom, which was partly planted with maize and barley. When the crop had been eaten up, and shortage of food was enhanced by grazing sheep, a sudden mass emigration of mice and voles (*Microtus californicus*), which here, too, irrupted simultaneously, took place in November 1926, followed later on by two other surges of mice. The mice invaded farmhouses 16 km apart. Hall (1927) estimated the original population to have had a density of 200 000 mice per ha, surpassing all the astronomical figures presented earlier! Such estimates may sound incredible, but some earlier sceptics have admitted at least to a correct order of magnitude (cf. Crowcroft 1966, p. 115).

In Australia, house-mouse plagues have a long history, and they are still continuing (Ryan and Jones, 1972). Hamar (1977) described a murid outbreak (6 species involved) on a Danube island, Insula Mare, with some 55 000–60 000 ha of maize and sunflower on recently meliorated fields. Generally, it seems that outbreaks of true mice usually occur on some kind of ecological island, i.e. on a patch of favourable habitat surrounded by a habitat type 'alien' to the species concerned.

The fauna of tropical Africa is rich in endemic murid rodents. Yet deleterious outbreaks appear to be a recent phenomenon, or at least poorly documented in the older literature. Taylor (1968) described rat outbreaks in Kenya in 1951 and 1962, the last-mentioned extending from northern Tanzania to Gezira in the Sudan. In Kenya, the main targets of rodent injury were wheat and barley. According to Taylor (1968) eight rodent species were involved, the most important being *Mastomys natalensis, Arvicanthus niloticus* and *Rhabdomys pumilio*. Both *Mastomys* and *Arvicanthus* are widely distributed in Africa.

Taylor (1968) suggested that the 1962 outbreak was linked with unusually heavy rainfall during the preceding rainy season, which, in turn, had delayed the harvest and caused luxuriant weed growth. Profiting from the unharvested crops and weed seed, the rats prolonged their breeding season (and survived better?) and, consequently, exhibited high densities at the very start of the breeding season. Later on, Taylor and Green (1976) provided evidence in support of this hypothesis. It was also corroborated empirically by the occurrence of similar weather conditions before another severe rodent plague experienced in Gezira, Sudan, in 1975–76 (Beshir *et al.*, 1976). In this case, too, *Arvicanthus* and *Mastomys* were the principal pests.

Even if the proximal causes of the Gezira plague were weather factors, the ultimate reasons must be sought in the recent changes in the agricultural system. Plans for converting the traditionally cotton-growing Gezira into the granary of Africa were accompanied by the introduction of new crops, such as wheat, rice and peanuts. At the start of the recent plague, shortcomings of the mechanical cropping system meant that large quantities of the crop lay for long periods in the fields, and food was thus provided for the rats. The local pest control unit was also unprepared for large-scale counteraction. Despite the treatment of 1.2 million hectares, pre- and post-harvest losses of wheat varied from 15 to 70 per cent, and those of groundnut from 15 to 30 per cent, according to the locality (Beshir *et al.*, 1976).

Taylor's (1968) climatic hypothesis is further supported by recent experiences from the Sahel. There a serious rodent plague broke out in 1975–76 in response to heavy rainfall following two years of extreme drought. As in Gezira, changes have occurred with cultivation systems, such as the introduction of irrigation and new crops. The main targets of damage in the Sahel were rice, sorghum, millet, cassava and peanuts, and the principal pests, once again, *Arvicanthus* and *Mastomys*, except in peanut, where gerbils (*Tatera*) made the trouble.

When tropical food production is considered as a whole, the central rodent problem is the widespread damage to rice. This problem has recently been tackled by several research units sponsored by international and bilateral development aid schemes. The best known of these units, the Rodent Research Center, Los Banos, Philippines (Fall and Sanchez, 1975, including a list of publications), has developed a scheme for assessment of rodent damage in rice (Lavoie *et al.*, 1970), which has since been recommended elsewhere.

Estimates of rice crop losses based on the Philippine national scheme (Anon. 1976), or on other surveys aiming at objective averages, have seldom produced figures for average damage higher than 5–6 per cent (e.g. Rodent Research Center, 1971; Funmilayo and Akande, 1977), although in individual paddies damage may be much heavier. It should be remembered, however, that the figures above do not include damage to young tillers immediately after planting, which many authors (Lavoie *et al.*, 1970; Wood, 1971; Funmilayo and Akande, 1977) consider most important. Yield comparisons between paddies under effective rat control and those without any control measures have produced surprisingly great differences (Wood,

1971, Greaves *et al.*, 1977). Furthermore, it is worth noting that, on a country-wide scale, a 5–6 per cent portion taken by rats would have been of great importance to people living below the limit of adequate daily energy rations.

Rodent species plundering ricefields mostly belong to the genus *Rattus*. The most widely distributed species in South East Asia is *R. argentiventer,* known as the 'rice-field rat'. In the Philippines the same name is given to *R. rattus mindanensis*, although several other species, like *R. norvegicus* and *R. exulans* also participate in the damage. In India and Pakistan, the dominant rice-field pest is the lesser bandicoot rat, *Bandicota bengalensis,* but the total array of rodent pests is wide (cf. Greaves *et al.*, 1977). In Africa, *Mastomys, Arvicanthus* and *Thryonomys* are the most important pests of rice fields, in Central America *Sigmodon*. The list is not exhaustive, but extensive enough to show that rodent problems are likely to evolve wherever rice is grown.

6.2.5 Rodent damage to tropical cash crops

In many tropical countries, the national economy is dependent on the cultivation and export of certain cash crops, such as sugar cane, cocoa and coconuts. Therefore, it is not surprising that in some of these crops rodent problems have been more thoroughly investigated than in the crops necessary for the people's subsistence. Rodent species involved in injury usually belong to the genus *Rattus*.

In Hawaii, sugar cane has a normal growth cycle of about two years (Hood *et al.*, 1970; Teshima, 1970). During the first 9 months or so, a new plantation is usually not yet rat-infested. The rat population is then concentrated in the gulches nearby, and the rats, headed by *R. exulans*, only occasionally make excursions into the cane-fields. At the age of about one year, the canopy of the cane reaches a height of 5 metres and stalks lodged near the ground provide good shelter; then *R. exulans* tends to move into the plantation, where it stays and feeds, the other species still making forays from the gulches. Rat damage is most extensive in maturing cane, more than 15 months old. At harvest 20 per cent to 40 per cent of the stalks may be found to be injured, and 30 per cent of these are dead (Hood *et al.*, 1970). Direct consumption by rats is, however, only the start of the damage – even at high population densities the rats cannot greatly reduce the 300 tonne/ha yield. The main problem is secondary infestation by pathogenic fungi and bacteria, with consequent reduction of the sugar content of the cane (Teshima, 1970).

Modern harvesting by heavy machines is fatal to rat populations in the canefields (Nass *et al.*, 1971). About 70 per cent of the animals are killed directly by the machines, and the rest are exposed to predators and other accidents after the harvest. By means of radiotelemetry, Nass *et al.* (1971) estimated that only 10 per cent of the canefield rats find new harbourage in the gulches. But these survivors are ready to start a new cycle.

Rodents cause heavy losses in sugar cane wherever their depredations have been studied (cf. Jackson 1977, Table 1); e.g. in Hawaii the losses were estimated at U.S. $4.5 million/year (Teshima, 1970). Similar figures are obtained for damage to other tropical cash crops, such as cocoa and coconuts (Jackson, 1977). However, a series of critical studies by Williams (1974a, b, 1975) on coconut damage in Fiji throw new light on these estimates.

Rats (*R. rattus, R. exulans*) damage coconuts by gnawing holes in the immature nuts, which then fall at the latest one week after the attack. Surveys on rat damage have usually been based on counting these fallen nuts, a procedure whose validity was the subject of Williams's study. He was able to show at least two mechanisms which compensate for the apparent losses, namely, an increase in the number of female flowers, which initiated a delayed compensatory effect, and a decrease in immature nutfall due to causes other than rat attack. Williams (1974a) arrived at a 'realistic' estimate of about 50 per cent compensation in terms of copra yield. On making corresponding adjustments to the survey counts, he estimated final losses per palm per year as 1.3–2.7 nuts (Williams, 1974b), a rate at which he considered rat control measures hardly economic at prevailing copra prices (Williams, 1975). These results do not imply that rat damage to coconuts is never economically significant; but, as Williams's work shows, there is a real need for loss assessment schemes extending beyond the mere scoring of apparent damage.

Although public health aspects are dealt with separately in Chapter 5, a brief bridging of rodent damage and disease transmission is appropriate here. Laird (1963) showed that the incidence of mosquitoes transmitting filariasis was proportional to the number of rat-gnawed coconuts used by the vector, *Anopheles polynesiensis*, as oviposition sites. Consideration of this aspect would warrant control measures against 'coconut rats' even when damage was below economic threshold set by copra prices. Another example is a high incidence of leptospirosis among workers in sugar cane and rice fields (Taylor, 1972), which is suspected to be linked with the density of rat populations in these damp habitats. Economic and public health aspects of rodent control could and should more often be taken into consideration together.

6.2.6 Rodent damage to stored products and structures

Rodent damage in the field and destruction of stored products are separated by a diffuse borderline: the harvested crop may be left on the spot for drying, stored temporarily in ricks or corresponding structures, or moved into the farmer's own store nearby. In this section I shall discuss post-harvest losses from the farmer's store onwards.

In tropical countries, the same murids damage farm stores and growing crops, although there may be minor exceptions to this rule. Thus, in Africa, *R. rattus* or *R. norvegicus* seldom occur in growing crops, but are commensal pests of stored products, as they tend to be in temperate regions. In open-air grain stores, shounas, much utilized in the arid region of Africa, *R. rattus* is apparently the dominant rodent pest. Among endemic African rats, *Mastomys* is as common in farm houses as it is in the field.

The main problems in farmers' stores all over the world are structural deficiencies (Jackson, 1977). Still, such problems are mainly encountered in the tropics, because a relatively larger proportion of the whole crop is stored in the farms. Fully mechanized systems of cropping cereals in the temperate zone do not leave much for rodents. The grain intended for human consumption is often marketed immediately or stored in rat-proof silos on the farms, while the grain used for animal feed is stored less carefully. Heavy rat infestations on farms in most European countries indicate that such stores are accessible everywhere.

A new type of rat problems at farm level has been introduced by ready-made fodder mixtures. In normal farm conditions, these are readily accessible to rats and mice and are preferred by these pests owing to their high nutrient content. The vicious circle is complete: the rats are attracted and sustained by the availability of high-grade feed, and the accessibility of this preferred fodder aggravates the baiting problem when attempts are made to poison the rats. Such problems are most crucial in industrialized animal farms, as shown by the following example from Hungary (Papocsi, 1974): The State Farm Babolna, with its 10 000 ha of cultivated land, was converted into a giant poultry and pig production unit in the 1960s. With the transition to concentrated and vitaminized fodder production for the animals, the rats also obtained an optimally balanced diet which led to an unforeseen increase in their numbers. Afterwards, as a result of a successful control campaign, the rat population during the outbreak was estimated at 80 000 head, which consumed some 2500 kg of fodder/day. The direct losses were estimated as an equivalent of 0.5 million U.S. $/year, but the main problem was the increased risk of *Salmonella*-type epidemics, which could both endanger animal production and interfere with the export of brood animals. Once again, it was demonstrated that economic and health problems are closely inter-related. Similarly, this example can be used to emphasize that rationalization and mechanization of any process in agricultural production is just as likely to create new types of rodent problem as to remove old ones.

Seldom has there been any serious attempt at quantifying storage losses due to rats and mice. It is methodologically difficult to obtain objective estimates on direct losses, and indirect damage owing to rats and mice as disease vectors is even more complicated to estimate. This situation produces a vicious circle again: applications for economic support for improvement of rodent control practices often fail, simply because the decision-makers at local or governmental level cannot be convinced without figures on losses and prospective gains.

A cue to the nature and extent of rat damage in granaries was provided by Barnett (1951). He exposed one ton of sacked wheat for 12–28 weeks to small enclosed populations of the Norway rat. The loss in weight was 4.4 per cent of the exposed wheat, while 70.4 per cent of the grain was fouled and had to be cleaned before use. Total monetary loss was about 18 per cent of the original value of the grain and sacks, and was mainly due to the damage to the sacks.

This example turns attention to indirect losses other than those due to disease transmission, namely, destruction of package materials and other structures. In this respect the house mouse, for example, is a worse pest than would be expected from its small size and low food consumption. On most occasions when I have been called in to help with mouse problems, the population itself has not been proportional to the nuisance produced. Thus, in a bakery producing a popular speciality of bread a problem arose because the mice preferred the same type of bread as the human customers. One loaf for mice, a thousand for the market would have been acceptable to the owner, but the trouble was that all thousand loaves had to be checked carefully before transportation to pick out the one selected by the mouse for its early breakfast.

Damage done by rodents is not limited to food packages; injury may be done to materials and structures that have no rational relation whatsoever to the rodent's vital needs. Economically the most important of such damage is injury to various types of cable and electric wiring (Becker, 1960; Connolly and Landstrom, 1969). The problem has been aggravated by the introduction of plastic as a coating material, although the rodents are also capable of gnawing lead, aluminium and other metals softer than their own teeth. Especially in the field where the damage is usually done by burrowing species, like *Arvicola* or the pocket gophers, it is difficult and time-consuming to localize the damage. Cable ruptures due to rodents have sometimes caused severe disturbances, e.g., to international flight and train connections. It would be frightening to contemplate the consequences of a free-ranging rat gnawing at electric wiring in a long-distance jumbo-jet above the Atlantic! Nonetheless, such a situation is by no means impossible in our time.

6.3 Principal means of combating small mammal damage

6.3.1 Introduction to control problematics

There is no better introduction to the control problematics of harmful small mammals than to recall the quotation from Elton (1942) in the introductory Section. A claim that Elton's words about 'Cures' are as valid now as they were a generation ago may dismay those who have devoted decades of their lives to rodent control. In fact, such a claim is grossly exaggerated, yet more true than would appear at first sight. To back this argument, a short historical account may be useful:

The pre-Eltonian 'Cures' were bacterial preparations, like the 'Danysz virus', or poisons, such as strychnine, zinc phosphide, barium carbonate, and arsenious oxide. Some of these are still in use, as are some other even more poisonous chemicals (thallous sulphate, sodium fluoracetate and other compounds of fluorine). Zinc phosphide, because of its low price, is one of the major rodenticides even today. The use of bacterial preparations has been abandoned with one notable exception, the USSR (Bykovski, 1975).

The first major innovation in rodenticide technology since Elton's (1942) review was the introduction of warfarin in the early 1950s, followed by a series of other anticoagulant rodenticides. The anticoagulants made a real breakthrough in the control of commensal rodents and were considered a panacea till the appearance and spread of warfarin resistance in the 1960s. Currently, the problem of anticoagulant resistance has been solved, at least temporarily (Lund, 1977).

An effective remedy against microtines was for a short spell the use of chlorinated hydrocarbons, *viz.* endrin, in the late 1950s and early 1960s. Endrin was, however, soon banned due to its potential hazards to non-target wildlife. Increased environmental concern has similarly caused withdrawal or stoppage of the development of several other effective but potentially dangerous control chemicals.

Like the technical development of 'Cures', the organizational status of rodent control operations today suffers from many shortcomings and setbacks. As in the past, the 'Cure' still tends to be found too late, i.e., the control is symptomatic rather than preventive. Notwithstanding, the rural community nowadays, at least in temperate regions, is seldom confronted by the menace of famine due to a rodent outbreak. However, the reason for this advance is generally not improved control measures but rather structural changes in agriculture, which have incidentally affected the rodent populations, sometimes creating new problems (cf. Section 6.2) but mostly alleviating old ones.

In what follows I shall not discuss the various rodent control agents in a conventional way; reviews of that type are to be found elsewhere (e.g. Brooks, 1973; Gratz, 1973; Myllymäki, 1975a; Lund, 1977). Instead, I shall try to describe topical problems associated with the application of prospective control techniques and, finally, to tie these elements together into an integrated rodent control programme.

6.3.2 Habitat manipulation and related methods as intentional rodent control measures

Ecologically minded scientists tend to assign priority to habitat manipulation as a long-term solution to rodent problems. Let us first assume that this is sound in principle. The next step is to look for habitat or landscape characteristics that create conditions favourable for rodent plagues and, correspondingly, for those that tend to prevent outbreaks. Here viewpoints differ: some serious students (e.g. Frank, 1956; Hansson, 1975a; Stenseth, 1977) criticize the creation of monocultures and swear by the beneficial effects of increased habitat heterogeneity, whereas others (e.g. Taylor, 1968) claim that one of the most important reasons for the problems is the patchiness of the landscape. Both may be partly right, although most of the examples cited above (Section 6.2) seem to favour the latter view.

In a predominantly agricultural landscape, the feature common to most rodent plague situations has been the abundance of permanent grassy verges, or the high proportion of such refuges and their proximity to the crop area. The smaller the patches under crops, the closer are the refuges and higher the risk. Creation of monocultures tends to minimize the ratio between the perimeter verges and the crop area, or the 'refuge quota'. Only exceptionally the pest is able to establish self-sustaining populations on the monocultures alone (e.g. the European hamster, *see* Section 6.2.3). At its best, removal of the verges virtually solves the problem, even though the overall mosaic character of the landscape remains unchanged. This is the principal means by which plagues of *M. arvalis* have been prevented in the Netherlands and Belgium (van den Bruel, 1969).

I discussed above (Section 6.2.2) the ways in which agricultural and forestry practrices favoured *M. agrestis*. The reverse side of the coin is that rationalization and mechanization of field management have considerably decreased the field vole's chances of inhabiting cultivated fields; a case in point is presented in Figure 6.2. (cf. also Myllymäki, 1970): Assuming an initial situation in which all the 100 ha or so of fields on the estate in question had open ditches, the fields included more than 100 km of verges that voles could inhabit, corresponding to an area of about 15 ha. The pastures of roughly equal area were intensively grazed and not suited to *M. agrestis*. Having their permanent refuges within and around the crop lots the voles, when abundant, inflicted damage to the cereals and hayfields. In

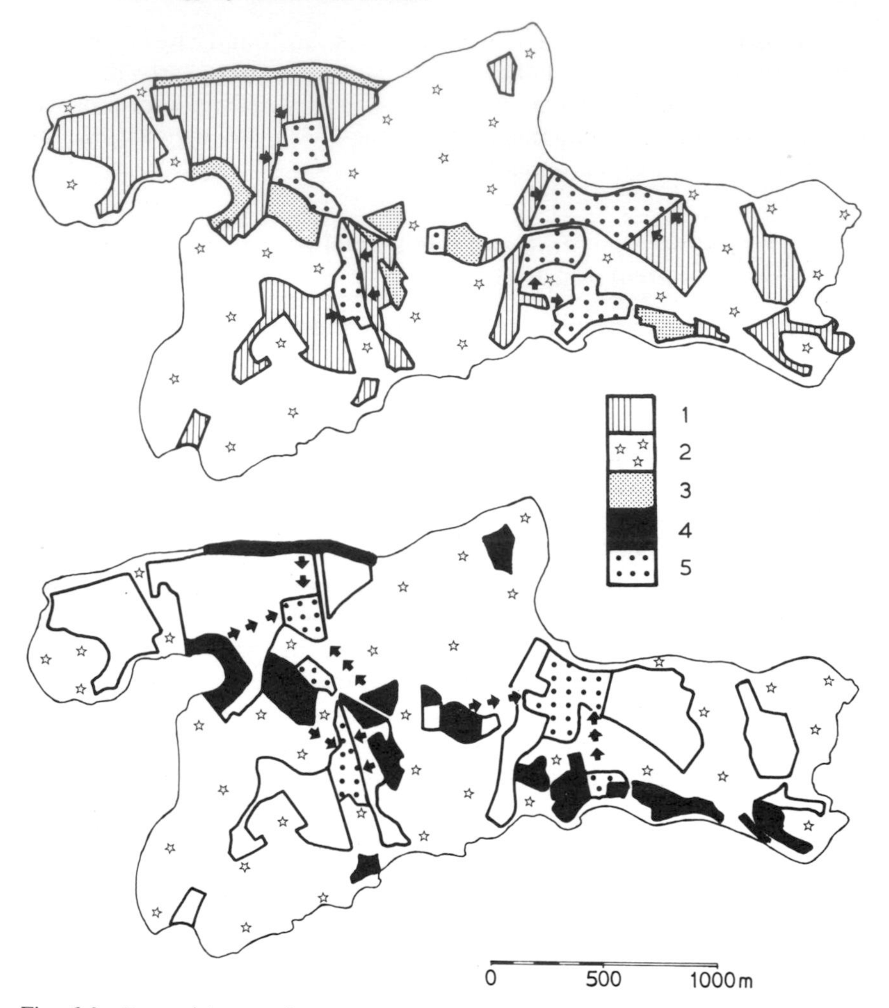

Fig. 6.2 Recent changes in the distribution and numbers of field vole habitats in a mixed agricultural landscape in southern Finland. *Upper map:* situation before the drainage reform, mechanization of field work and abandonment of dairy farming, *lower map:* the present situation (more details in the text). Explanations: (1) fields (hatched: open ditches, blank: drainage pipes); (2) mature forest; (3) grazed pastures; (4) former pastures, abandoned and mainly forested; (5) orchards and nurseries. Supposed migration of *M. agrestis* into orchards indicated by arrows.

winter population invaded orchards and nursery lots, destroying apple trees and other garden plants.

Let us then consider a 'reform', the main constituents of which were replacement of open ditches with drainage pipes, introduction of combines and herbicides, and slaughtering of cattle. The field perimeter refuges were reduced by some 80–90 per cent, and the field vole virtually lost close contacts with field crops. Weed control further decreased its chances to inhabit cereal fields. But, in terms of area, habitats suitable for *M. agrestis* were doubled rather than reduced; this was due to abandonment of earlier pastures and formation of clear-cut reforestation lots on rich forest soil. As a consequence, a minor vole problem, damage to field crops, was reduced by unintentional application of the principles of ecological control, but the principal problem, the threat to orchards and nurseries, continued unabated. Moreover, a new problem was created by the planting of saplings of forest trees on the newly-created grassland patches.

The case history above is useful also in illustrating the role of surface treatment in preventing damage to fruit trees. At the time of the actual study (cf. Myllymäki, 1970), the orchards and nurseries, about 20 ha in all, were scattered in separate lots and intermingled with other components of the landscape mosaic. The capacity of the orchards themselves to support propagating population of *M. agrestis* was minimal. During the winter 1968–69 it was actually shown that the voles moved into the orchards under, or above, the snow (Myllymäki, 1970, Fig. 5), and destroyed practically all the trees in one of them. The field vole is nomadic outside the breeding season (Myllymäki, 1977c), and winter invasions into orchards are the rule rather than the exception wherever there is a permanent snow cover. I would even venture to suggest that the same is true of *M. pennsylvanicus* in the northern parts of the U.S.A. and Canada. Consequently, the role of surface clearance is often over-emphasized (e.g. Davis, 1977), or virtually irrelevant in most geographical areas where tree-barking is a problem.

Intentional application of the removal of grass was reported by Green and Taylor (1975) from Kenya. With one exception, *Mastomys*, the existing rodent species proved very sensitive to removal of the vegetative cover. This finding, though based on treatment on an experimental scale only, is encouraging enough to warrant similar measures all over the tropics, where the dominant pests are murid rodents and the landscape is characterized by a high refuge quota. The means of managing the problems are thus necessarily not difficult in theory. In practice the problem is predominantly psychological and social: how to persuade farmers that the best way to prevent outbreaks is to abolish unproductive habitats even before any damage is visible – a task that has been proved difficult also in the developed countries (cf. Bouyx, 1967).

Whether ecological control could reasonably be used in forest regeneration areas, has been studied experimentally by Larsson (1977). Preliminary trials show that burning of slash before planting may retard grass growth and is thus argued to prevent damage by *M. agrestis* during the early history of a plantation. Whether this finding has practical implications, depends, as in the case of the orchards, on the proximity to sources of invasion by the field vole. Saplings are vulnerable until the age of about 10 years or so, and will thus experience approximately two outbreak situations even after a delayed invasion of grassy vegetation. On seed-eating small mammals fire does generally not have even the temporary deterrent effect that it exerts on herbivorous species (Tevis, 1956).

Authors advocating removal of the vegetation as a rodent control measure usually base their arguments on increased exposure of the pest population to predation. Indeed, in regions where migrating predatory birds gather in great numbers (Green and Taylor, 1975), this effect may be paramount. The situation is different in the northern latitudes, where there are often few predators during the breeding season, and in winter the rodents, i.e. voles, are well protected against most of the resident predators. However, if predation merits discussion as a prospective ocmponent of a rational rodent control programme, it is mostly relevant in connection with habitat manipulation only.

In connection with the reasoning that monocultures may create rodent problems, Frank (1956) and Hansson (1975a) argue that increased habitat heterogeneity increases predation pressure. A recent study on total impact of predation on the continental vole population by Ryszkowski *et al.* (1973) revealed that joint pressure by all predators was low during the population increase of *M. arvalis* and increased only in response to the peak density of the prey. Similarly, predation pressure may be high at high hamster densities in Hungarian monocultures, i.e., again in response to the prey population rather than to the habitat type.

Introduction of alien predators has been utilized as an intentional rodent control measure; among the best known examples are the introduction of weasels (*Mustela* spp.) to some islands in the North Sea and the Pacific, and of the mongoose (*Herpestes auropunctatus*) to Jamaica, Puerto Rico, the Hawaiian Islands, etc. (for the original references, *see* Wodzicki, 1973, and Myllymäki, 1975a). In most cases such measures have met with initial success but later on secondary problems have arisen wherever the situation has been observed long enough. Consequently, Wodzicki (1973) strongly opposed further introduction of weasels or monitor lizards (*Varanus indicus*) to the Pacific islands; [*see* also the general discussion by Howard (1967)].

An alternative to predation, competitive displacement of a pest rodent by another, less harmful species, was proposed by Taylor (1975) as an idea worth exploring. In the literature, there is no shortage of reports on negative species interaction, or competitive exclusion, but as regards practical solutions based on this idea, we are at the stage of guesswork. Moreover, the proposal involves all the risks connected with the introduction of alien predators.

Another way in which a pest rodent may be attacked selectively is by influencing its food chain. Keith *et al.* (1959) sprayed gopher-infested rangelands with a selective herbicide, 2,4-D, which killed the dicotyledonous plants favoured by *Thomomys*, and observed a subsequent decrease in gopher numbers. Provided that the application of herbicides is more acceptable than that of rodenticides, a similar procedure could probably be used also against other herbivorous rodents.

6.3.3 Exclusion methods and rodent deterrents

Where the borderline should be drawn between habitat manipulation and exclusion methods is sometimes a matter of convenience. Depriving rats of access to household refuse by placing the garbage in rat-proof bins is, presumably, classified as habitat alteration, while prohibiting the same rats with mechanical barriers from invading granaries may be taken as an example of exclusion techniques. This example also indicates the main field of application of exclusion methods, namely, for preventing post-harvest crop losses and damage to industrial products.

There is no doubt that urban rat populations can be brought down to, and maintained at, a tolerable level predominantly by strict care of environmental hygiene, including rat-proofing (Brooks, 1974; Myllymäki, 1974a). However, solving the rat problem in the field involves more than finding the best technique; it has psychological, social and organizational aspects, and, consequently, is seldom fully effective.

In the food-processing industry, mechanical barriers are less easily applied in practice than might be expected; an indication of the validity of this claim was provided by Dykstra (1966) in terms of the high incidence of rodent 'signs' in milled products. In such situations application of poison has obvious drawbacks, and this has led to the construction of special rodent deterring devices, such as ultrasound generators. Greaves and Rowe (1969) showed that the efficacy of these devices was questionable, at best they delayed reinfestation of buildings where the rats had been removed earlier by some other means. Much effort has also been devoted recently to the development of rat-proof packages. One such product consists of a multilayer polyethylene cover with a repellent chemical layer in between (Tigner, 1966).

In farm conditions garbage plays a secondary role in supporting rat populations; the main source of rat feed is the farm storage of cereals and animal fodder. In old farm buildings it is difficult to exclude rats by mechanical means, and even more difficult to prevent the entry of the house mouse. However, mechanical post-harvest treatment of cereals has greatly diminished rodent attacks, wherever cereal is produced above the immediate household needs of the farmer. In developing countries, this unintentional improvement of rodent control has had very little influence at the small-farmer level. A recent review by Jackson (1977), however, shows that

Fig. 6.3 Typical signs of *M. agrestis* in snow (left), and injury on an apple tree after a winter with much snow. The aluminium foil was installed to protect the trunk against vole gnaw but the measure failed in unusual snow conditions.

almost everywhere in the tropics there are indigenous rat-proof storage solutions which would considerably decrease post-harvest losses to rodents if practised generally.

In growing crops, mechanical exclusion is economic only in exceptionally valuable stands, such as coconut trees (e.g. Williams, 1975), apple trees and other fruit trees (Myllymäki 1970, 1977a), or grafts of forest trees in seed orchards (Myllymäki 1977a, b). The materials used include aluminium foil, wire netting, and hard plastic, all involving relatively high costs of both material and labour. Additionally, almost every time mechanical barriers have been used, there have been difficult problems of application. Williams (1975), for instance, reported that in coconut trees less than 15 m high a hanging frond often gave rats access to the tree above the aluminium band set at 3–4 m height. In Finnish seed orchards of Scots pine, solid materials (aluminium foil, hard plastic) encouraged attacks by insect pests and fungi beneath the collar, and in apple trees the problem has been that the trunk is generally too short for the tree to be fully protected by the collar in the case of a thick snow cover (Fig. 6.3).

One possible solution, fencing of the whole plantation to be protected with fine-mesh wire netting, has been practised in some Swedish seed orchards. But for one reason or another, high populations of *M. agrestis*, with consequent damage, have been found inside these exclosures, established at astronomical cost.

One extension of the exclusion technique are chemical repellents. Repellents are less used against rodents than against game damage, which may depend both on a more critical attitude by rodent research workers and on the rodent's special ability to gnaw off the repellent-treated surface without allowing the repellent to come into contact with the sensory organs of the mouth. At the time being, there are no unambiguously effective and non-phytotoxic surface repellents against tree barking by voles and other rodents causing similar damage. Attempts at applying highly poisonous chemicals, as crimidin or endrin, as repellents on tree stems have likewise failed to produce adequate results (Myllymäki, 1975a, 1977a). In contrast, some success has been obtained by coating seed with the same or similar poisons or repellents, like the combination of endrin-thiram-latex or the repellent 'R-55' (e.g. Radvanyi, 1973).

Repellents are also used against rodent attacks on cables; for instance, the above-mentioned 'R-55' has been used for this purpose, apparently with some success. Because of its penetrating smell, 'R-55', like most other repellents, cannot be applied for protecting packages containing foodstuffs.

Related to the repellent approach is the search for 'natural repellency' in the plant itself (Hansson, 1975b). Plant pathology and applied entomology provide several examples of crop strains resistant to certain pest organisms, while rodent research has not yet reached a single solution along this line. It is true that some varieties of apple trees, or provenances of forest trees are less susceptible to rodent damage than others in choice situation, but the same strains may be severely injured when standing alone. According to empirical experience, the strains liked by voles also possess properties (rapid growth, good quality fruits, etc.) that are desirable from man's point of view, and this may discourage the breeding expert from developing varieties resistant to voles.

One further difficulty in resistance breeding would be the fact that the attractiveness of a given plant to a rodent is not solely due to its hereditary properties, but may be greatly modified by soil factors, fertilizers applied, etc., in other words, by the chemical composition of the cambium (or some other plant part attacked). This, in turn, suggests theoretical chances of influencing certain elements of the cambium by means of fertilizers, or their additives, i.e. systemic repellents (Myllymäki, 1970, 1975a; Hansson,

1975b). These should, of course, not exert any adverse effect on the growth pattern or other desirable properties of the plant – a demand likely to lead to a vicious circle again. In contrast to Hansson (1975b), however, I do not see any special need for stressing the 'natural', i.e. non-synthetic, origin of the prospective substances to be used.

6.3.4 Reductional control methods

In this section I shall discuss control methods implying direct attack on the target population, i.e. means of killing the existing animals or influencing their vitality or reproductive performance. Rodent toxicants are conventionally divided into two main categories, acute and chronic rodenticides. This classification principally concerns bait poisons; additionally, some insecticides are used against rodents as surface sprays, and fumigants for gassing rodents in their burrows. Chemo-sterilants and microbial preparations can also be discussed in this context, because their mode of application – baiting – is similar to that of rodenticides proper.

The simplest method of chemical rodent control is blanket surface spraying or dusting of the vegetation inhabited and grazed by the target species, typically a herbivorous microtine. The animals then have no choice between treated and untreated food items, and the potential repellent properties of the chemical play no role. The effect is directly predictable from the specific toxicity of the poison and the behaviour of the species concerned.

The short spell when this control method predominated in Europe was, as regards efficacy, the most advanced era in the control of microtine pests. Owing to criticism of all use of chlorinated hydrocarbon compounds, endrin, the most effective surface spray, was soon withdrawn in most European countries (cf. Myllymäki, 1977a). Some less effective substitutes, like endosulphan and manocrotophos, are relatively widely used in some East European countries against *M. arvalis*.

Fumigants are used against burrowing rodents and moles. Several gas generators have been used; in Europe the principal gas used is hydrogen phosphide (van den Bruel and Bollaerts, 1960), in the U.S.A. cyanide (Brooks, 1973). Both fumigants are applied as powders, granulates or tablets which disintegrate when coming into contact with soil moisture, freeing the effective substances, phosphine or cyanogas. Reports on the efficacy are somewhat variable; however, these gasses are usually more effective than rodenticide baits against burrowing rodents.

Fumigants can also be used against insectivorous moles, although the results have not always been fully satisfactory (Lund, 1976). The applicability of fumigants is restricted by two additional factors, the extreme hazardousness for the field workers and the amount of manual work required.

Gassing with carbon dioxide (dry ice) as a means of controlling house mice in corn ricks and storage rooms was introduced by Southern (1954). The method is useful, though rather expensive, also for mice control in cold storage rooms.

Poison baiting is the traditional, and principal, method for combating pest rodents. The effect on the target animal is due to the combined action of two different factors, the toxic properties of the active ingredient and the attractiveness of the carrier material and the poison itself. Not infrequently, there are difficulties in balancing bait palatability and efficacy. An ideal rodent bait has not yet been devised.

As regards acute rodenticides, palatability problems are often aggravated by the taste of the active ingredient, or by a phenomenon called bait shyness, which means stoppage of feeding owing to symptoms initiated by the intake of a sublethal dose of the poison. An increase in concentration of a repellent active ingredient may lead to a decrease in bait acceptance (cf. Myllymäki, 1975a, p. 321) and hence to a lower percentage kill. On the other hand, there is at least one recent example in which an increase in crimidin concentration in a vole bait to a level sufficient to kill the animal at the first intake increased the percentage efficacy of the treatment. Many authors (e.g. Brooks, 1973) stress the importance of the latent period from the first take to the appearance of the first symptoms as the principal means of overcoming bait shyness. Generally speaking, the palatability of acute rodenticides is greatly overrated; a typical example is classification of the palatability of zinc phosphide as 'good' (e.g. Brooks 1973, Table 6), although this compound is sometimes used as a rodent repellent (Bykovski, 1975).

Chronic rodenticides are usually readily accepted at the concentrations used in rat baits. The fact that their effect *a priori* implies continued feeding imposes heavy demands on the carrier material, which must compete successfully with all other food sources available to the target species. This competition is aggravated in farm conditions and in the food industry. A recent survey and series of experiments in the Finnish countryside, for example, showed that radical bait innovations are necessary for essential improvement of farm rat control (Myllymäki, unpubl.). High farm infestation levels reported from many other countries provide circumstantial evidence that the problem is universal. Industrial bait production inevitably means a compromise between the biological needs and the technical and commercial aspects involved. Too often the latter carry more weight. What is unfortunate also is that 'commercialized rodent control' is most widespread in less developed countries where quality control is virtually lacking. Very often a better level of control could be obtained with domestic carriers than with imported ready-made pellets. This criticism is not intended, however, to underrate industrial bait development in general; in fact, there is far too little serious activity in this field.

Because bait palatability problems are generally recognized, there have been various approaches towards overcoming them. Attempts to find universally applicable bait additives, attractants and feeding enhancers have so far failed, and another line of research, the microencapsulation technique (Greaves *et al.*, 1968), has not yet been perfected. Possible applications of the results of rodent pheromone research (Stoddart, 1974; Christiansen, 1976) have also attracted attention, but not a single practical solution is yet in sight. Bait formulation seemingly continues to be a matter of trial and error.

A demand for good-quality ready-made baits is aggravated in connection with mechanized control of field rodents. Distribution of rodent bait by means of fertilizer applicators, burrow-builders (cf. references in Turner *et al.*, 1973) or aircraft (Marsh, 1968; Panteleev, 1968) is already a commonplace practice. But the bait types so far broadcast are not sufficiently palatable and consequently not effective. Most experts agree that control of field rodents should be based on acute rodenticides, which should be applicable without prebaiting, a procedure traditionally recommended when these chemicals are used against commensal rodents. Prebaiting is generally not needed in connection with the use of chronic rodenticides which, in turn, implies continued availability of fresh bait. Therefore, chronic rodenticides have mainly been used against commensal rodents. Recently, this more or less established dichotomy has broken down in both fields, firstly due to the introduction of the principle of sustained baiting in the control of field rodents and secondly as a response to the problem of anticoagulant resistance.

The sustained baiting procedure was developed in Germany (Telle, 1969; *see* also Drummond *et al.*, 1977) for maintaining 'rat-free' urban areas. The method, based on the establishment of a great number of permanent bait stations where fresh bait is laid at regular intervals (one month or less), was recently introduced into the Philippines for control of rice-field rats (Anon., 1976). At first sight, such a solution seems economically unsound but, reportedly, observed increase in rice yield justifies continuation of the programme. Most recently, Radvanyi (1974) applied the sustained baiting procedure to the control of *M. pennsylvanicus* in forest plantations. The anticoagulant chlorophacinone used by Radvanyi (1974) had been recommended for microtine control even earlier (Giban, 1970), although the percentage kill obtained was far from sufficient (70 per cent or less).

The phenomenon of anticoagulant resistance has been a leading topic of international meetings on rodent control for a decade or more (of the innumerable reviews, *see*, e.g. Drummond, 1970). Resistance was first detected against warfarin, but later on it was shown that cross resistance against most available anticoagulants was a widespread phenomenon, which occurred both in the Norway rat and in the house mouse. Studies on the genetics of warfarin resistance revealed that it was due to an autosomal dominant gene. Long-term monitoring of resistant rat populations showed that the populations are maintained in a state of balanced polymorphism, because the heterozygotes have a selective advantage over the homozygotes (Greaves *et al.*, 1977). Attempts to eradicate resistant populations have therefore usually failed.

Although the problem of resistance was for a while obviously overemphasized in relation to other rodent control problems, this close attention led to the development of a series of new rodenticides (Lund, 1977). Some of these are also anticoagulants (diphenacoum), while others, like calciferol (Vitamin D), act chronically but are not anticoagulants and, finally, some are acute rodenticides (pyrimidil). Some of these last, *viz.* pyrimidil, probably means a partial return to acute poisons for controlling commensal rodents. Of the new anticoagulants, diphenacoum is effective against resistant rats. Vitamin D, also effective against warfarin-resistant rats, is a good example of a 'natural', or even essential, substance that may be highly poisonous when taken in excess. It is also the only one of the new compounds that is effective against microtines; still, owing to its chronic action and high price it will probably never come into wide-scale use in the field.

Preparations of *Salmonella* and other pathogens are, as stated above, out of date, and their use is strongly discouraged by the World Health Organization (WHO, 1967). In spite of this authoritative objection, *Salmonella* strains claimed to be specific to rodents are in large-scale use in the U.S.S.R. (Bykovski, 1975). Of the species controlled with these bacterial preparations, the most susceptible are *Microtus* spp., the percentage mortality being reportedly between 65 and 100. A well known example of the myxoma virus against the European rabbit shows that, theoretically, it is possible to find species-specific pathogens but, on the other hand, intentional search for such pathogens may take a life time. If occasionally successful, the continued control of a pest with such a specific pathogen requires continual testing of the virulence in order to detect the appearance of attenuated strains, or resistance in the host (cf. Fenner and Marshall, 1957).

In contrast to microbial control, the use of chemosterilants is generally considered a modern challenge, and readily accepted even by those who strongly criticize all use of poisons against vertebrate pests. A chemosterilant is intended to interfere with the target animal's reproductive capacity along physiological pathways. The interference may be induced by various means: by sterilizing one or both sexes permanently or temporarily, by preventing implantation of fertilized ova, by abortion, inability to lactate, neonatal interference with the sexual development of the offspring, etc. As with rodenticides there are application problems, i.e. the chemosterilant must be fed to the target animal in bait. Most chemosterilants are exceedingly unattractive to rodents; hence baiting problems are aggravated as compared with rodenticides. The chemosterilants are generally not very specific; hence their application in the field may be associated with hazards to non-target animals. Furthermore, it is not easy to explain to a practical man why he should await the effect on the offspring, when the pest animals could be killed immediately with the same or less effort by application of poison.

These reservations and objectives are severe, and it is hardly surprising that no practical control solutions have been reached. Marsh and Howard (1973) regarded as the most sensible mode of application the use of antifertility agents to prevent recovery of the population of rats after conventional poisoning in garbage dumps, recreational areas, etc. In a field experiment on Richardson's ground squirrel, Goulet and Sadleir (1974) obtained some interesting results with mestranol, a steroid female sterilant. The finding that this compound induced behavioural changes in the animals, calls for further experimentation. Summarizing, it could be said that chemosterilants appeared to be a promising control technique, but that this promise has not been borne out in practice.

Lastly, trapping as a reductional rodent control measure has been criticized as too laborious and ineffective. However, Shuyler and Sun (1974) have recently suggested that trapping can be used even on a practical scale where cheap labour is available and funds are not adequate for the main alternative, chemical control. In Europe, trapping has been applied with variable, and often questionable, success in connection with hamster and muskrat control programmes. In the Netherlands muskrat control is based solely on the activities of state-paid trappers. To be rational, such activities should be performed under permanent monitoring and supervision, as it is in the Netherlands (Doude van Troostwijk, 1977).

6.4 Towards integrated control programmes

6.4.1 Introduction

Rodent control programmes are burdened by a reputation for shortcomings and by worn-out clichés. These are most obvious in connection with urban rat control campaigns, which are generally the only type of rodent control action known to the average citizen. Rodent damage to crops is seldom recognized before it is too late for any preventive measures; hence the 'Cure' is mostly applied in emergency and without the necessary planning and precautions.

These practical man's shortcomings have often been sharply criticized by biologists. No doubt, there is room for criticism, but the biologists themselves may be guilty of half-truths and of dragging ideology into a subject that should be treated in a rational way. For example, biologists tend to advocate encouragement, or even management, of natural predators as a solution to rodent problems; the irrational nature of such suggestions is shown when the very same persons deny the influence of predation upon populations of 'useful' game at the same trophic level as rodents.

The following is a brief outline of the essential elements of a rational rodent control programme. In principle, all such programmes should be directed towards minimizing the damage but, in practice, reaching this goal usually involves reduction of the numbers of the target pest. However, as Howard (1967), for example, has emphasized, such selective action focused on one, or a few, species at a time has a much slighter impact on the functioning of the ecosystem than is usually produced by habitat manipulation.

6.4.2 Identification of the problem and the pest

The first step towards a rational control programme is to identify the pest species responsible for the damage. This is seldom a crucial problem in Europe or North America, where both the pest species and their typical ways of damaging crops are adequately known. But the situation is different in the developing world, *viz.* in Africa, where little is known about even the most widely distributed pests, *Mastomys* and *Arvicanthus*, whose taxonomic status is still obscure. Large-scale control operations, like that in Gezira, Sudan (*see* Section 6.2.4) have been implemented with only a vague notion of the species involved. Such treatments tend, of course, to be a complete failure.

The mere presence or absence of a given species does not usually prove that it is responsible for a certain type of damage. Feeding experiments, microscopical diet analysis (Hansson, 1970) or the use of radioactive tagging (Radvanyi, 1973; Myllymäki and Paasikallio, 1976) are useful aids to correct determination of the pest.

6.4.3 Determination of the control thresholds

Two kinds of information are needed for determination of the feasibility of a control operation: a reliable estimate of the density of pest populations and an estimate of the potential damage without control.

Determination of the densities of small mammals is a science in itself (cf. Smith *et al.* 1975; and Myllymäki, 1975b). What is needed, however, for planning control programmes is not exact density values but reliable population indices. The Small Quadrat Method (SQM), developed and used for small mammal surveys in Scandinavia (Myllymäki *et al.*, 1977), will serve as an example of a trapping method that is simple but accurate enough for such a purpose. The merits of index trapping are obvious compared with mere reading of rodent signs. The latter method may produce useful population indices when field work is conducted by highly dedicated persons, i.e., by the scientists themselves (Hayne and Thompson, 1965), but as soon as surveillance in the field is handed over to technicians, there are far too many temptations to find shortcuts. From my own experience, for instance, I know that indices based on counting 'active' hamster burrows, the only surveillance method used in practice, is largely nonsense, both because of the way such counts are made in the field and becaue background studies are missing (cf. Section 6.2.3).

For some species, e.g. the Norway rat, which are not easily trappable, sign counts are the only feasible prcedure for large-scale surveys (Myllymäki, 1974a).

Accurate assessment of damage suffers from lack of adequate methods (for a general review, *see* Taylor, 1971). In the case of exceptionally valuable objects, such as fruit trees or grafts of forest trees in seed orchards, a reasonable approach may be to make complete inventories (cf. Myllymäki, 1977a, b). Usually, however, one should devise some kind of sampling scheme. The main obstacles to representative surveys are economic, but technical difficulties also arise when attempt is made to measure compensatory responses of the crops (for examples and detailed discussion, *see* Sections 6.2.1 and 6.2.5). The Philippine rice loss assessment scheme (Anon., 1976) is a good example of crop loss assessment applied on a country-wide scale.

Final determination of the economic threshold of a given control operation means interrelating the prevailing population indices with the estimates of potential losses. But this goal can be adequately realized only if forecasts on forthcoming population development can be made at least on a short-term basis (i.e. for at least a few weeks or months).

The need for population forecasts is generally agreed, but the methods so far used have been largely guesswork. It seems likely that in arid areas, for example in Africa, rodent populations increase in response to unusual rainfall (Section 6.2.4), a situation which, if true, would render forecasts relatively simple. This interdependence of climatic factors and rodent populations is not yet scientifically established. Climatic models have been developed for the temperate zone also (e.g. Spitz, 1977) but, as in the case of the arid region, the general applicability of these models still needs confirmation. Recent work based on the Scandinavian survey material (Myllymäki *et al.*, 1977) has revealed that at least prediction of the outbreaks of *M. agrestis* is more complex than would be expected from the studies referred to above.

6.4.4 Selection of control measures

The choice of an actual control method is often governed by economic considerations or by the fact that one of the potential alternatives, usually the 'non-poisonous' method, must be regarded as a long-term solution, while, e.g. rodenticides may be needed for coping with the topical problem.

The urban rat problem is a good example of where the use of rodenticides on the one hand and exclusion methods and environmental sanitation on the other are largely intercompensatory. There is sufficient evidence (e.g. Brooks, 1974; Myllymäki, 1974a) that in urban settlement areas the principal means of lowering rat populations is sanitation, but the difficulties of managing the human component of the rat problem have led to the development and large-scale use of an alternative approach based on sustained baiting with rodenticides (Telle, 1969; Drummond *et al.*, 1977). Both are reported to give similar results, but the 'poisonous' alternative is easier to manage, although not less expensive. An integrated compromise would be that suggested by Myllymäki (1974a) on the basis of experiences in a medium-size city (Tampere, Finland) and calling for continuing supervision of environmental hygiene in living quarters supplemented with a limited number of permanent bait stations at 'strategic' sites (food industry, traffic terminals, etc.).

A corresponding example of a field rodent control programme provided by Myllymäki (1977b) concerned a centralized damage control scheme for all seed orchards of forest trees in Finland. Here, the initial, and at first sole, control method against the main pest, *M. agrestis*, was treatment with endrin. This treatment was highly effective against the prevailing autumn population of the pest but, in addition to the environmental hazards, had another serious limitation: it was virtually ineffective against winter invasions of the pest. Therefore, mechanical guards were recommended as the chief means of combating the field vole. Implementation of this recommendation delayed, due both to high investment costs and to the extreme sensitivity of the conifer grafts to the materials tested. Finally the problem was solved by ecologically acceptable means, fitting all the grafts with guards of wire netting. Subsequently, there is no further need for the continued surveillance which formed an essential part of the original programme.

In the great majority of cases, control programmes against field rodents are based on application of poison, *viz.* poison baiting. In tropical countries, especially, non-poisonous alternatives are exceptional, although not ruled out (Green and Taylor, 1975). Thus, the choice primarily concerns which active ingredients and bait materials to use. The main properties and potential of the rodenticides used at present were discussed in Section 6.3.4. As far as bait material is concerned, the best results could often be obtained with domestic carriers.

6.4.5 Follow-up of the efficacy of control operations

Lack of follow-up of the effects of a control treatment on a target population is usually the weakest point in an integrated rodent control programme. Omission of this final stage may have grave consequences. Practically all pest rodents have a well-developed capacity to reproduce and increase in numbers in response to the population dilution caused by natural or artificial predation. Thus, it was a commonplace observation during the old-time rat control campaigns that the population, when reduced, say, to 20 per cent of the original numbers, soon returned to the original level, and a new campaign was needed at latest after one year. The round estimate of a half-year recovery period by the African rodents after a corresponding decrease (Gratz and Arata, 1975) coincides well with this empirical evidence from temperate latitudes.

Compensatory mechanisms have been studied in detail in microtine populations, for example, by Morris (1970, 1972) and Myllymäki (1974b). Morris (1970) produced experimental evidence in support of Howard's (1967 and later papers) suggestions of the beneficial effects of predation (*sensu lato*; sublethal dosing of a poison in this particular case) on the vigour of the populations. Further empirical evidence on the beneficial effects of the 'management' of pest rodent populations by means of insufficient percentage kills were given in Section 6.2.1. From the point of view of agriculture, the trapping of European hamsters for fur may also be regarded as pest management on a sustained yield basis.

Whenever there are no realistic chances for decreasing the capacity of the environment to maintain the pest or for applying exclusion methods, the best way to counteract these compensatory processes seems to be to aim at as high an initial percentage kill as possible and/or sustained baiting.

A final point worth mentioning in this context is the need for monitoring programmes that will rapidly detect any resistance phenomena. After the detection of warfarin resistance in rats, monitoring programmes were established in Denmark, Great Britain and the U.S.A., at least. Some of these programmes have now been terminated owing to the development of new compounds which are effective against warfarin-resistant rats. This may be economically feasible, but one should continually be ready to establish new monitoring surveys as soon as new problems arise. In the case of warfarin resistance, unpreparedness to meet the problem greatly delayed the countermeasures.

6.4.6 Need for international co-ordination of efforts

During 1977, the Food and Agriculture Organization of the United Nations (FAO) was approached by 40 countries for advice or help in rodent control problems (Shuyler, pers. comm.). The problems are increasing from year to year, while simultaneously the capacity of at least the FAO to give concrete assistance is tending to decrease. This weakness of a central international organization is partly compensated by bilateral development aid programmes of national agencies. An obvious drawback in implementing assistance programmes is the almost complete lack of co-ordination and, hence, the overlap of efforts. Also, despite the agreement of experts at innumerable meetings about the necessity for close contacts between groups dealing with agricultural damage and those dealing with the public health aspects of rodent control (cf. WHO, 1974; EPPO, 1977), combined action by FAO and WHO has been minimal in the field. There is a wide gap between the recommendations and the practical implementation of these recommendations.

Co-ordinative action is similarly needed among the donor countries themselves. As an example, since World War II, working groups in some 25–30 laboratories have been engaged in studies on the ecology and control of *M. arvalis* in Europe. With good reason it can be asked whether this research has been worth the outlay. At least in terms of practicable solutions to the pest problem there is room for doubt. In spite of excellent individual studies, there are as yet no scientifically based and generally acceptable schemes for assessment of damage, determination of economic control thresholds or fully efficient control measures. Discussions in recent expert meetings have clearly shown that most of these defects could be remedied simply by efficient and co-ordinated use of existing data.

So as not to leave the reader with a too pessimistic view, it should be added that some important advances have been achieved, for example by the action of the working groups of the European and Mediterranean Plant Protection Organization (EPPO). These activities include a set of published guidelines for rodenticide tests (EPPO, 1975); these guidelines will in all probability prove useful even outside the sphere of EPPO, *viz.* in the developing countries. Initial steps have also been undertaken to co-ordinate studies on forecasting outbreaks of *M. arvalis* and other microtines, as well as for co-ordinated control of the muskrat in Western Europe. Progress may seem slow, but that is a feature common to all international action.

6.5 References

Anon. (1958) The Oregon meadow mouse irruption of 1957–1958. *Federal Cooperative Extension Service, Oregon State College, Corvallis, Oregon, Bulletin,* 88 pp.

Anon. (1976) Rat control in rice fields. *The Philippine Recommends for Rice – 1976,* 4 pp., Laguna, Philippines.

Barnes, V.G. (1973) Pocket gophers and reforestation in the Pacific northwest: a problem analysis. *Special Scientific Report – Wildlife,* **155**, 1-18.

Barnett, S.A. (1951) Damage to wheat by enclosed populations of *Rattus norvegicus, Journal of Hygiene,* **49**, 22-25.

Becker, K. (1960) Über die Beschädigung kunststoffisolierter Leitungen durch Nagetiere. *Elektrotechnische Zeitschrift – B,* **12**, 311-314.

Bernard, J. (1977) Damage caused by the rodents Gerbillidae to agriculture in North Africa and countries of the Middle East. *EPPO Bulletin,* **7**, 283-296.

Beshir, S.A., Abbas, H. & Bashir, S. (1976) Rat outbreak and control in the Sudan with special reference to the Gezira Scheme. *First Afro-Asian Vertebrate Pest Congress, Cairo 1976,* (MS).

Bodenheimer, F.S. (1949) Dynamics of vole populations in the Middle East. *Israel Scientific Research Council, Jerusalem,* pp. 19-25, 48-50.

Bouyx, L. (1967) The economic incidence of field vole control in France. *EPPO Publications, Series A,* **41**, 93-97.

Brooks, J.E. (1973) A review of commensal rodents and their control. *Critical Reviews in Environmental Control,* **3**, 405-453.

Brooks, J.E. (1974) A review of rodent control programs in New York State. *Proceedings: Sixth Vertebrate Pest Conference, March 5-7, 1974, Anaheim, California,* pp. 132-141.

Bruel, W.E. van den (1969) Le campagnol des champs *Microtus arvalis* Pallas etat du probleme en Belgique. *Parasitica,* **25**, 117-151.

Bruel, W.E. van den & Bollaerts, D. (1960) Mise au point du dispositif utilisé pour détruire au moyen de l'hydrogène phosphoré les mammifères dissimulés dans des terriers profonds. *Proceedings of the IVth International Congress of Crop Protection Hamburg, 1957,* pp. 1335-41.

Bykovski, V.A. (1975) Selective action preparations for protecting plants against field rodents. *VIII International Congress of Plant Protection,* Vol. II, pp. 80-84.

Christiansen, E. (1976) Pheromones in small rodents and their potential use in pest control. *Proceedings: Seventh Vertebrate Pest Conference, March 9-11, 1976, Monterey, California,* pp. 185-191.

Connolly, R.A. & Landstrom, R.E. (1969) Gopher damage to buried cable materials. *Materials Research and Standards,* **9**, 13-18.

Crowcroft, P. (1966) *Mice all over.* London: G.T. Foulis & Co Ltd.

Davis, D.E. (1977) Advances in rodent control. *Zeitschrift für Angewandte Zoologie,* **64**, 193-211.

Doude van Troostwijk, W.J. (1977) Monitoring musk-rat control in the Netherlands. *EPPO Bulletin,* **7**, 415-421.

Drummond, D.C. (1970) Variation in rodent populations in response to control measures. In *Variation in Mammalian Populations,* eds. Berry, R.J. & Southern, H.N., pp. 351-367. London: Academic Press.

Drummond, D.C., Taylor, E.J. & Bond, M. (1977) Urban rat control: further experimental studies at Folkestone. *The Environmental Health Officers Association, Monograph Series,* **85**, pp. 265-267.

Dykstra, W.W. (1966) The economic importance of commensal rodents. *Seminar on Rodents and Rodent Ectoparasites, Geneva, October 24-28, 1966.* Mimeo., pp. 9-12.

Elton, C. (1942) *Voles, mice and lemmings. Problems in population dynamics.* Oxford: Clarendon Press.

EPPO (1968) Report of the International Conference on the Musk-Rat, *Ondatra zibethica* L. (1968) *EPPO Publications, Series A,* **47**, 88 pp.

EPPO (1975) Guide-lines for the development and biological evaluation of rodenticides. *EPPO Bulletin,* **5**, 1-49.

EPPO (1977) Report of the Joint FAO/WHO/EPPO Conference on Rodents of Agricultural and Public Health Concern. *EPPO Bulletin*, **7**, 151-554.

Fall, M.W. & Sanchez, F.F. (1975) The Rodent Research Center, *PANS*, **21**, 206-212.

Fenner, F. & Marshall, I.D. (1957) A comparison of the virulence for European rabbits (*Oryctolagus cuniculus*) of strains of myxoma virus recovered in the field in Australia, Europe and America. *Journal of Hygiene*, **55**, 149-191.

Frank, F. (1956) Grundlagen, Möglichkeiten und Methoden der Sanierung von Feldmausplagegebieten. *Nachrichtenblatt des Deutschen Pflanzenschutzdienstes*, **10**, 147-157.

Funmilayo, O. & Akande, M. (1977) Vertebrate pests of rice in South-western Nigeria. *PANS*, **23**, 38-48.

Giban, J. (1970) Experiments on the practicability of control methods against the continental vole, *Microtus arvalis* (Pallas). *EPPO Publications, Series A*, **58**, 73-80.

Gorecki, A. (1977) Energy flow through the common hamster population. *Acta Theriologica*, **22**, 25-66.

Goulet, L.A. & Sadleir, R.M.F. (1974) The effects of a chemosterilant (mestranol) on population and behaviour in the Richardson's ground squirrel (*Spermophilus richardsonii*) in Alberta. *Proceedings: Sixth Vertebrate Pest Conference, March 5-7, 1974, Anaheim, California*, pp. 90-100.

Gratz, N.G. (1973) A critical review of currently used single-dose rodenticides. *Bulletin of the World Health Organisation*, **48**, 469-477.

Gratz, N.G. & Arata, A.A. (1975) Problems associated with the control of rodents in tropical Africa. *Bulletin of the World Health Organisation, 52*, 697-705.

Greaves, J.H., Choudry, M.A. & Khan, A.A. (1977) Pilot rodent control studies in rice fields in Sind, using five rodenticides. *Agro-Ecosystems*, **3**, 119-130.

Greaves, J.H. *et al* (1968) Microencapsulation of rodenticides, *Nature*, **219**, 402-403.

Greaves, J.H. *et al*. (1977) Warfarin resistance: a balanced polymorphism in the Norway rat. *Genetical Research*, **30**, 257-263.

Greaves, J.H. & Rowe, F.P. (1969) Responses of confined rodent populations to an ultrasound generator. *Journal of Wildlife Management*, **33**, 409-417.

Green, M.G. & Taylor, K.D. (1975) Preliminary experiments in habitat alternation as a means of controlling field rodents in Kenya. *Ecological Bulletins/NFR*, **19**, 175-181.

Hall, E.R. (1927) An outbreak of house mice in Kern County, California. *University of California Publications in Zoology*, **30**, 189-203.

Hamar, M. (1977) Mechanized control of Muridae in isolated foci. *EPPO Bulletin*, **7**, 541-550.

Hansson, L. (1970) Methods of morphological diet micro-analysis in rodents. *Oikos*, **21**, 255-266.

Hansson, L. (1975a) Effects of habitat manipulation on small rodent populations. *Ecological Bulletins/NFR*, **19**, 163-173.

Hansson, L. (1975b) Natural repellence of plants towards small mammals. *Ecological Bulletins/NFR*, **19**, 213-219.

Hansson, L. & Zejda, J. (1977) Plant damage by bank voles (*Clethrionomys glareolus* Schreber) and related species in Europe. *EPPO Bulletin*, **7**, 223-242.

Hayne, D.W. & Thompson, D.Q. (1965) Methods for estimating microtine abundance. *Transactions of the Thirtieth North American Wildlife and Natural Resources Conference*, March 8, 9 and 10, 1965, 393-400.

Higuchi, S. (1976) Damage caused by vertebrates, especially by voles, in forests in Japan. *Proceedings: XVI IUFRO World Congress,* Division II, pp. 485-490.

Hood, G.A., Nass, R.D. & Lindsay, G.D. (1970) The rat in Hawaiian sugarcane. *Proceedings: Fourth Vertebrate Pest Conference, March 3-5, 1970, West Sacramento, California,* pp. 34-37.

Howard, W.E. (1967) Biocontrol and chemosterilants. In *Pest Control: Biological, Physical and Selected Chemical Methods,* eds. Kilgore, W.W. & Doutt, R.L., pp. 343-386.

Howard, W.E. & Childs, H.E. (1959) Ecology of pocket gophers with emphasis on *Thomomys bottae mewa. Hilgardia,* **29**, 277-358.

Jackson, W.B. (1977) Evaluation of rodent depredations to crops and stored products. *EPPO Bulletin,* **7**, 439-458.

Keith, J.O., Hansen, R.M. & Ward, A.L. (1959) Effect of 2,4-D on abundance and foods of pocket gophers. *The Journal of Wildlife Management,* **23**, 137-145.

Laird, M. (1963) Rats, coconut, mosquitoes and filariasis. *Tenth Pacific Science Congress, Honolulu, 26 Aug. – 6 Sept. 1961,* pp. 535-542.

Larsson, T.-B. (1977) Small rodent abundance in relation to reforestation measures and natural habitat variables in northern Sweden. *EPPO Bulletin,* **7**, 397-409.

Lavoie, G.E., Swink, F.N. & Sumangil, J.P. (1970) Destruction of rice tillers by rats in relation to stages of rice development in Luzon. *Philippine Agriculturist,* **54**, 175-181.

Lund, M. (1976) Control of the European mole, *Talpa europaea. Proceedings: Seventh Vertebrate Pest Conference, March 9-11, 1976, Monterey, California,* pp. 125-130.

Lund, M. (1977) New rodenticides against anticoagulant-resistant rats and mice. *EPPO Bulletin,* **7**, 503-508.

Maksimov, A.A. (1977) *Tipy vspysek in prognozy massovogo razmnozenija gryzunov.* Novosibirsk: 'Nauka', Sibirskoje otdelenije. (in Russian).

Marsh, R.E. (1968) An aerial method of dispensing ground squirrel bait. *Journal of Range Management,* **21**, 380-384.

Marsh, R.E. & Howard, W.E. (1973) Prospects of chemosterilant and genetic control of rodents. *Bulletin World Health Organisation,* **48**, 309-316.

Meylan, A. (1977) Fossorial forms of the water vole, *Aricola terrestris* (L.), in Europe. *EPPO Bulletin,* **7**, 209-221.

Mizushima, S. (1976) On damage of apple trees by the red-backed vole. *The Bulletin of Hokkaido Prefectural Agricultural Experiment Stations,* **34**, 59-66.

Morris, R.D. (1970) The effects of endrin on *Microtus* and *Peromyscus.* I. Unenclosed field populations. *Canadian Journal of Zoology,* **48**, 695-708.

Morris, R.D. (1972) The effects of endrin on *Microtus* and *Peromyscus.* II. Enclosed field populations. *Canadian Journal of Zoology,* **50**, 885-896.

Myllymäki, A. (1970) Population ecology and its application to the control of the field vole, *Microtus agrestis* (L.) *EPPO Publications, Series A,* **58**, 27-48.

Myllymäki, A. (1974a) A programme for the extermination of rats at minimal cost in Tampere, Finland. *Beihefte der Zeitschrift für angewandte Zoologie,* **3**, 155-159.

Myllymäki, A. (1974b) Experience from an unsuccessful removal of a semi-isolated population of *Arvicola terrestris. Proceedings of the International Symposium on Species and Zoogeography of European Mammals, November 22-26, Brno,* pp. 376-387.

Myllymäki, A. (1975a) Control of field rodents. In *Small Mammals:*

their productivity and population dynamics, eds. Golley, F.B., Petrusewicz, K. & Ryszkowski, L., Ch. 14, pp. 311-338. Cambridge University Press.

Myllymäki, A. (1975b) Rodent surveillance and the prediction of rodent outbreaks. *Ecological Bulletins/NFR,* **19**, pp. 275-282.

Myllymäki, A. (1977a) Outbreaks and damage by the field vole, *Microtus agrestis* (L.), since World War II in Europe. *EPPO Bulletin,* **7**, 177-207.

Myllymäki, A. (1977b) A program for control of damage by the field vole, *Microtus agrestis* (L.), in seed orchards of forest trees. *EPPO Bulletin,* **7**, 523-531.

Myllymäki, A. (1977c) Intraspecific competition and home range dynamics in the field vole *Microtus agrestis. Oikos,* **29**, 553-569.

Myllymäki, A., Christiansen, E. & Hansson, L. (1977) Five-year surveillance of small mammal abundance in Scandinavia. *EPPO Bulletin,* **7**, 385-396.

Myllymäki, A. & Paasikallio, A. (1976) Scots pine seed depredation by small mammals, as releaved by radioactive tagging of the seeds. *Annales Agriculturae Fenniae,* **15**, 89-96.

Nass, R.D., Hood, G.A. & Lindsey, G.D. (1971) Fate of Polynesian rats in Hawaiian sugarcane fields during harvest. *Journal of Wildlife Management,* **35**, 353-356.

Nechay, G., Hamar, M. & Grulich, I. (1977) The common hamster (*Cricetus cricetus* L.); a review. *EPPO Bulletin,* 7, 255-276.

Panteleev, P.A. (1968) *Population ecology of water vole and measures of control.* Moscow: Nauka (in Russian; English summary).

Papocsi, L. (1974) Das Entrattungssystem von Babolna. *Beihefte der Zeitschrift für angewandte Zoologie,* **3**, 155-159.

Poljakov, I. Ja. Ed. (1976) Rasprostranjenije glavnjeisih vrjeditjeljei selskohozjaistvennyh kultur v SSSR i effjektivnost borby s nimi. Leningrad: VIZR (in Russian).

Prakash, I. (1976) Rodent pest management – principles and practices. *CAZRI Monograph,* **4**, 28 pp.

Radvanyi, A. (1973) Seed losses to small mammals and birds. *Direct Seeding Symposium, Timmins, Ontario, September 11-13, 1973,* 1339, 67-75.

Radvanyi, A. (1974) Small mammal census and control on a hardwood plantation. *Proceedings: Sixth Vertebrate Pest Conference, March 5-7, 1974, Anaheim, California,* pp. 9-19.

Rodent Research Center, College, Laguna, Philippines (1971) *Annual Report,* 80 pp.

Rowe, F.P., Taylor, E.J. & Chudley, A.H.J. (1963) The numbers and movements of housemice (*Mus musculus* L.) in the vicinity of four corn-ricks. *Journal of Animal Ecology,* **32**, 87-79.

Ryan, G.E. & Jones, E.L. (1972) A report on the mouse plague in the Murrumbidgee and Coleambally irrigation areas, 1970. *N.S.W. Department of Agriculture,* mimeo. 60 pp.

Ryszkowski, L. & Myllymäki, A. (1975) Outbreak of *Microtus arvalis* (Pall.) and other microtine rodents in central and eastern Europe. *Ecological Bulletins/NFR,* **19**, 57-64.

Ryszkowski, L., Goszczynski, J. & Truszkowski, J. (1973) Trophic relationships of the common vole in cultivated fields. *Acta Theriologica,* **18**, 125-165.

Santini, L. (1977) European field voles of the genus *Pitymys* McMurtrie and their damage in agriculture and forestry. *EPPO Bulletin,* **7**, 243-253.

Shuyler, H.R. & Sun, R.F. (1974) Trapping: a continuous integral part of a rodent control programme. *Proceedings: Sixth Vertebrate Pest Conference, March 5-7, 1974, Anaheim, California,* pp. 150-160.

Smith, M.H. *et al.* (1975) Density estimations of small mammal populations. In *Small Mammals: their productivity and population dynamics,* eds. Golley, F.B., Petrusewicz, K. & Ryszkowski, L., Ch. 2, pp. 25-53. Cambridge University Press.

Southern, H.N. (1954) Carbon dioxide fumigation for house mice. In *Control of Rats and Mice,* eds. Chitty, D. & Southern, H.N., Vol. 3, pp. 199-215.

Spitz, F. (1968) Interactions entre la vegetation epigee d'une luzerniere et des populations enclose ou non enclose de *Microtus arvalis* Pallas. *La Terre et la Vie,* **3**, 274-306.

Spitz, F. (1977) Dévelopment d'un modèle de prévision des pulluations du campagnol des champs. *EPPO Bulletin,* **7**, 341-347.

Stenseth, N.C. (1977) Forecasting of rodent outbreaks: models and the real world. *EPPO Bulletin,* **7**, 303-315.

Stoddart, D.M. (1974) The role of odor in the social biology of small mammals. In *Pheromones,* ed. Birch, M.C., Ch. 15., pp. 297-315. Amsterdam: North-Holland Publishing Company.

Taylor, J.C. (1970) The influence of arboreal rodents on their habitat and man. *EPPO Publications, Series A,* **58**, 217-223.

Taylor, K.D. (1968) An outbreak of rats in agricultural areas of Kenya in 1962. *East African Agricultural and Forestry Journal,* **34**, 66-77.

Taylor, K.D. (1971) Assessment of losses due to rodents. *Crop Loss Assessments Methods. FAO Manual on the evaluation and prevention of losses by pests, diseases and weeds.* ed. L. Chiarappa. Oxford. **3.2.1**., 1-3.

Taylor, K.D. (1972) Rodent problems in tropical agriculture. *PANS,* **18**, 81-88.

Taylor, K.D. (1975) Competitive displacement as a possible means of controlling commensal rodents on islands. *Ecological Bulletins/NFR,* **19**, 187-194.

Taylor, K.D. & Green, M.G. (1976) The influence of rainfall on diet and reproduction in four African rodent species. *Journal of Zoology, London,* **180**, 367-389.

Telle, H.-J. (1969) 12 Jahre grossräumige Rattenbekämpfung in Niedersachsen – ein kritischer Rück- und Ausblick. *Schriftenreihe des Vereins für Wasser-, Boden- und Lufthygiene,* **32**, 131-159.

Tertil, R. (1977) Impact of the common vole, *Microtus arvalis* (Pallas) on winter wheat and alfalfa crops. *EPPO Bulletin,* **7**, 317-339.

Teshima, A.H. (1970) Rodent control in the Hawaiian sugar industry. *Proceedings: Fourth Vertebrate Pest Conference, March 3, 4, 5, 1970 West Sacramento, California,* pp. 38-40.

Tevis, L. (1956) Effect of a slash burn on forest mice. *Journal of Wildlife Management,* **20**, 405-409.

Tigner, J.R. (1966) Chemically treated multiwall tarps and bags tested for rat repellency. *Journal of Wildlife Management,* **30**, 180.

Turner, G.T. *et al.* (1973) Pocket gophers and Colorado mountain rangeland. *Colorado State University Experiment Station, Bulletin* 554S, 90 pp.

Wieland, H. (1973) Beiträge zur Biologie und zum Massenwechsel der Grossen Wühlmaus (*Arvicola terrestris* L.). *Zoologische Jahrbücher, Abteilung für Systematik, Ökologie und Geographie der Tiere,* **100**, 351-428.

Wijngaarden, A. van (1957) The rise and disappearance of continental vole plague zones in the Netherlands. *Verslagen van Landbouwkundige Onderzoekingen,* **63**.15, 29 pp.

Williams, J.M. (1974a) The effect of artificial rat damage on coconut yields in Fiji. *PANS,* **20**, 275-282.

Williams, J.M. (1974b) Rat damage to coconut in Fiji. Part 1. Assessment of damage. *PANS,* **20**, 379-391.

Williams, J.M. (1975) Rat damage to coconuts in Fiji. Part II. Efficiency and economics of damage reduction methods. *PANS,* **21**, 19-26.

Wodzicki, K. (1973) Prospects for biological control of rodent populations. *Bulletin World Health Organisation,* **48**, 461-467.

Wolf, Y. (1977) The levant vole, *Microtus guentheri* (Danford et Alston, 1882). Economic importance and control. *EPPO Bulletin,* **7**, 277-281.

Wood, B.J. (1971) Investigations of rats in ricefields demonstrating an effective control method giving substantial yield increase. *PANS,* **17**, 180-193.

WHO (1967) Joint FAO/WHO Expert Committee on Zoonoses. Third Report. *World Health Organisation, Technical Report Series*, **378**, 127 pp.

WHO (1974) Ecology and control of rodents of public health importance. *World Health Organisation, Technical Report Series,* **553**, 42 pp.

Ecology of bats

7

JIRÍ GAISLER

Information on the ecology of bats is included already in the classical monographs by Eisentraut (1937), Allen (1939), Ryberg, (1947) and Kuzyakin (1950) as well as in books summarizing data on the bat faunae of certain regions (Abelentsev *et al.*, 1956; Verschuren, 1957; Rosevear, 1965; Villa, 1966; Barbour and Davis, 1969; Kingdon, 1974). Problems of the ecology of bats are the subject of a whole monograph by Brosset (1966) and considerable parts of those by Wimsatt (1970) and Slaughter and Walton (1970). In addition, round 3200 papers on the ecology of bats were published during the past three decades (Krzanowski, in litt.). Due to the limited space allotted to this chapter, only an outline of the problems can be given. Data and papers which could not be utilized are not considered less valuable.

7.1 Habitat

7.1.1 Structural and functional requirements

The ecological niche occupied by bats is the result of a number of adaptations of which flight capacity, nocturnal feeding activity and the habit of hiding in caverns and crevices may be considered the basic ones. While the latter does not pertain to certain species, particularly among Megachiroptera, it is apparently a basic characteristic of the whole order (Jepsen in Wimsatt, 1970). Morphological and physiological mechanisms cannot be described in detail but the functional complexity of the whole system should be pointed out. Thus, the adaptation to flight is closely connected with the adaptation to motor activity at limited or no light as well as with that to moving and hanging in shelters. Hence, modifications of the motor apparatus are realized in connection with special neuro-sensory mechanisms and further modified by, e.g., trophic specialization. The types of echolocation utilized have been found to be correlated, to some extent, with the way of flying, and with feeding strategies (Novick, 1958; Griffin, 1958; Airapetiantz and Konstantinov, 1974).

The maintenance of homeostasis and survival requires a successful interaction of the above and other mechanisms with environmental factors. A number of feedback loops can be found in these relationships. The very wings which permit long range flight increase, at the same time, the body surface considerably. Since they are, moreover, strongly vascularized bats face problems of heat loss which do not occur in other mammals (Lyman in Wimsatt, 1970). Unusual energy expenditure occurs even in foraging and spatial dispersal. Therefore, special adaptations have developed in the metabolism of bats which effect the saving of energy. These adaptations are due to a special type of thermoregulation. However, external factors, especially microclimatic ones, are required for their realization. Most bats show very marked requirements for specific shelters during periods of rest. Of primary importance are shelters of nursery colonies, being the bearers of the reproduction potential of the bat populations. The presence or absence of such shelters appears to be capable of limiting the diversity and abundance of bat communities in extensive areas (Humphrey, 1975).

7.1.2 Ecological factors: control through ambient influence

The ecological phenomena which regulate the life cycles and numbers of bats are likewise complex in nature and hence are difficult to categorize. Climatic factors seem to be of primary importance for bats in colder regions and biotic factors for those in the tropics (Eisentraut, 1947; Krzanowski, 1969). The importance of temperature for the natural regulation of bat numbers in temperate regions is paramount. Recent studies (Heithaus *et al.*, 1975; Vaughan, 1976) point out that climatic changes, especially the pattern of alternating wet and dry seasons, can have the same controlling effect in the tropics as temperature in the temperate zone. Climatic factors, together with the soil substrate, affect the vegetational cover on which the food supply depends. There are very few regions where the climate is virtually 'stable' and where biotic factors are of primary importance. The major factor in these regions may be the interspecific competition both within the Chiroptera (Tamsitt, 1967) and between bats and other animals (Fenton and Fleming, 1976). Among ecological factors, very important is periodic oscillations of light; darkness is the main timing system of bats' circadian activity rhythms.

At present, the life cycles and numbers of bats are significantly affected by anthropogenic factors, an analysis of which can be found in all monographs cited in the introduction. Their major positive effect is that many bat species can find shelters in human buildings; their negative effects are connected with the transformation of ecosystems, environmental pollution, and disturbance and/or killing of bats in their shelters.

7.1.3 Foraging strategies

Flight patterns

The principle of bat flight is basically known from the analyses of high-speed photographs of flying bats (Eisentraut, 1936; Hartman, 1963; Norberg, 1970; Vaughan in Wimsatt, 1970; Pennycuick, 1972). Bats were found to be capable of mastering different flight patterns from gliding to hovering and their wings were found to beat in synchrony, except in such specific situations as making turns. During a normal wing-beat cycle, the movement of wings is downward and forward in the downstroke, and upward and backward in the upstroke. Characteristic of bats is their ability to vary the camber of the wing. The uropatagium may have special function in flight and in certain species it participates in seizing the prey (Webster and Griffin, 1962); its absence or feeble development in many families show, however, that it is not indispensable for flying.

Many authors, including the present one, studied the relative size and shape of bats' wings. These data are less adequate to the needs of an ecological survey although they reveal some information about the correlations of structure and function. The reader is referred to the paper by Findley *et al.*, (1972) in which a statistical evaluation of the morphological properties of wings of 136 bat species representing 15 families is presented. According to these authors, distinctive combinations of the variables studied characterize groups of bats that presumably have similar flight modes.

Although selection for modification of flight structure in bats seems to operate within narrow limits (Lawlor, 1973), several styles of flight can be distinguished. Vaughan (in Wimsatt, 1970) distinguishes five:

(1) Insectivorous bats (Emballonuridae, Chilonycterinae, Natalidae, Thyropteridae, Vespertilionidae) – slow, manoeuverable flight with the ability to remain continuously on the wing when foraging.

(2) 'Flycatcher' bats (Rhinolophidae and certain Megadermatidae, Nycteridae, Phyllostomatidae and Vespertilionidae) — very slow, highly manoeuverable flight, not remaining continuously on the wing when foraging, sometimes resembling the feeding habits of flycatchers.

(3) Nectar-feeding bats (Macroglossinae, Glossophaginae, Phyllonycterinae) – slow flight near vegetation and hovering, perhaps not continuously on the wing when foraging.

(4) Frugivorous bats (most Pteropodidae and some Phyllostomatidae – enduring straight flight, not highly manoeuverable, not feeding on the wing.

(5) Fast fliers (Molossidae and some Vespertilionidae as *Nyctalus* and *Miniopterus*) – rapid, enduring flight, less manoeuverable but with the ability to travel long distances during the foraging periods. *Tadarida brasiliensis,* a typical representative of this group, may reach altitudes of over 3000 m and a speed of 105 km per h when flying to its foraging grounds, as ascertained by Williams *et al.*, (1973) using radar and direct observations from a helicopter. This case is exceptional, however. Bats as a whole are characterized by slow, manoeuverable and enduring flight, by which they differ from most birds.

Activity rhythms

The foraging activity of most bats occurs at night. As it is difficult to study the course of this activity, most authors have confined themselves to studying its beginning and end. The current method of study involves using photocell detectors with infra-red beams placed at the exit chambers or holes from the roosting sites. The detectors are coupled to automatic recorders registering the interruptions of the beams by bats flying out or in. The main source of information is in the time relations between the first exit flight and the last return flight, or first exit – last exit and first return – last return, defining the duration of the activity and its various phases. Most studies of this type have been conducted in Europe (DeCoursey and DeCoursey, 1964; Nyholm, 1965; Böhme and Natuschke, 1967; Laufens, 1972; Voûte *et al.*, 1974). An example of the results obtained is given in Fig. 7.1.

The beginning and end of the activity was found to vary, in the course of the year, rather synchronously with the time of sunset and sunrise. The most conspicuous deviation from the current scheme is the shift of activity to afternoon hours, as ascertained, for instance, in the colony of *Myotis mystacinus* recorded in Fig. 7.1 b. In this case cold nights in Finland, which prevented the bats from obtaining food in spring and autumn, caused the pattern change (Nyholm, 1965). Diurnal foraging activity under similar situations was reported in several species throughout the Palaearctic and Nearctic regions and even in the tropics (Brosset, 1966) where it cannot be explained by the effect of low temperatures. Such cases are rather scarce, however.

Another conspicuous phenomenon is the considerable interval between the first and the last return flight, observed, for instance, in the colony of *Myotis dasycneme* in the Netherlands (Fig. 7.1 a). The first return flight is often followed by alternating exits and returns of many individuals of the colony, which may last for several hours. We observed the same phenomenon in colonies of various European species of Vespertilionidae and Rhino-

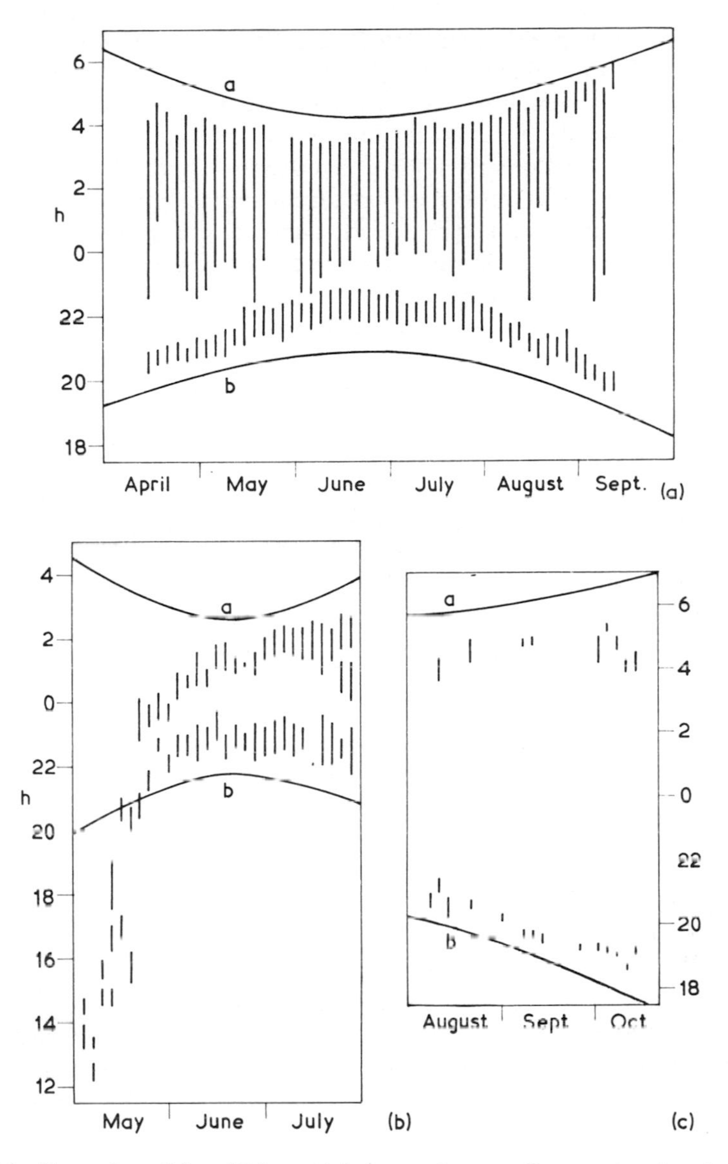

Fig. 7.1 Examples of bat flight activity records according to the time intervals between the first and last exit and first and last return flights (vertical bars). Explanations: ordinate, time (h); upper curves (a), times of sunrise; lower curves (b), times of sunset; (**a**) summer colony of *Myotis dasycneme* (after Voûte *et al.*, 1974); (**b**), summer colony of *Myotis mystacinus* (after Nyholm, 1965); (**c**), two individual *Myotis nattereri* (after Laufens, 1972). (Adapted and very schematized.)

lophidae. According to Laufens (1972) and Voûte *et al*. (1974), this occurs especially at the time of nursing the young.

An essential problem is provided by the relation of the endogenous circadian timing system to the L:D schedule and its changes. As pointed out by DeCoursey and DeCoursey (1964), light-sampling at the light-to-dark transition is the chief means of synchronizing an endogenous activity rhythm to the daily light cycle. Light-sampling precedes the proper exit flights and consists either in descending and waiting near the exit hole or in short exit flights followed by instant returns. Certain observations tend to show that light-sampling does not occur in bats roosting in tree holes and small cavities, but it is unknown to what extent light can be perceived by the roosting bats in such cases. Through individual registration of marked members of a small colony of *Myotis nattereri*, Laufens (1972) found that individual bats differ in the duration of activity, the first ones to fly out being the last to return, and *vice versa*, which corresponds with the general idea of endogenous activity rhythms. Besides the major synchronizer, L:D, the activity can be affected by some arythmic factors as sudden changes in ambient temperature, shower, etc.

Another approach towards activity studies is based on nocturnal observations and netting outside the roosting places. Nearly all such studies indicated that the activity, quantified by the number of observed or netted bats, is greater during the first half of the night (Aellen, 1962; Mutere, 1969; O'Farrell and Bradley, 1970; Kunz, 1973; Gaisler, 1973; Davis and Dixon, 1976). In some species, a smaller second peak of activity after midnight or before dawn was observed. Summer activity at hibernacula and in different temperate, subtropical and tropical habitats was described in a series of papers by Fenton. This author has developed an automated ultrasonic sensing system to detect the echolocating cries of bats (Fenton *et al*., 1973) which operates as a bat traffic counter. Kunz and Brock (1975) consider monitoring to be more satisfactory than netting or trapping; however, if the species under study must be identified with certainty, it is advisable to combine both methods (Fenton and Morris, 1976). The hitherto cited results pertain mostly to insectivorous bats. Of the non-insectivores, the vampire *Desmodus rotundus* has been studied in detail. Most individuals of this bat forage only once nightly during the darkest part of the night, that is, either before the moon rises or after it sets (Wimsatt, 1969; Greenhall *et al*., 1971; Crespo *et al*., 1972). Their activity patterns are also strongly affected by the activity of prey and an inverse relationship exists between vampire foraging and cattle grazing periods (Turner, 1975).

The data, obtained by various methods, suggest that the activity rhythms of bats inhabiting the same environments are not identical and that even minor differences can reflect partitioning of food and other resources (Fenton, 1974).

Feeding grounds

Bats forage in various environments from all types of forest ecosystems to city centres. They are absent only from polar regions, alpine zones of high mountains and over open sea. In general, bats seem to be more active along watercourses and in diversified habitats than over fields and other 'open' habitats or inside the densest forest stands (Brosset, 1966; Fenton, 1974).

To date, precise investigations on natural feeding behaviour have been made with only a few bat species. Nyholm (1965) studied the foraging activity of *Myotis mystacinus* and *M. daubentoni* in southern Finland. The bats first forage in forests, and from August onwards, over meadows and shores of lakes. The mean area of individual feeding grounds in forests is 240 m^2 (*M. mystacinus*) and 420 m^2 (*M. daubentoni*). Open feeding grounds are used by the whole colony and are 5450 m^2 (*M. mystacinus*) or 4580 m^2 (*M. daubentoni*). The mean distance from the shelter of the colony to the feeding ground is 187 m for *M. mystacinus* and 236 m for *M. daubentoni*. In forest habitats, these bats fly higher than in the open field.

Vaughan (1976) studied the activity of *Cardioderma cor* in the bushland in southern Kenya with the use of a night viewing device. These bats roost in hollow baobab trees and at night each individual occupies an exclusive foraging area. These territories are 8 m up to 1.2 km away from the daytime roosts. Foraging is characterized by long periods spent perching and listening for terrestrial prey with brief and quick flights to capture prey, usually covering less than 25 m. Seasonal shifts in foraging behaviour were observed associated with the pattern of alternating wet and dry seasons.

The two examples given above show, on the one hand, considerable differences in feeding strategies among bats and, on the other hand, their capability of changing feeding grounds and feeding behaviour during the year. Other authors studying the foraging activities of whole bat communities (Kunz, 1973; Bateman and Vaughan, 1974; Fenton, 1975) have ascertained seasonal changes in nightly dispersal behaviour. While some species were encountered in only a few habitats, others were widespread. In some cases, higher levels of flight activity were found near buildings than over adjacent habitats. The temporal and spatial components of bat activity may both be important strategies in the reduction of competition within sympatric associations of bats.

7.1.4 Daytime shelters during the active life

Classification

The most detailed and best substantiated classification of shelters is contained in the monograph by Verschuren (1957). It is based on three

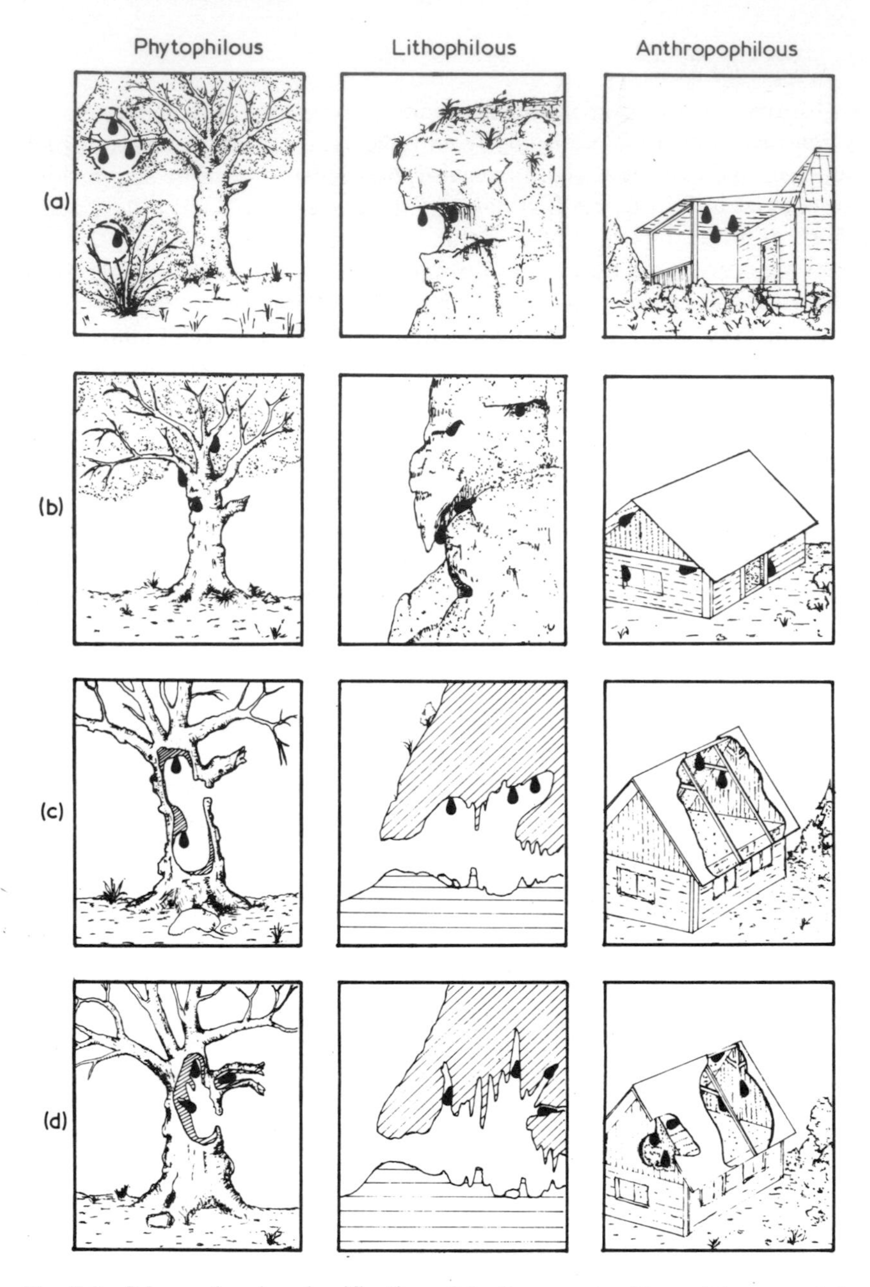

Fig. 7.2 Scheme showing classification of daytime roosts of bats: (a), external, free; (b), external, in contact; (c), internal, free; (d), internal, in contact. (Modified after Verschuren, 1957, Brosset, 1966 and Greenhall & Paradiso, 1968).

criteria: A, by the degree of isolation from the macro-habitat: (1) external, (2) internal; B, by the degree of bodily contact with the roost: (1) free, (2) in contact; C, by the environment of the micro-habitat (= roost): (1) phytophilous, (2) lithophilous, (3) anthropophilous. This classification was elaborated by subsequent authors (Rosevear, 1965; Brosset, 1966; Villa, 1966) and its scheme is outlined in Fig. 7.2. A systematic review of diurnal retreats of bats was presented by Dalquest and Walton (in Slaughter and Walton, 1970). These authors enlarged the classification by additional aspects illustrated by such terms as permanent/semipermanent/temporary roosts, or occurrence in aggregations as solitary bats. At the same time, they pointed out problems encountered when a static system of classification is applied to a dynamic process. In fact there are numerous transitions between the individual groups and certain species cannot be classified by the above schemes. Nevertheless, one can roughly say that the relative number of species that are internal and in contact increases in distances away from the equator and decreases towards the equator. Irrespective of latitude, bats that utilize caves or man-made structures tend to be more gregarious, both interspecifically and intraspecifically.

Free roosting sites

Nearly all free-roosting species are tropical. The group comprises, first of all, most members of Megachiroptera and, in tropical America, certain species of Phyllostomatidae. However, free-roosting species can be found in other families as well. Most frequently, free roosting sites are found in trees and/or shrubs where the bats hang themselves both among the foliage and freely on branches, trunks or even aerial roots. An interesting type of roost is provided by young, rolled-up leaves of bananas or other plants, used as shelter by *Pipistrellus nanus*, *Myotis bocagei* and *Thyroptera tricolor*. South American bats of the genera *Artibeus* and *Uroderma* show an exceptional behaviour, in which they gnaw through parts of leaves which then form a sort of roof over the roosting bats (Brosset, 1966). *Rousettus aegyptiacus* and a few other species can roost in entrances to caves or buildings (e.g., ancient ruins), and such places also can be considered free roosting sites. Occasionally, certain bats can also roost free in colder regions. This phenomenon is normal for five North American species of the genus *Lasiurus* which spend the daylight hours hanging in the foliage of trees or in the interior of the clumps of *Tillandsia* which festoon the trees (Barbour and Davis, 1969).

Tree and other natural holes and crevices

All big families of the order, except Pteropodidae, include many species

Fig. 7.3 Various types of bat boxes designed by many authors may serve as daytime shelters for certain species roosting in trees. The boxes are occupied mainly by small groups or individual bats and like tree holes, they are very frequently changed. (Photo by P. Eleder, scale 1:12.)

which roost in tree holes or under loosened areas of bark, and this type of roosting is common in both warm and temperate regions. For comparison, of the 38 bat species studied by Verschuren (1957) in the Garamba National Park (Zaïre), about 45 per cent roost in tree holes; of the 76 species in the palearctic region, this way of roosting is typical of about 26 per cent. The number of species roosting in trees is naturally proportional to the abundance of trees; hence, these species are absent from many arid regions. In these regions, certain species of the genera *Tadarida, Pipistrellus, Otonycteris* or *Antrozous* roost in rock crevices or under big boulders. Vaughan and O'Shea (1976) studied in detail the roosting ecology of *Antrozous pallidus* in Arizona and found that these bats roosted in crevices responsive to changes in ambient temperature. Even in woody areas, however, individual bats can use rock crevices as their daytime shelters.

Underground cavities: natural and man-made

Underground cavities harbour bats from all families. In cold regions, however, this stable environment is of greater importance as hibernating quarters than daytime shelters. So far, nobody has drawn the northern and southern limits of the zone in which bats form summer nursing colonies in caves; roughly estimated, the limits approximate the mean annual isotherm of 10°C. In colder regions, only single males and non-breeding females roost in caves and galleries during the non-hibernating phase of life. Of the families containing many species, the Pteropodidae show the lowest affinity to this environment. There is little difference ecologically between natural and man-made underground cavities; whether they are frequented by bats is governed by other attributes, such as size, situation and number of entrances, air currents, temperature and relative humidity. These features are treated in detail in the spelaeological literature. The precise characteristic of roosting is often specific. Extremes are provided by 'fissure bats' (hiding in crevices) and 'space bats' (hanging freely); for particulars, *see* Brosset (1966), Gaisler (1966) and Slaughter and Walton (1970). The numbers of bats found in underground cavities can be very high and on the average they are higher than in other types of shelters. The greatest colonies are known from caves in the south of the U.S.A. and in Mexico which may harbour several millions of *Tadarida brasiliensis* (Barbour and Davis, 1969). Huge colonies of bats are also known from the tropics and it is possible that summer aggregations of certain bats in caves represent the greatest societies a mammal species can form.

Buildings

A large number of originally lithophilous and phytophilous bat species have

Fig. 7.4 A typical summer nursing colony in a loft of an old building. In the present case, it is a mixed colony of two large European *Myotis* species: *M. myotis* and *M. blythi;* the arrow denotes one of the young-of-the-year. (Photo by Dr V. Hrabě, scale 1:10.)

adapted to utilize buildings. The adaptation process to roosting in buildings would appear to have not yet ended. Some species visit buildings only occasionally whereas others prefer this environment to natural shelters. The bats show remarkable plasticity in occupying a wide range of various types of human constructions. We have observed in Czechoslovakia that certain species (e.g., *Eptesicus serotinus*) can occupy quite modern urban houses within 5 years after their construction. By the variety of their roosting sites in buildings, bats probably surpass all other vertebrates (Fig. 7.4). Brosset (1966) distinguishes 12 major types of bat roosts in buildings, not including the cellars and similar underground spaces. The shelters in buildings offer considerable variety of ecological conditions, which in fact has been utilized by certain species for regular changes between two or more roosts according to external climatic conditions and intrinsic physiological state (Gaisler, 1963).

Permanent and temporary shelters

Probably no shelter is 'permanent' in the sense that the same individuals would keep to it throughout their lives. Generally, however, such shelters are considered permanent when they are occupied by the bats for several consecutive months in connection with a certain phase of their annual life cycle (reproduction, hibernation, etc.). Besides, there are many additional shelters that are used for one or a few days or weeks; such shelters are denoted as temporary, transitory, secondary, etc. Dalquest and Walton (in Slaughter and Walton, 1970) also pointed out the existence of night roosts used for rest stops during nightly feeding forays. At present, the problem of temporary shelters is currently under intense investigation by bat ecologists. It appears that most species possess such shelters irrespective of the preferred type of daytime roost or geographic origin. We cannot deal here with the results of the various studies in detail; in general, it may be stated that the temporary shelters belong to three types, *viz*., (1) localities ecologically similar to summer roosts; (2) localities ecologically similar to winter roosts; and (3) localities dissimilar to any of the two main roost types. The latter category comprises, for instance, mass 'invasions' of *Pipistrellus pipistrellus* to peculiar places of buildings in some Central European towns (Hůrka, 1966; Roer, 1974). An example of utilizing a series of temporary shelters by a bat population is given in Figure 7.12.

7.1.5 Hibernation

The problems of hibernation are closely connected with thermoregulation and metabolism in bats. This field comprises a large number of papers and the reader is referred to the very good summaries by Davis (hibernation) and Lyman (thermoregulation and metabolism), both in Wimsatt (1970). Bats show exceptional variability in the control of body temperature. They enter torpor more readily and arouse to the active state more quickly than other hibernators of comparable size. Bats are peculiar in many other respects concerning thermoregulation, and it has been proposed to transfer bats from the group of heterotherms to a separate group of 'chiropterotherms'. This term will probably never gain widespread acceptance, but it points to the adaptive and specialized, rather than primitive, character of bat thermoregulation (cf. Davis loc. cit., Lyman loc. cit., Dwyer, 1971; Kulzer, 1972; McNab, 1974).

Not all bats hibernate. The large Megachiroptera are true homeotherms;

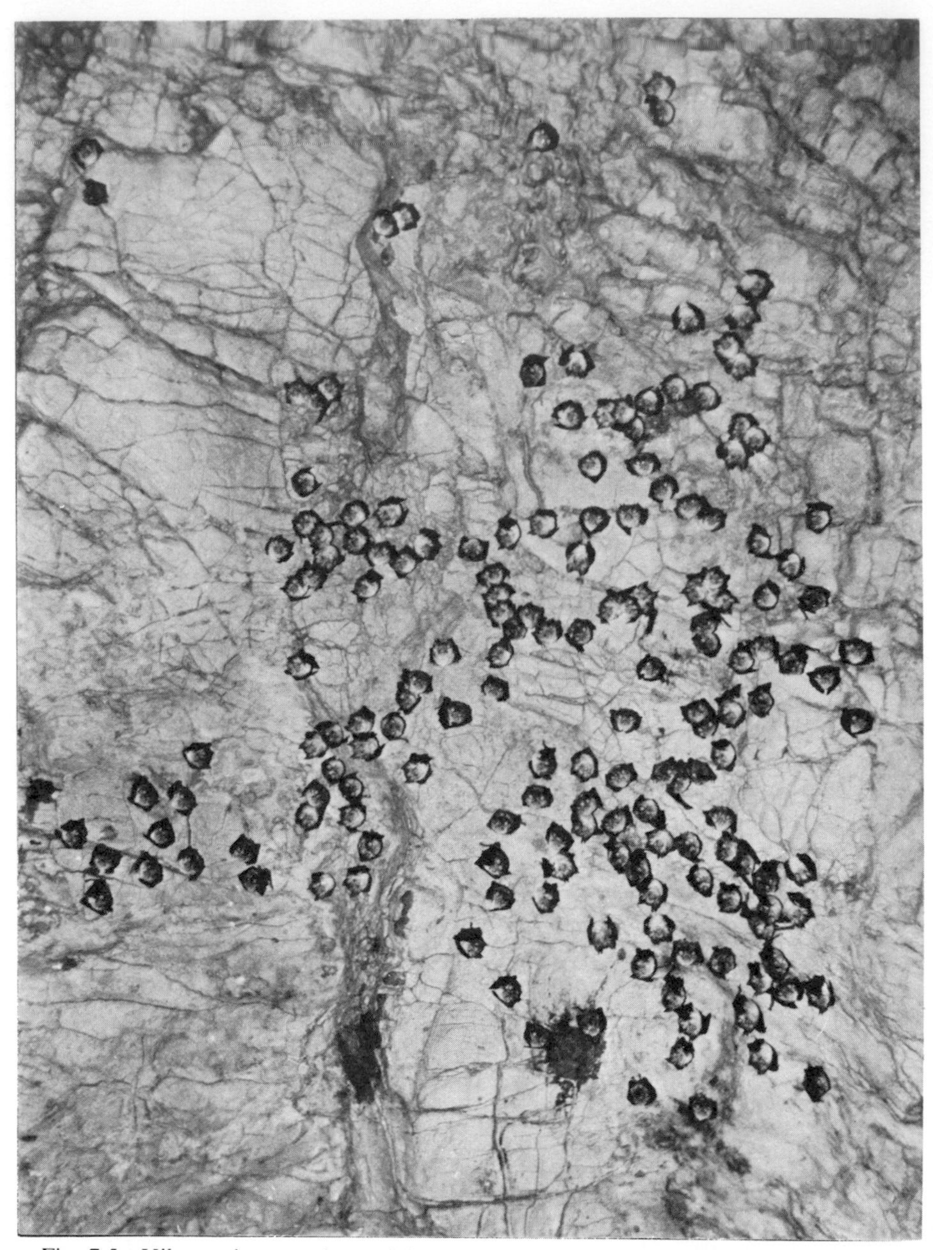

Fig. 7.5 Hibernating members of the genus *Rhinolophus* hang freely on the ceiling of their hibernaculum and individuals mostly do not touch one another. The picture shows a winter colony of *R. euryale* in a natural cave. (Photo by Dr V. Hrabě, scale 1:12.)

they are capable of maintaining constant body temperature and incapable of active hypothermia (Kulzer, 1965; Lyman loc. cit.). Concerning the tropical Microchiroptera, there is great variation in temperature regulation with some of the species or families (e.g., Megadermatidae) seeming to be nearly homeothermous, and others (e.g., Desmodontinae*) showing little ability to regulate their body temperature. All temperate zone Microchiroptera are 'deep' hibernators (Figs. 7.5 and 7.6). They enter lethargy, or torpor, not only during the winter season but even during the day, usually when they return from feeding forays. Thermoregulation of this type is the most pronounced in Vespertilionidae and Rhinolophidae. As far as one can conclude from available information, species of Vespertilionidae maintain the capability of entering dirunal lethargy even in the subtropics and tropics. In a limited extent, this is true of Rhinolophidae; particularly the members of Hipposiderinae do not enter lethargy as a rule. The temperature regulation of Molossidae appears to form a transition between that in truly tropical bats on the one hand, and the temperate zone bats on the other. Those that are limited to the tropics have some diurnal fluctuations of body temperature, the species that enter the temperate zone undergo a daily torpor, but prolonged hibernation is questionable.

The hibernating quarters could be classified similarly as the daytime roosts. There are, of course, no free roosting sites. Underground cavities and buildings are the most common winter quarters, especially in the coldest regions inhabited by bats. According to Dwyer (1971, caves serve mainly as hibernacula in regions with annual mean temperatures 2–12°C. They are dominated by bats that enter torpor readily. In regions with annual mean temperatures 12–22°C, the caves are dominated by bats which rarely hibernate although they are able to enter torpor voluntarily. Certain typically cave-dwelling bats behave variously according to the temperatures prevailing in their range; thus, *Miniopterus schreibersi* is a deeply hibernating species in Central Europe (Gaisler, unpublished observations) whereas in warmer regions it can behave as an obligatory or non-hibernating species (Dwyer, 1966).

The limit of the possible utilization of tree holes as hibernacula is shifted to low latitudes compared to caves, galleries and buildings. For instance, *Nyctalus noctula* can hibernate in hollow trees in the Netherlands (Sluiter and Heerdt, 1966) but cannot do so in the central regions of the European part of the U.S.S.R. where the species lives in summer (Strelkov, 1969). According to the latter author, the northern limit of this species' hibernating in tree holes approximates the mean January isotherm of −2 to −4°C.

*Throughout this chapter, the classification and nomenclature of Chiroptera by Koopman and Jones (in Slaughter & Walton, 1970) is adopted. Contrary to former alignment, Desmodontinae are not regarded as a separate family. Subsequently, two new families not recognized by Koopman and Jones (loc. cit.) are established, *viz.*, Mormoopidae (Smith, 1972) and Craseonycteridae (Hill, 1974) resulting in the total number of 17 recent bat families.

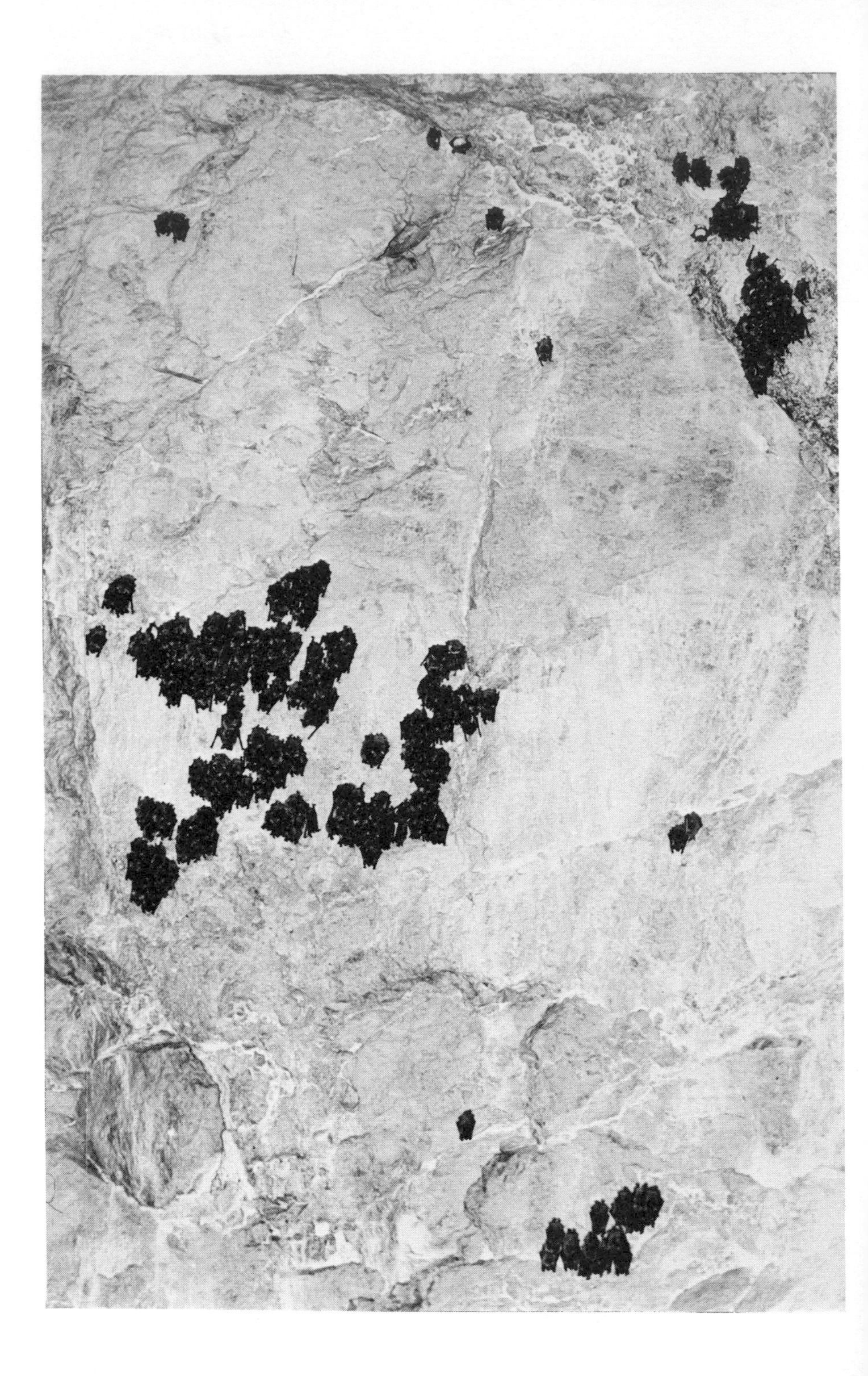

Fig. 7.7 Among other man-made structures, World War II underground fortresses are used as hibernacula by bats. In the mountain regions of Central Europe, species adapted to hibernate under low ambient temperature – such as *Barbastella barbastellus* and *Eptesicus nilssoni* – are common there. (Photo by the author, scale 1:60.)

Numerous American and European authors have studied the selection of hibernating sites particularly in underground cavities which are most accessible for examination (Fig. 7.7). The most detailed study in this field is by Bezem *et al.* (1974) who studied some characteristics of the hibernating locations of bats in the artificial caves in southern Netherlands. The authors conclude that the choice of a hibernation site in the cave is similar to the choice of a site in a summer roost, but it shows no connection with flying and hunting habits. Hibernation sites are most frequently on or in the walls

---

Fig. 7.6 Hibernating members of the genus *Myotis* often form closely aggregated groups. The picture shows a mixed winter colony of *M. myotis* and *M. blythi* in a natural cave. (Photo by J. Rys, scale 1:20.)

Table 7.1 *The distribution of feeding categories of bats by the zoogeographic regions and taxonomic pertinence. Modified after Wilson (1973) to whose paper the reader is referred as for the computation of individual criteria. In cases marked with an asterisk* the importance values are considered unrealistic as the evidence of natural feeding on fish in Palearctic bats is questionable.*

Faunal Region	Criterion		Feeding category					
			Insect	Fruit	Nectar	Carn	Fish	Blood
Nearctic	Number of:	Families	3	1	1	–	1	1
		Specialized genera	12	–	2	–	–	1
		Contributing genera	2	3	–	–	1	–
	Importance value		89.0	1.0	3.8	0	3.7	2.4
Neotropical	Number of:	Families	7	1	1	1	2	1
		Specialized genera	33	22	15	3	1	3
		Contributing genera	15	3	–	1	1	–
	Importance value		53.7	30.2	13.3	1.8	0.9	1.4
Palearctic	Number of:	Families	6	1	1	–	1	–
		Specialized genera	19	1	–	–	–	–
		Contributing genera	–	–	1	–	1	–
	Importance value		96.0	0.7	0.7	0	*2.6	0

Table 7.1 (*cont.*)

Faunal Region	Criterion		Feeding category					
			Insect	Fruit	Nectar	Carn	Fish	Blood
Ethiopian	Number of:	Families	8	1	1	1	1	—
		Specialized genera	29	10	3	1	—	—
		Contributing genera	3	1	2	—	1	—
	Importance value		80.1	13.8	5.5	0.3	0.3	0
Oriental	Number of:	Families	8	1	1	1	2	—
		Specialized genera	30	20	2	1	—	—
		Contributing genera	4	1	3	—	2	—
	Importance value		76.1	19.5	3.5	0.6	0.9	0
Australian	Number of:	Families	6	1	1	1	2	—
		Specialized genera	24	7	4	2	—	—
		Contributing genera	3	4	3	—	2	—
	Importance value		57.7	35.1	6.4	1.0	0.3	0
Total	Number of:	Families	15	2	2	2	2	1
		Specialized genera	82	53	23	6	1	3
		Contributing genera	23	7	3	1	2	—
	Importance value (mean)		75.5	16.7	5.5	0.6	*1.5	0.6

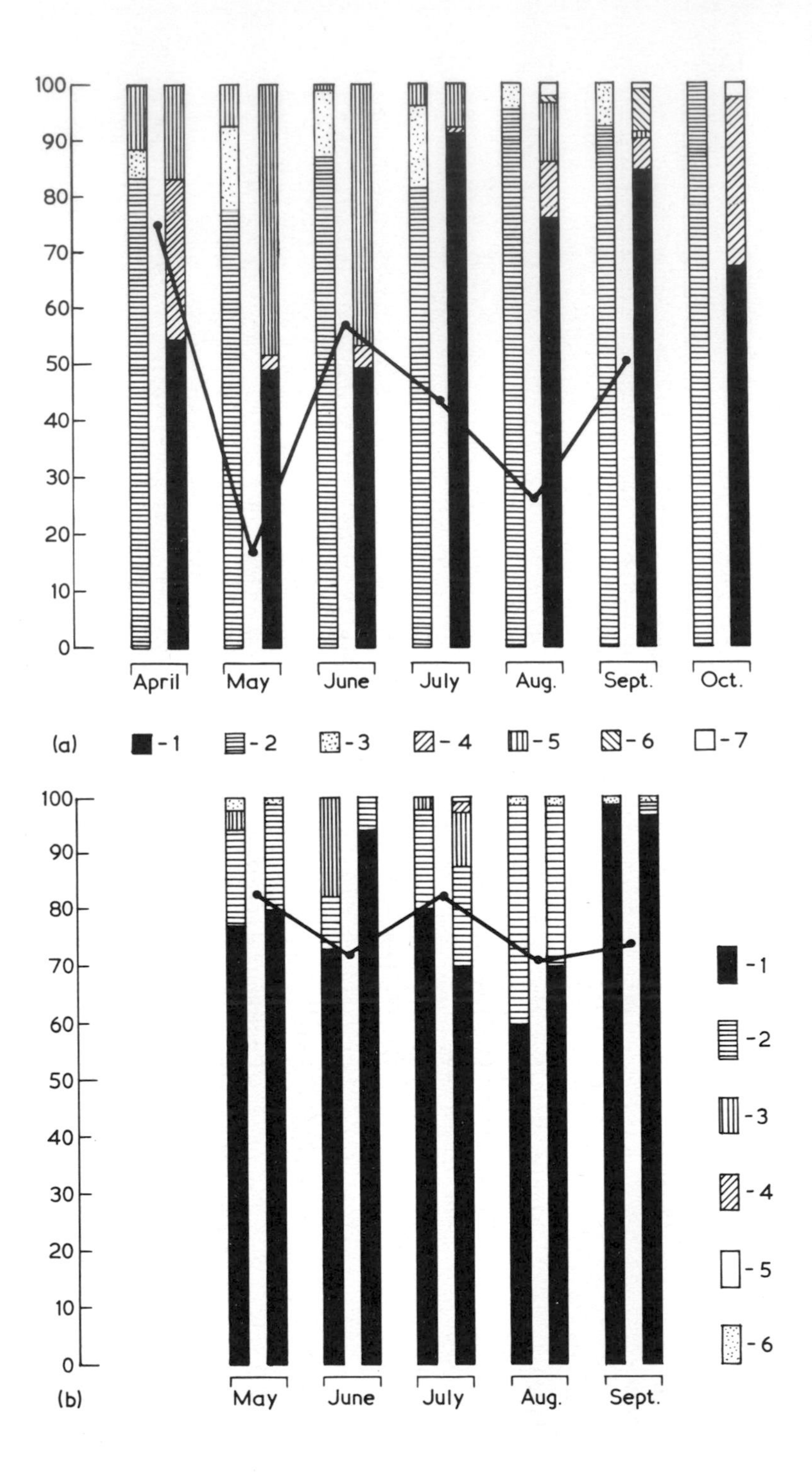

100
90
80
70
60
50
40
30
20
10
0
April
May
June
July
Aug.
Sept.
Oct.
(a)
-1
-2
-3
-4
-5
-6
-7
100
90
80
70
60
50
40
30
20
10
0
(b)
May
June
July
Aug.
Sept.
-1
-2
-3
-4
-5
-6

and ceilings. Cases of bats hibernating in protected situations in the floor of caves and mines have infrequently been observed in North America and Europe (Roer and Egsbaek, 1966; Davis in Wimsatt, 1970).

No mammals remain in hibernation during the whole hibernation season but all arouse from time to time. In bats, this problem has been investigated in detail by Daan (1973) whose investigation involved individual marking, automatic recording of intracave and extracave flights, and assessment of the frequency of movements between hibernation sites by short-interval searches. The author found that the frequency of intracave movements showed great seasonal variation, being relatively high in autumn and spring, and low in midwinter. The frequency of movements was positively correlated with ambient temperature. Weight loss of the bats was rapid in the first half of hibernation and occurred at a slower rate in the second half.

7.2 Diet

7.2.1 Food specializations

The diversity of trophic specializations within the order Chiroptera enticed many authors to review the matter (Eisentraut, 1951; Brosset, 1966; Glass in Slaughter and Walton, 1970; Wilson, 1973). The main feeding categories or trophic roles are as follows: insectivore, frugivore, nectarivore, carnivore, piscivore, and sanguivore. For completeness' sake, one should add that there are also species consuming centipedes, scorpions, crustaceans, etc.; the share of such items in the diet of the whole order is negligible, however. A review of the distribution of the various trophic categories is given in Table 7.1. Gillette (1975) presented a probable explanation of the evolution of feeding strategies in bats. He sees the starting point in generalized insectivory which evolved through specialized insectivory and food-source duality to alternate feeding strategies.

Insectivores

Insectivory is even at present the most widespread trophic specialization in bats and hence it has received the greatest attention. The basic methodical

Fig. 7.8 Examples of results of analyses of food habits in two insectivorous bats: (a), *Plecotus austriacus* and (b), *Myotis myotis*. Explanations: ordinate, percentage representation of food items. – (a): in each month, first column relates to discarded indigestible parts of insects; second column, to faeces; 1, Lepidoptera; 2, Noctuidae; 3, Lepidoptera minus Noctuidae; 4, Diptera; 5, Coleoptera; 6, Heteroptera; 7, other insect orders; heavy curve, representation of Noctuidae in the catch in a light trap. – (b): in each month, first column relates to stomach contents; second column, to faeces; 1, Carabidae; 2, Scarabeidae; 3, Coleoptera minus Carabidae and Scarabeidae; 4, Diptera; 5, Lepidoptera; 6, Araneidea; heavy curve, representation of adult Carabidae in the catch in pitfall traps. (Based on the original by Z. Bauerová.)

problem is associated with the identification of food items which is difficult due to the fineness of mastication and the high digestion rate. The most frequent methods include (1) analyses of remains of the prey (wings or similar fragments discarded by bats); (2) analyses of stomach contents; (3) analyses of faeces. The first method is the easiest but available with only a few species, e.g. of the genus *Plecotus* (cf. Roer, 1969). Even in these bats, however, the information on the diet, obtained in this way, is incomplete; a comparison with data obtained by faecal analyses carried out in our laboratory has shown that important food items are not discarded by bats and therefore not found under the feeding places (Fig. 7.8a).

The second method provides good information on the identity of the predator and on individual food consumption but can be employed only if the stomachs are examined shortly after the end of feeding activity. Since the relative difficulty of determining the kind and number of arthropods eaten from faecal materials is not greater than from stomach contents, the third method is the most frequently used at present. Bats netted during feeding activity are held over-night to collect their faeces, and then released. An excellent summary of these problems is provided by Fenton (1974). This author also states the mean food consumption of insectivorous bats as between 0.12 and 0.24 g food per gram of body mass per day. While earlier authors merely listed the taxa represented in the diet, the current workers attempt to express the relationships between the predator (bat) and prey (insect) communities. As the work of Ross (1967), Black (1972, 1974) and Hussar (1976) shows, moths and beetles generally contribute the greatest part to diets of temperate insectivorous bats; most bats can be classified as either moth or beetle strategists. Additional insect orders represented in the diet of bats include Hymenoptera, Diptera, Homoptera, Hemiptera, Neuroptera, Trichoptera, Orthoptera and Odonata. Diptera may strongly predominate in the diet of smaller bat species (Sologor and Petrusenko, 1973; Belwood and Fenton, 1976), and Ephemeroptera in that of species foraging over water bodies. Competition for food is mostly avoided through different foraging strategies and partial specialization to different size and/or taxonomic groups of insects. Fenton and Morris (1976) used a 'black light' to attract insects and found that bats were significantly most active during periods when the light was on and insects were aggregated over it. This opportunistic feeding, however, is not incompatible with selective feeding and both strategies can be used by the same species according to the local ecological conditions. Particular note must be made of bats collecting prey from the soil surface. This behaviour had long been known in the North American *Antrozous pallidus* (cf. Barbour and Davis, 1969) and was discovered and thoroughly investigated in the European *Myotis myotis* by Kolb (1959, 1973). The share of food items collected from soil surface by the latter species varies with the seasons, but in general Carabidae predominate, many species of which are rarely or never on the wing (Fig. 7.8b). Collecting prey from the ground is known to occur in tropical bats as well.

The diet of tropical insectivorous bats has mostly been studied simultaneously with other trophic specializations and the published reports frequently refer also to feeding grounds. McNab (1971) compared the trophic structure of several tropical bat faunae and found that insectivorous species account for 50–78 per cent. The two most important parameters of food partitioning are the type of food and food particle size. Whitaker and Black (1976) studied the diet of 10 species of cave bats from Zambia and found that their food habits were similar to those of some north temperate faunae. As far as the taxonomic diversity of prey exploited is the measure of the width of the ecological niche of the exploiting species, that of the tropical bats seems to be narrower than that of temperate ones.

Other food specialists

As shown in Table 7.1, the remaining trophic specializations are invariably confined to only one or two families of bats. Frugivory (feeding on fruit) and nectarivory (feeding on flowers, nectar, and pollen) are rather widespread specializations which occur in two ecologically parallel groups, the Old World Pteropodidae and the New World Phyllostomatidae. Most members of the large family of Pteropodidae feed exclusively on fruit. Specialized nectarivores are found namely among members of the subfamily Macroglossinae but occasional visits to flowers (of trees) is known in some other fruit bats as well (Kock, 1972). Many fruit bats feed on such plants of economic importance as banana, pawpaw and mango (Jones, 1972; Mutere, 1973).

The food of the family Phyllostomatidae has been systematically described by Gardner (in Baker *et al.*, 1976). This family displays a wide variety of food preferences, and relatively few species are restricted to a specific dietary regime. Frugivory is often coupled with nectarivory and sometimes also insectivory and carnivory. Carroliinae and Stenoderminae are the most pronounced frugivorous subfamilies; Glossophaginae are specialized flower nectar feeders. The relationships between resource utilization and foraging patterns in some species of this family have been studied by Heithaus *et al.* (1975). Floral resources were found to be seasonally abundant and competition for them was minimal, whereas fruit resources were more evenly available through the year and competition for them was more important in determining the species diversity of bats. Small species of bats fed on resources of high abundance, whereas large species utilized resources that were patchy in time and space. Information on specialized nectarivorous habits of bats is summarized by Howell (1976) who also deals with problems of pollination and chiropterophily of plants.

The remaining trophic roles are of minor importance. The carnivorous species feeding on other bats, birds, small terrestrial mammals and lizards are represented by the Old World Megadermatidae and New World Phyllostomatinae. All carnivorous species are tropical; many of them partly feed on arthropods as well. The peculiar New Zealand species, *Mystacina tuberculata,* which normally feeds on arthropods, may occasionally feed on carrion (Brosset, 1966).

Piscivory is practised by the genus *Noctilio* of the family Noctilionidae. *Noctilio leporinus* feeds mainly on freshwater fish but has been observed fishing in salt water as well (Goodwin and Greenhall, 1961). Another American species, *Myotis vivesi*, feeds mainly on marine fish and crustaceans which it collects along sea coasts and gulfs (Villa, 1966). Further *Myotis* spp. of the subg. *Leuconoë,* which have elongate feet, offer clues to the origin of fishing behaviour in bats (Dwyer, 1970) but none of them can be considered a specialized piscivore.

Three single species genera of the subfamily Desmodontinae are the only bats – and the only terrestrial vertebrates – specialized to feed exclusively on blood. They are restricted to the neotropical region, only one of them having penetrated into the south of the nearctic. Owing to the great epidemiological importance the most abundant species, *Desmodus rotundus*, has been studied in great detail. It attacks a large number of mammals and, under laboratory conditions, even other vertebrates, but most importantly it feeds on domestic cattle (Greenhall, 1972; Turner, 1975).

7.2.2 Water intake

Bats have frequently been observed to drink by skimming close to the surface of water and lapping it up while in flight, but detailed information on their drinking behaviour is scarce. Rosevear (1965) suggests that insectivorous bats show higher consumption of fluids than do frugivorous ones. Roer (1970) demonstrated that even insectivorous bats can be adapted to life in arid regions to such an extent that they can do without water for several months if they find enough insects. On the other hand, Greenhall (1972) found that vampire bats required to drink; a curious observation in view of their highly fluid diet. Greenhall also observed the bats licking dew and drinking rain water from puddles, and this behaviour may be common in many other bat species.

7.2.3 The position of bats in food chains

Since insectivorous bats are by far the most numerous, I shall concentrate on this trophic group. Although some information is available on predator-prey relationships (Black, 1974; Fenton, 1974), it appears that so far nobody has described the position of bats in view of the concepts of trophic levels and food chains.

Insectivorous bats are secondary consumers or first order predators. It is thought that they participate in the consumption of insect biomass to a very small extent, at least during the peak of the growing (temperate zone) or rainy (tropics) season. At that time they cannot essentially affect even that part of the insect community which has practically no other predators than bats. Their special ecological niche enables them to be isolated from predators. Although Gilette and Kimbrough (in Slaughter and Walton, 1970) list a large number of vertebrate predators which occasionally prey on bats, cases of occurrence of larger numbers of bats in the diets of carnivores are scarce. Thus it appears that the food chain involving bats is rather short, as it is 'prematurely' terminated by decomposition. Dead bodies of bats and notably their faeces are the trophic bases of certain decomposers. This part is of particular importance in the ecosystems of many caves where bat guano is the primary source of nutrition for a large number of organisms (Horst, 1972). Thus it is probable that as regards the energy flow through ecosystems, bats occupy a special position which could be tentatively denoted as a 'lateral trophic chain'. Verification of this concept will require greater application of the synecological approach not only of bats together with their trophic base, but also with additional secondary consumers of similar size.

7.3 Reproduction

7.3.1 Sexual cycles

The main features of the sexual cycle of bats have been known for over fifty years. A modern ecological approach to the problems involved, however, dates from the study by Pearson *et al*., (1952). A synopsis of the major findings can be found in Brosset (1966), Wimsatt (1969), Saint Girons *et al*., (1969) and Carter (in Slaughter and Walton, 1970).

The male

In a large majority of hibernating bats, particularly of the families Vespertilionidae and Rhinolophidae, spermatogenesis begins shortly after the end of hibernation and culminates in late summer; in the northern hemisphere this is in August or September. It is followed by a rapid regression of testes but the spermatozoa are stored in the cauda epididymidis throughout winter. Males retain their capacity of inseminating females throughout the winter and in certain species inhabiting northern regions copulations are more

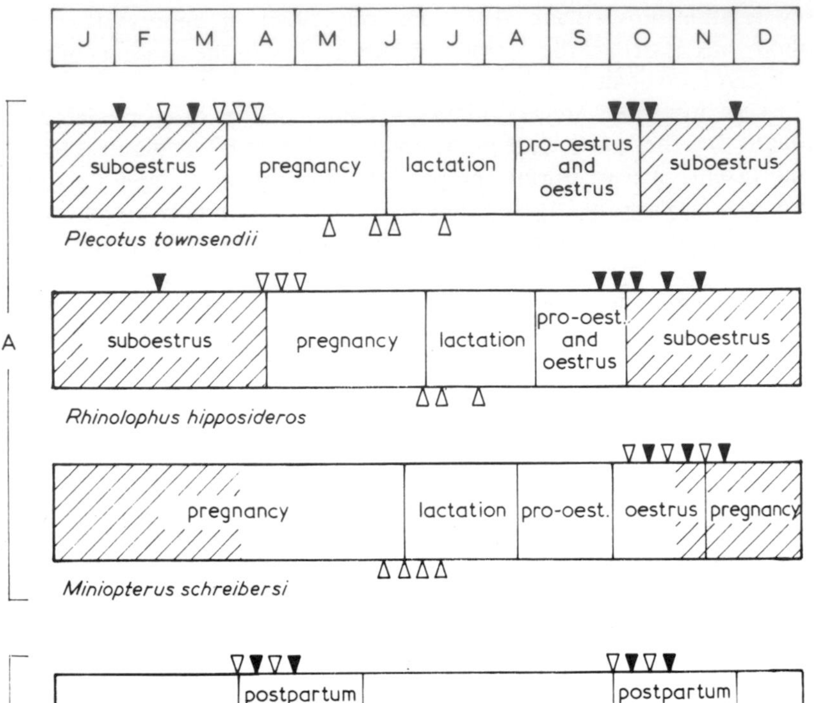

J F M A M J J A S O N D
suboestrus
pregnancy
lactation
pro-oestrus and oestrus
suboestrus
Plecotus townsendii
suboestrus
pregnancy
lactation
pro-oest. and oestrus
suboestrus
Rhinolophus hipposideros
A
pregnancy
lactation
pro-oest.
oestrus
pregnancy
Miniopterus schreibersi

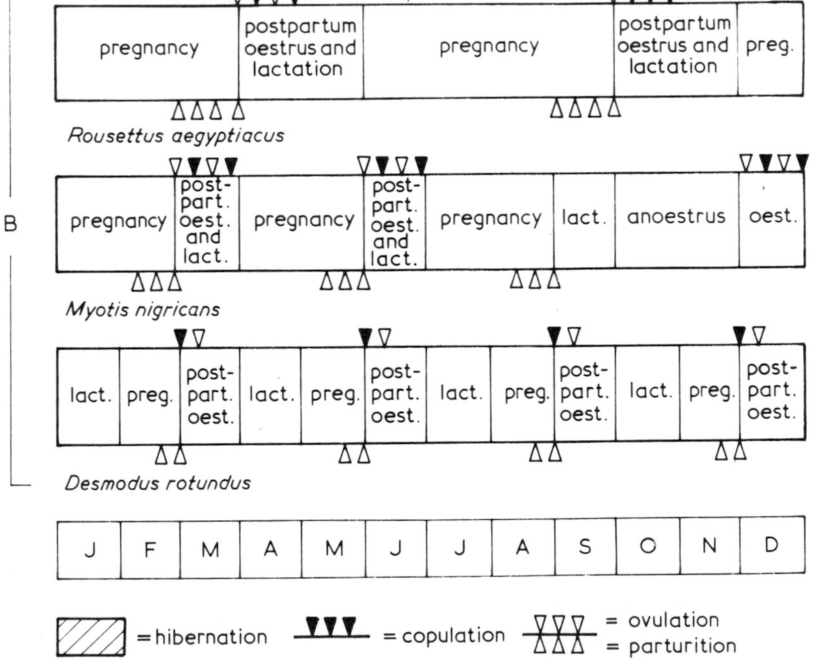

pregnancy
postpartum oestrus and lactation
pregnancy
postpartum oestrus and lactation
preg.
Rousettus aegyptiacus
B
pregnancy
post-part. oest. and lact.
pregnancy
post-part. oest. and lact.
pregnancy
lact.
anoestrus
oest.
Myotis nigricans
lact.
preg.
post-part. oest.
lact.
preg.
post-part. oest.
lact.
preg.
post-part. oest.
lact.
preg.
post-part. oest.
Desmodus rotundus
J F M A M J J A S O N D
= hibernation
= copulation
= ovulation
= parturition

frequent in winter than in autumn (Strelkov, 1962). The problem of hormonal control of prolonged male sexual activity is complex and its analysis is beyond the scope of the present chapter. The highest levels of androgens are usually synchronous with spermatogenesis, but at least in some species these hormones continue to be produced by the testes during winter (Racey and Tam, 1974). The accessory glands are fully developed during most of the year, but show involution between the end of hibernation and the middle or end of summer. In the genus *Rhinolophus* there is a special unpaired glandula urethralis which undergoes a pronounced seasonal cycle and participates in the formation of the vaginal plug (Gaisler, 1966).

In non-hibernating bats, spermatogenesis is either seasonal or aseasonal. The males of many species having cyclical spermatogenesis store spermatozoa for a considerable time and in Vespertilionidae sperm storage appears to be associated with heterothermy (Racey *et al.*, 1975). Males of other tropical species store spermatozoa only temporarily (Carter in Slaughter and Walton, 1970). It appears that in many non-hibernating tropical bats copulations occur shortly after maximum spermatogenesis, as in most other mammals.

The female

The female sexual cycle of hibernating bats is still more peculiar than that of males. The time elapsing between insemination and parturition can be very long and this prolongation is attained in various ways. The most common is through delayed fertilization which occurs in most Vespertilionidae and all hibernating Rhinolophidae. Viable spermatozoa are stored in the female reproductive tract for a period up to six months and it is only after the end of hibernation that ovulation, fertilization and gestation take place. Racey (1975) has summarized the problems of the prolonged survival of spermatozoa in bats and demonstrated that the lining epithelia of the sperm-storing organs secrete substances which may be of nutrient value to the sperm. Gestation in these species lasts two and a half or three months, and lactation one or two months. There is a very short (if any) anoestrus followed, in the late summer and autumn, by oestrus which is prolonged throughout the winter as suboestrus or 'submaximal oestrus' (Carter in Slaughter and Walton, 1970). There is some evidence of the hormonal stimulation of the

---

Fig. 7.9 Examples of breeding cycles of selected bat species: A, hibernating; B, non-hibernating. Research areas and references pertaining to the species: *P. townsendii,* USA, California (Pearson *et al.*, 1952); *R. hipposideros*, Czechoslovakia (Gaisler, 1966); *M. schreibersi,* France (Brosset, 1966); *R. aegyptiacus,* Uganda (Mutere, 1968); *M. nigricans*, Panama (Wilson & Findley, 1970); *D. rotundus*, Neotropical America (Wilson, 1973). First and last line indicate months. The timing is only approximate, especially in *D. rotundus* with aseasonal polyestry.

prolonged female oestrus but the whole mechanism is not yet clear. The female sexual cycle is controlled by external factors which may markedly affect its timing (Wimsatt, 1969; Racey, 1977).

Another phenomenon, long known to occur in *Miniopteus schreibersi* in Europe and Australia and discovered rather recently in American bats of the genera *Macrotus* and *Artibeus* (Bleier, 1975), is delayed development. In this case, fertilization of the egg follows copulation and the blastocyst develops and implants in the uterine endometrium, but development ceases or proceeds very slowly. Since insemination takes place in autumn as a rule, and parturition in early summer, pregnancy lasts 8 to 9 months.

A still longer period of pregnancy sometimes results from delayed implantation of the blastocyst. This phenomenon, however, appears to occur only in tropical bats, e.g. the African pteropid, *Eidolon helvum* (Mutere, 1967). Thus, a prolongation of the period between insemination and parturition may occur even in non-hibernating bats. Although the knowledge of the female sexual cycle in tropical bats is limited for the time being, it appears that in most of them copulation, fertilization and pregnancy follow without interruption. Parturition can be followed by either an anoestrus or a postpartum oestrus.

Breeding cycles (Fig. 7.9)

All temperate zone bats are monoestrous and produce young once a year, at the beginning of the period of maximum food supply. Wilson (1973) grouped the reproductive behaviour of neotropical bats in four categories, *viz.*, (1) aseasonal polyoestry; (2) seasonal polyoestry; (3) bimodal polyoestry; and (4) seasonal monoestry. Similarly, Anciaux de Faveaux (1973) distinguished, in African bats, (1) a polyoestrous type with uninterrupted reproduction; (2) a polyoestrous type with seasonal reproduction (two cycles a year); (3) a monoestrous type. Basically, both classifications agree and can be applied to tropical bats in general. The types of breeding cycles of tropical bats do not appear to be primarily confined to different families. Thus, the African families Pteropodidae, Vespertilionidae, Molossidae and Emballonuridae comprise species belonging to all groups. Brosset (1966) and Anciaux de Faveaux (loc. cit.) also attempted at defining 'l'équateur biologique', determining the boundary between species showing the 'boréal' type cycle and those showing the 'austral' type one. Obviously, this 'équateur' does not coincide with the geographic equator and differs in different species or even populations.

7.3.2 Effect of reproduction

Number of young

Contrary to the majority of small mammals, bats are mostly monotocous, i.e., they produce one young per pregnancy. As far as is known, one young per pregnancy is produced by all species of Rhinopomatidae, Emballonuridae, Noctilionidae, Nycteridae, Megadermatidae, Rhinolophidae, Phyllostomatidae, Natalidae and Molossidae (Brosset, 1966; Carter in Slaughter and Walton, 1970). This is also true of most species of Pteropodidae, of which at least two commonly have twins, however. Also Vespertilionidae most frequently have one young but the number of species with two young per litter is relatively large here. Twins are common in the genera *Nyctalus, Pipistrellus* and *Vespertilio* and certain species of *Eptesicus* and *Myotis*. According to Christian (1956), *Eptesicus fuscus* has twins in the eastern United States, but has only a single young in the western U.S.A. Differences in this respect have also been ascertained between different populations of the European species *Nyctalus noctula* and *Pipistrellus pipistrellus*, the frequency of two embryos or young increasing approximately from the west to the east and from the south to the north. The incidence of twins, however, is nowhere a hundred per cent; it varies, e.g. in the European part of the U.S.S.R. around 50–75 per cent (Panyutin, 1970; Rachmatulina, 1971).

Three embryos per female have occasionally been found in several species. Such cases, however, are very rare except in the North American bats of the genus *Lasiurus*. These bats can probably bear even more than three young. Barbour and David (1969) state, for *L. borealis*, an average of 3.2 embryos per female, and Hamilton and Stalling (1972) observed two cases of a female with 5 clinging young.

Sexual maturity

The available information on the onset of sexual maturity is rather controversial. In particular, it is difficult to define the onset of sexual maturity in females with delayed fertilization; also, insufficient knowledge of the age of an individual may lead to certain misinterpretations concerning the age at sexual maturity of both male and female bats. It appears from the published reviews (Brosset, 1966; Carter in Slaughter and Walton, 1970) that most bats attain sexual maturity during the second year of life. This has been documented, above all, in a number of American and European species of the genus *Myotis*. Dinale (1968) believes that the onset of sexual maturity is positively correlated with body size and supports his view by observations from among members of the genus *Rhinolophus*. According to that author, *R. ferrumequinum*, the largest European species, attains sexual maturity only when 3–4 years old. Although the latter date may be overestimated, the onset of sexual maturity later than two years of age cannot be ruled out, particularly in certain large bats in the tropics.

The onset of sexual maturity during the first year of life has been evidenced for the females of the following hibernating species: *Plecotus townsendii* (Pearson *et al*., 1952), *Tadarida brasiliensis* (Barbour and Davis, 1969), *Nyctalus noctula* (Cranbrook and Barrett, 1965; Racey and Kleiman, 1970), *Rhinolophus hipposideros* (Gaisler, 1966), *Pipistrellus pipistrellus* (Rachmatulina, 1971) and *Myotis velifer* (Kunz, 1973). Also the males of most of these species attain sexual maturity at less than one year of age, but male *P. townsendii* and *R. hipposideros* enter breeding only during their second year of life. On the other hand, male *Plecotus auritus* attain sexual maturity during their first, and females during their second year of life (Stebbings, 1966). The share of yearlings in breeding is rarely a hundred per cent and frequently it is very low; for instance, our observations in Czechoslovakia show that of yearling females, about 80 per cent enter breeding in *N. noctula* but only 15 per cent in *R. hipposideros*. In this respect, different populations of the same species can show appreciable differences. Maturation during the first year of life is perhaps more widespread than is supposed, and very likely it occurs in many tropical species. So far it has been demonstrated in only one species, *Myotis nigricans* (Wilson, 1971).

Reproductive rate

Besides the litter size and age at sexual maturity, the breeding rate of females and juvenile mortality are indices of breeding success. Since no general information is available, mention should be made of the results of several studies, although their data include various aspects and may be subject to observation errors. According to Pearson *et al*. (1952), a colony of 100 female *Plecotus townsendii* produces about 45 young females each year. According to Davis *et al*. (1962), virtually all female *Tadarida brasiliensis* coming north to Texas bear young every year, and according to Herreid (1967), the juvenile mortality in the same species during the first two months of life is 1.3 per cent. On a broadly theoretical basis, Bezem *et al*. (1960) calculated juvenile mortality rates during the first six months of life in certain European *Myotis* spp. to be 10–50 per cent. According to Panyutin (1970), up to 50 per cent of young of the woodland bat species inhabiting the European part of the U.S.S.R. die before attaining independence. According to Rachmatulina (1971), 72–100 per cent of adult female *Pipistrellus pipistrellus* reproduce in various parts of its range. Roer (1973) found the juvenile mortality of *Myotis myotis* to vary between 2.2 and 43.1 per cent, depending upon climatic factors. Kunz (1974) summarized data on early mortality in *Eptesicus fuscus* as follows: prenatal mortality, 10–34 per cent; mortality from birth to weaning, 7 per cent; and mortality from birth to one year, 40 per cent.

In polyoestrous species, the natality per unit time is higher. Adult female *Myotis nigricans* can produce young three times per year (Wilson and Findley, 1970) and in *Eonycteris spelaea* more than 50 per cent of the adult females are pregnant or lactating at any time of the year (Beck and Lim, 1973). In tropical environments, however, the juvenile mortality rate is higher (Brosset, 1966) so that the annual reproductive output can be similar. Gathering up all the available information on the effect of reproduction in Chiroptera, one may assume that the annual population increment varies between 0.5 and 1.5 living individuals per one adult female. As pointed out by Herreid (1964), this is an exceedingly low rate compared to other mammals of similar size.

7.4 Population

7.4.1 Population size

Most of the information on population sizes of individual bat species is based on observations of groups in shelters. Such groups are currently denoted as colonies. As mentioned already, of greatest functional importance among them are aggregations of females bearing and rearing young, which Eisentraut (1937) denoted as Wochenstuben and which are called 'maternity', 'nursing' or 'breeding' colonies in the English literature. Additional types of bat colonies are the adult (involving both sexes), male, juvenile, mating, wintering (hibernating), or transient (Dwyer, 1966; Gaisler, 1966). Not all species form all types of colonies. In many of them, the males hide individually or only a small number of them are mingled with the females. There are a few species which do not form colonies at all (Brosset, 1966).

Various methods have been used to estimate the size of a bat colony, including visual census, random sampling combined with area estimates, photographic recording, etc. Certain authors based their computations on mark-and-recapture data and used the simple Lincoln index or derived and more complicated statistical procedures (Bezem *et al.*, 1960; Tinkle and Milstead, 1960; Constantine, 1967; La Val, 1973). These studies resulted in estimates of population size in selected summer and/or winter quarters, but yielded little information concerning actual population densities. While it is very difficult to estimate population densities of bats the author of this chapter considers any attempts to do so very useful, provided that they are based on good knowledge of the ecology of the respective species.

Table 7.2 gives a survey of such data considered sufficiently reliable by

Table 7.2 *Population density estimates (n per ha.) for several bat species. In cases where the authors give more than one value, a mean was computed or the more representative data were chosen. As the sizes of areas under study are not always stated in the respective papers they cannot be given here. In the cases of* L. borealis *(Iowa),* P. pipistrellus *(Sibiu),* N. noctula *(Sibiu) and* T. tricolor *(Puntaneras) the values relate to small areas of local concentration of the species.*

Species	Density	Country	Method of estimating	Reference
Plecotus townsendii	0.01	USA, California	Sampling in buildings and mines	Pearson *et al.*, 1952
Plecotus townsendii	0.02	USA, Kansas & Oklahoma	Sampling in caves and over streams	Humphrey and Kunz, 1976
Miniopterus schreibersi	0.03	Australia, NS Wales	Sampling in caves and mines	Dwyer, 1966
Myotis lucifugus	0.1	USA, New England	Sampling in buildings	Barbour and Davis, 1969
Lasiurus borealis	2.5	USA, Lewis (Iowa)	Sampling in trees and visual observation	Barbour and Davis, 1969
Tadarida brasiliensis	1.4	USA, Texas	Trapping and netting, mostly in caves	Barbour and Davis, 1969
Tadarida brasiliensis	0.8	USA, Texas, N. Mexico, Oklahoma, Kansas	Trapping and netting, mostly in caves	Packard and Mollhagen, 1973
Myotis mystacinus	0.1	USSR, central zone	Sampling in trees, nest boxes and buildings	Panyutin, 1970
Plecotus auritus	0.1	USSR, central zone	Sampling in trees, nest boxes and buildings	Panyutin, 1970

Table 7.2 (*cont.*)

Species	Density	Country	Method of estimating	Reference
Vespertilio murinus	0.5	USSR, central zone	Sampling in trees, nest boxes and buildings	Panyutin, 1970
Pipistrellus nathusii	1.1	USSR, central zone	Sampling in trees, nest boxes and buildings	Panyutin, 1970
Pipistrellus pipistrellus	1.1	USSR, central zone	Sampling in trees, nest boxes and buildings	Panyutin, 1970
Pipistrellus pipistrellus	3.0	Rumania, camping site near Sibiu	Visual observation and netting	Gaisler, unpubl.
Nyctalus noctula	2.0	Rumania, camping site near Sibiu	Visual observation and netting	Gaisler, unpubl.
Nyctalus noctula	0.7	USSR, central zone	Sampling in trees and nest boxes	Panyutin, 1970
Nyctalus leisleri	0.1	USSR, Voronezh Res.	Sampling in trees and nest boxes	Panyutin, 1970
Pteropus mariannus	0.6	USA, Guam Island	Visual census on trees	Perez, 1973
Thyroptera tricolor	21.9	Costa Rica, Puntaneras	Sampling in rolled leaves	Findley and Wilson, 1974
Eptesicus serotinus	0.2	Czechoslovakia, Brno	Visual census of hunting bats	Gaisler, unpubl.

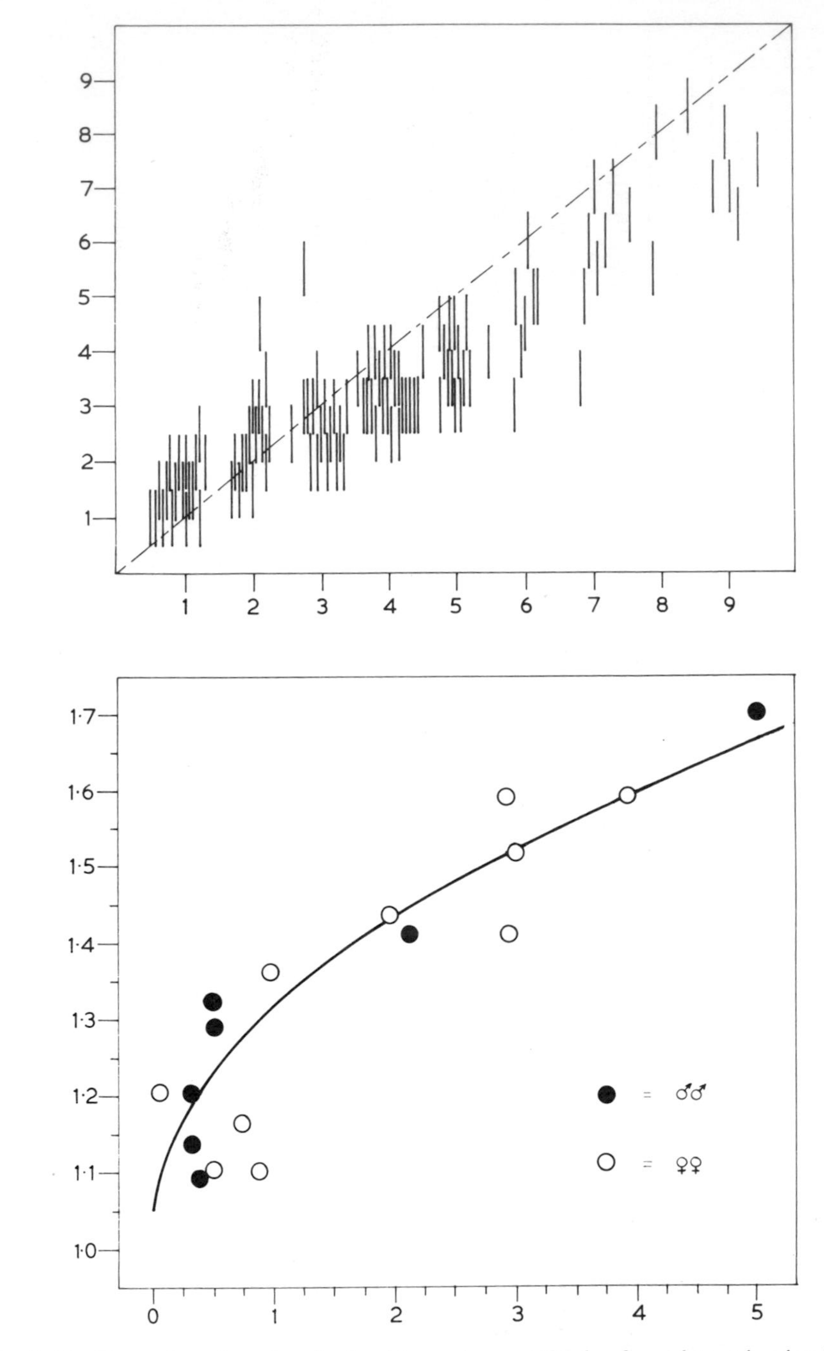

Fig. 7.10 Examples of results obtained by various methods of age determination in bats: A, *Myotis myotis*; B, *Nyctalus noctula*. Explanations for A: ordinate, age

the present author. All numbers have been converted to the standard unit of area used in population studies on small terrestrial mammals. The available information suggests that bat population densities range between 0.01 and 20.0 individuals per ha. Everybody who has observed – and, especially, netted – bats in warm regions knows that the densities there are much higher than in temperate regions. The greater community size in some tropical habitats depends, however, on the larger number of species sharing the same habitat.

In conclusion, it is necessary to emphasize that the pattern of dispersion of most bat populations differs from that of most other small mammals. Bats are not only highly concentrated in their shelters but their general population densities vary in a geographically patchy, rather than clinal, fashion (Humphrey, 1975). The regions of relative concentration may be surrounded by areas of very scarce occurrence. The infuence of density-dependent factors on bat population dynamics cannot be ruled out (Mills *et al.*, 1975) but it is probably smaller than in terrestrial mammals.

7.4.2 Population structure

Sex ratio

The secondary sex ratio, calculated by studies on foetuses, neonates and suckling young, is 1:1 for most species (Dwyer, 1966; Gaisler and Klíma, 1968; Panyutin, 1970; Sluiter *et al.*, 1971; Kunz, 1973; Mills *et al.*, 1975). Although significant differences have been observed in sufficiently large samples of neonates or juveniles of certain bat species (Wilson, 1971, La Val, 1973), they do not exceed ± 10 per cent as a rule and sometimes occur in only one population, while in others a balanced sex ratio is found.

The literature on the sex ratio among adults (i.e. tertiary sex ratio) is very extensive, apparently because the sex ratio is the most easily determined population parameter. Situations in which samples from all seasons reveal statistically balanced sex ratio, such as in *Miniopterus schreibersi* (Dwyer, 1966) are rather scarce. Since the clustering behaviour is, as a rule, more pronounced in females, this component predominates in most samples taken from daytime shelters. In hibernating bats, one could expect that both sexes would be equally represented in the wintering quarters, but even here the sex ratio is often biased, especially in favour of males (Davis, 1959; Tinkle and Milstead, 1960; Gaisler, 1975). Many authors have attempted to

---

estimated by tooth wear (years); abscissa, age determined by banding results, all individuals having been banded as juveniles. Explanations for B: ordinate, dry weight of right eye (mg); abscissa, number of dentin layers on cross-section through right upper canine. (Based on the originals by, (A) I. Horáček; (B) J. Dungel and J. Gaisler).

explain this phenomenon. Factors considered include differential preference for different quarters, differential mortality, or differential migrality between the sexes. The influence of these factors has actually been observed for certain species (Davis, 1966) but the problem of the extent to which the sexually biased samples are 'real' has not been solved. Surprisingly enough, a biased sex ratio has also been ascertained in certain non-hibernating tropical bats, the samples of which were partly or entirely obtained by mist-netting (Lim, 1966; Arata and Vaughn, 1970). Considering all available information, one must admit that the sex ratio is not equal in all bat populations although real cases of an unbalanced ratio may be rarer than indicated.

Age structure and longevity

By their size, coloration, presence or absence of cartilaginous growth zones in long bones, and macroscopic signs of sexual activity, bats can be divided into juveniles, subadults, and adults; or juveniles, yearlings, and adults (Dwyer, 1966). Regarding the longevity of bats, as revealed by banding studies, the category of adults comprises a rather long period of time. For this reason, additional age criteria have been sought. The most frequently used ones include tooth wear, dry mass of eye lens, and number of layers of dentine or cement observed in a cross section through a tooth. The former two criteria have been subject to criticism (Hall *et al.*, 1957; Barbour and Davis, 1969). The latter criterion seems to bring the most reliable results (Christian, 1956; Klevezal and Kleinenberg, 1967; Rachmatulina, 1971; Lord *et al.*, 1976). According to Rachmatulina (pers. comm.), the method has been verified on large sample of material of marked individuals of known age and it was confirmed that the number of appositional layers of dentine directly indicates the age of bats in years, at least in the hibernating species. Certain aspects of the applicability and comparability of the three methods mentioned above are illustrated in Fig. 7.10.

The age structure of bat populations has also been ascertained by calculation from marking-recapture data. In Table 7.3, the available information on the age structure of several species is summarized, based on mark-recapture data or on age determination by dentine layers. As expected, the lower age groups up to 3–5 years of age are the most frequented ones, including more than 50 per cent of the samples. The time interval between them and the highest group (life span) is unusually large in most species, indicating that a small part of the population attains very great age.

The maximum known life span is 26 years in the European species *Rhinolophus ferrumequinum* (Sluiter *et al.*, 1971) and 24 years in the North American *Myotis lucifugus* (Griffin and Hitchcock, 1965). From regions of mass marking, the maximum life span of around 20 years is known for

Table 7.3 *Age structure in some bat populations. The data of the first three authors are based on the marking-recapture method; those of the remaining two, on age determination by the number of dentin layers (for particulars see the text). Age groups 1, 2, etc. comprise 0.1–1,0, 1.1–2.0, etc. years. The values of per cent distribution and maximum life span are rounded to the nearest integer; the latter were updated. The representation of categories over 13 years (if any) is not given.*

Species	n	Sex	Age in years and per cent distribution													Life span	Reference
			1	2	3	4	5	6	7	8	9	10	11	12	13		
Plecotus townsendii	1500	♀	20	16	13	10	8	7	5	4	3	3	2	2	1	19	Pearson *et al.*, 1952
Rhinolophus hipposideros	1717	♂, ♀	36	23	15	9	6	4	2	2	1	–	–	–	–	18	Bezem *et al.*, 1960
Myotis mystacinus	1828	♂, ♀	25	19	14	10	8	6	4	3	2	2	1	1	–	19	Bezem *et al.*, 1960
Myotis emarginatus	1608	♂, ♀	21	17	13	10	8	6	5	4	3	2	2	1	1	16	Bezem *et al.*, 1960
Myotis daubentoni	920	♂, ♀	20	16	13	10	8	7	5	4	3	3	2	2	1	20	Bezem *et al.*, 1960
Myotis dasycneme	668	♀	30	21	15	10	7	5	3	2	2	1	–	–	–	16	Sluiter *et al.*, 1971
Myotis blythi	88	♀	39	18	6	5	9	3	6	4	–	3	2	2	1	13	Rachmatulina, 1971
Pipistrellus kuhli	62	♀	64	24	10	2	–	–	–	1	–	–	–	–	–	8	Rachmatulina, 1971
Eptesicus serotinus	46	♀	39	6	11	6	9	9	6	2	–	4	2	4	–	12	Rachmatulina, 1971
Miniopterus schreibersi	143	♂, ♀	33	24	22	12	5	1	1	2	–	–	–	–	–	16	Rachmatulina, 1971
Desmodus rotundus	217	♂, ♀	39	16	6	8	8	5	2	2	3	3	2	1	3	17	Lord *et al.*, 1976

several additional species (Barbour and Davis, 1969; Roer, 1971). According to Herreid (1964), warm climate forms do not appear to have a shorter maximum life span than their temperate counterparts. Thus it appears that within the order Chiroptera age is not correlated with metabolic rates. Rachmatulina (1971) believes that species bearing one young show a life span twice as long as those bearing two young. Negative correlation between the age and reproductive capacity within the bats may really exist; it certainly does when bats are compared with other small mammals.

7.4.3 Population changes

Mortality and survival

The development of a population is known to be affected by natality, mortality and migrality. Since natality of bats was discussed sub 7.3.2, only the two remaining parameters will be discussed here. Gilette and Kimbrough (in Slaughter and Walton, 1970) summarized in great detail the causes of chiropteran mortality. Many species of vertebrate and even some invertebrate predators of bats are known but most of them prey on bats only exceptionally. Also disease and fatal accidents act rather sporadically, although they may result in occasional mass mortality. The authors do not state mortality rates of the various species but point out the relatively low mortality rate of the whole order.

Starting with the work of Eisentraut (1949), several authors calculated survival rates and mean life expectancies based on the marking and recapture method. Recently, Humphrey and Cope (1977) pointed out the great analytical difficulties in the literature on bat demography resulting from the fact that most data are based on unknown-age cohorts of bats sampled during hibernation. Only the average adult survival rate at the prevailing population age structure then has clear and direct biological meaning. Table 7.4 gives examples of survival rates of several bat populations calculated in this way. Fortunately, some unambiguous survival data on bats are available from census work and from recapture of marked animals of known age. In most cases, survival rates are low during the first year of life (0.1 – 0.5), high during the subsequent 3 – 5 years (0.4 – 0.8) and then decreasing, often more rapidly in females than in males (Dwyer, 1966; Davis, 1966; Goehring, 1972; Mills *et al.*, 1975; Humphrey and Cope, 1977).

Summarizing the above information, around 30 per cent of juveniles and 60 per cent of adults of the species examined will survive annually. These species form a rather uniform ecological group – all of them are internal, hibernating, monoestrous, etc. The survival rates of species showing different ecological properties are unknown; lower values may be expected in polyoestrous and polytocous species.

Table 7.4 *Mean annual rate of survival and mean expectation of life (years) of certain bats in unknown-age cohorts, resulting from the 'traditional' processing of the marking-recapture data (cf. Table 7.5). Modified after Humphrey & Cope (1977).*

Species	Sex	Years after marking	Percent survival	Mean expect. of life	Reference
Myotis mystacinus	♂, ♀	1–7	75.2	3.5	Bezem *et al.*, 1960
Myotis emarginatus	♂, ♀	1–6	70.1	2.8	Bezem *et al.*, 1960
Myotis daubentoni	♂, ♀	1–4	80.0	4.5	Bezem *et al.*, 1960
Myotis dasycneme	♂, ♀	1–5	66.7	2.8	Bezem *et al.*, 1960
Rhinolophus hipposideros	♂, ♀	1–5	56.7	1.8	Bezem *et al.*, 1960
Rhinolophus euryale	♂	1–4	83.0	–	Dinale, 1968
Miniopterus schreibersi	♂, ♀	1–7	70.0	3.5	Dwyer, 1966
Plecotus auritus	♀	1–6	70.0	4.0	Stebbings, 1966
Pipistrellus subflavus	♂	1–8	57.0	–	Davis, 1966
Pipistrellus subflavus	♀	1–12	75.9	–	Davis, 1966
Pipistrellus subflavus	♀	12–13	46.3	–	Davis, 1966
Eptesicus fuscus	♂	1–17	76.0	–	Goehring, 1972
Eptesicus fuscus	♀	1–16	81.7	–	Goehring, 1972
Myotis lucifugus	♂	1–11	77.1	–	Humphrey & Cope, 1977
Myotis lucifugus	♀	1–13	85.7	–	Humphrey & Cope, 1977

Migrality

Contrary to many other population parameters, migrality has been more thoroughly investigated in bats than in other small mammals. After some observational evidence of migration, Allen, Mohr, and Griffin in America, and Eisentraut in Europe started to mark bats for later recognition. Starting in the 1930s the method has gradually developed a mass character and at present bats are being marked not only in Europe and North America but also in South America, Australia and several countries of Asia and Africa. The number of marked bats probably exceeds a million; in *Tadarida brasiliensis* alone around 170 000 individuals have been marked (Cockrum, 1969). The application of metal bands to the forearm is the most frequently used marking method. Other methods involve ear tagging, colour marking, punch-marking, radioactive tagging, and radio tracking. Their synopsis as well as a review of extensive literature on the results of bat marking has been provided by Griffin (in Wimsatt, 1970). The experience of the European workers has been summarized by Roer (1960, 1971); that of the Soviet workers by Strelkov (1969) and Strelkov and Kuzyakin (1974).

Many authors discussed the problem of disturbance connected with the marking and methods are being sought either to improve the classical banding technique (Bonaccorso *et al.*, 1976) or to induce new techniques (Buchler, 1976). While it is difficult to judge the degree of distortion of the behaviour under study, the present accumulation of records undoubtedly enables some insight into the problems of chiropteran migrality.

In the temperate zone bats, summer daytime shelters are rarely identical with hibernacula. The pattern of alternating summer and winter quarters usually involves movements and this 'interseasonal' migrality is the most obvious population change in bats. Although it is rarely possible for a migratory flight to take place precisely between the time of banding and recovery, it is possible to make inferences on the character of the movement from the timing of banding. The shorter the time between recovery and banding or a former recovery, the more significant the observations.

By examining the length and direction of the movements observed, bat species can be divided into several groups, the extremes being stationary and migratory (Strelkov, 1969; Roer, 1971). Stationary species, such as the members of the genera *Rhinolophus, Plecotus* and *Antrozous,* show little interseasonal migrality and disperse more or less radially to distances at most 50 km. Migratory species above all, the members of the genera *Nyctalus, Vespertilio, Lasiurus, Lasionycteris, Tadarida, Miniopterus* and certain *Pipistrellus* spp., migrate to distances up to about 1000 km. The records are 1600 km for *N. noctula* and *P. nathusii* (Strelkov, 1969), and 1300 km for *T. brasiliensis* (Cockrum, 1969). In some species, circumstantial evidence exists for even longer migrations. Only some populations of the above species, namely those living near the northern limit of their ranges, perform unidirectional migrations comparable to those of certain birds; the capability of long (multidirectional) movements, however, remains preserved even in other populations and is obviously a function of the flight patterns of these species. A third group comprises a large number of species, namely of the genus *Myotis*, which migrate to various distances depending on how far their summer daytime shelters are from their hibernacula; at most, up to 400 km. We have called them vagrant bats (Gaisler and Hanák, 1969) although they are of course no more vagrant than are the migratory ones. Certain populations of some vagrant species, e.g., *M. myotis, M. dasycneme, M. lucifugus*, show preference for certain directions or zones when migrating, others do not (cf. Fig. 7.11). Contrary to the migratory species, no vagrant species leave extensive areas of their summer range when migrating to hibernacula.

Recent investigations have shown that migrations between summer daytime shelters and hibernacula form only a minor part of the migrality of most species. Changes between shelters take place throughout the growing season and, to a smaller extent, even during winter. Migrality of this type has been witnessed in *M. schreibersi* (Dwyer, 1966), *R. hipposideros*

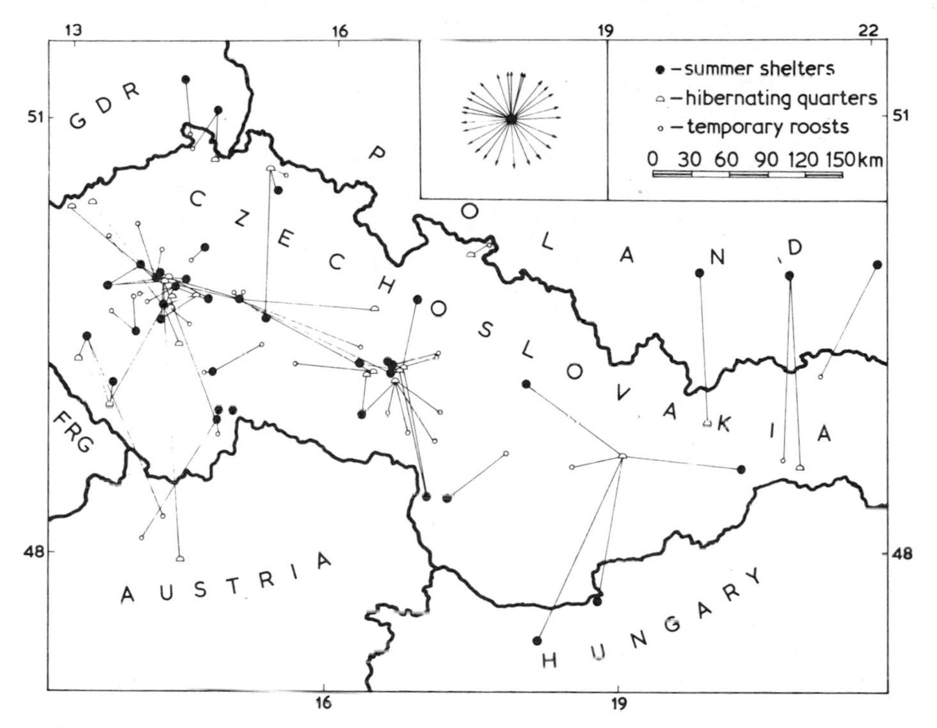

Fig. 7.11 Movements of typically vagrant populations of *Myotis myotis* in Central Europe. Inset: directions in which the members of the central Bohemian population disperse from summer nursing colonies to hibernation quarters. (After Gaisler & Hanák, 1969, supplemented.)

(Gaisler and Hanák, 1969) and *M. myotis* (Haensel, 1974), that is, in species belonging to all three above-mentioned categories. Various components of populations participate in these movements in various ways and the resulting pattern is too complicated to be analysed hereunder (cf. Fig. 7.12). The purpose of this 'intraseasonal' migrality may differ in different species and situations but there is no doubt that this activity is spontaneous and functional in connection with the annual life cycle. An interesting phenomenon belonging to this category is the nightly visitation of caves not related to roosting, as revealed by mist-netting (Hall and Brenner, 1968). This behaviour may be motivated trophically (Gaisler, 1975) but at the same time it appears to manifest a general affinity of bats to caves.

In spite of their migrality, many bats demonstrate strong loyalty to both summer home range and wintering quarter or a set of quarters. This is evidenced by recoveries in the place of banding, the number of which exceeds that of foreign recoveries several times. According to Tuttle (1976), this loyalty or philopatry appears to be a general phenomenon among bats. Multiple recaptures and roundtrip recoveries between summer and winter

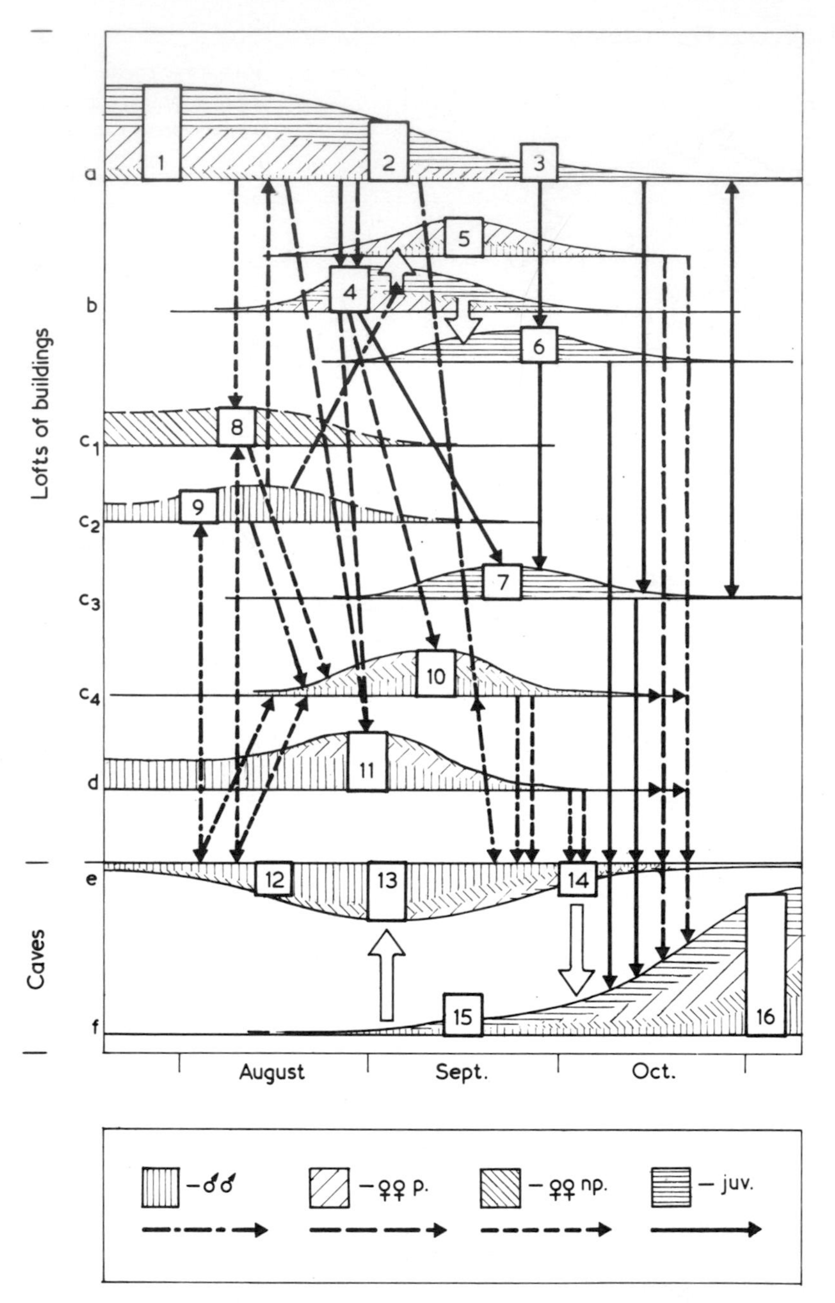

Fig. 7.12 Scheme showing disintegration of summer colonies and autumn movements in *Myotis myotis* (the same population as in Table 7.5). Explanations: a, shelters of summer nursing colonies; b, alternative shelters of summer colonies; c, temporary shelters of both sexes; d, summer shelters of males; e, community

quarters are highly significant indicators of philopatry. In *Myotis grisescens* and some other *Myotis* spp., philopatry attains around 90 per cent. Apparently, migrality and philopatry are the two sides of one and the same phenomenon and many apparent cases of disloyalty probably are the result of insufficient knowledge of a bat's entire pattern of annual movements.

The loyalty to shelters is also evidenced by the results of homing experiments, as summarized by Davis (1966). The longest recorded homing flight was 720 km in 34 days by 3 out of 51 adult female *Eptesicus fuscus*. Such cases are exceptional, however, and the percentage of recoveries decreases as a rule as the distance of migration increases. Numerous examples of successful homing has made it possible for the observed phenomenon to be interpreted from various points of view. The more impressive instances of homing suggest that many individuals actively choose the correct homeward direction (Griffin in Wimsatt, 1970). However, according to Wilson and Findley (1972), randomness in bat homing cannot be disproven. According to their hypothetical model the point at which 50 per cent of the animals home should mark the edge of familiar area. They estimated the familiar area in two tropical species: in *Myotis nigricans* as 13 km and in *Phyllostomus hastatus* as 27 km in radius.

The migrality of tropical bats is little known. In certain flying foxes there is indirect but strong evidence for flights of up to tens of kilometers (Brosset, 1966). Through spraying a fruit bat camp with the dye rhodamine red from a plane, Nelson (1965) observed a movement of about 115 km in *Pteropus poliocephalus*. In some flying foxes, food shortage is the probable stimulus for dispersal and transition to a nomadic way of life. In neotropical bats, the extent of movements may be positively correlated with body size (Fleming *et al.*, 1972; Heithaus *et al.*, 1975). Movements revealed by banding are mostly short, however, not exceeding several km. While long migrations appear to be uncommon in tropical bats, the frequency of their movements may be similar to that in temperate zone bats.

Population dynamics

The dynamics of bat populations, when measured by the fluctuations in numbers and changes in population structure, appears to be relatively low. Rapid changes in abundance of the gradation and depression type are unknown. Climatic and some other factors can cause mass mortality

---

netted at cave entrances; f, temporary and hibernating shelters in caves. 1–3, subsequently disintegrating summer nursing colony; 4, 6, 7, transient colonies of different types; 5, 10, 11, mating aggregations at different places; 8, solitary, nonparous females; 9, solitary males; 12–14, cave-visiting bat community (only at night); 15, 16, individuals and groups roosting in caves and the onset of hibernation. Open arrows indicate movements within one locality. (Based on the original by I. Horáček.)

(Gilette and Kimbrough, in Slaughter and Walton, 1970) but they rarely affect the whole population. As long as the numbers of bats increase or decrease, it is usually a long-term process. Cases in which the abundance is observed to increase are rare (Goehring, 1972; Rachmatulina, in Strelkov and Kuzyakin, 1974). The opposite is much more frequent. The decrease can often be explained by disturbance in roosts in connection with economic or tourist interests of man as well as with too frequent interference of investigators (Sluiter and Heerdt, 1957; Stebbings, 1969). The numbers of certain species, e.g., *Rhinolophus hipposideros*, have been observed to decrease locally so conspicuously that the decrease cannot be accounted solely to disturbing factors but to a general deterioration of the environment (Roer, 1972). Nevertheless, the abundance of most species appears relatively constant (Gaisler, 1975).

Considering population dynamics as a complex phenomenon we may infer its pattern from the various elements described in the preceding paragraphs. Although many questions are still unanswered, the essentials of population dynamics are known in a number of North American and European bat species as well as in the Australian *Miniopterus schreibersi* and tropical American *Desmodus rotundus*. To summarize the basic quantitative data in at least one species, the central European populations of *Myotis myotis* can be considered. The data are kindly supplied by I. Horáček (unpublished).

Most population parameters cannot be ascertained directly from observation data. Thus, for their quantification one must use indirect estimates, especially the changes in numbers or components of the population. To express these changes, the author used mark-recapture data, similarly to those used in earlier studies (Bezem *et al.*, 1960; Dwyer, 1966). The computations are based on a modification of the Petersen or Lincoln index:

$$N(t) \approx \hat{N}(t, t+\Delta t) = \frac{M(t) \,.\, S(t+\Delta t)}{R(t, t+\Delta t)}, \qquad (7.1)$$

where $N(t)$ is the actual population size at time t; $\hat{N}(t, t + \Delta t)$ is the population size estimated on the basis of catches obtained at time t and $t + \Delta t$; $M(t)$ is the number of bats marked at time t, $S(t + \Delta t)$ is the total number of bats captured at time $t + \Delta t$; and $R(t, t + \Delta t)$ is the number of bats marked at time t and recovered at time $t + \Delta t$. The applicability of $\hat{N}$ with respect to N was examined (at 0.966 probability level) by the relation

$$N \in (\hat{N} \pm \delta\hat{N}), \text{ where } \delta \text{ (the applicability factor)} = 1.5\sqrt{\frac{2(S-R)\,.\,(M-R)}{R\,.\,(S\,.\,M-R)}}. \qquad (7.2)$$

For simplicity's sake, the time characteristics were left out, being the same as in the equation (7.1).

In a natural population, the decrease in numbers is the result of mortality and emigration; the increase in numbers, then, is the result of natality and immigration. To estimate these population parameters, use was made of the fact that at $t \gg 0$, $\hat{N} \gg N$. This disparity can be expressed as

$$\frac{N(t)}{\hat{N}(t, t + \Delta t)} = P(t, t + \Delta t) \in (0, 1) \tag{7.3}$$

Then

$$P(t, t + \Delta t) = \frac{I(t, t + \Delta t)}{\hat{N}(t, t + \Delta t) - N(t)} =$$

$$\frac{N(t, t + \Delta t) - I(t, t + \Delta t)}{N(t)} = 1 - \frac{D(t, t + \Delta t)}{N(t)}, \tag{7.4}$$

where $I(t, t + \Delta t)$ is the population increment, and $D(t, t + \Delta t)$ is the decrease in number during time $(t, t + \Delta t)$.

The probability that the change in numbers is due to any component of migrality is lower, the greater the area whose inhabitants are considered to be a population. Dr Horáček found that in an area 1200 km^2, the variation in the numbers due to migrality does not exceed 5 per cent of the population size. Therefore, the effect of migrality on the variation in numbers may be neglected, which considerably simplifies subsequent considerations and computations. In practice, further simplification is attained by a stratification of the examined series by sex, age, etc.* In such cases, then, the value of $P(t, t + \Delta t)$ in the equations (7.3) and (7.4) can be considered directly to be the probability of survival $p(t, t + \Delta t)$. For a series of animals of equal age, N_i or $N_{i + k.\Delta t}$, where $k\Sigma(1, T)$ is an entity, with the uniform time interval $\Delta t = 1$ (year), it applies that

$$N_{i + T}(t + T) = N_i(t) \prod_{k = 1}^{T} p_{i + k.\Delta t}(k.\Delta t - 1, k.\Delta t). \tag{7.5}$$

Some of the results obtained by using the above procedures are summarized in Table 7.5. The population of *M. myotis* under study appears to be an open balanced system capable of self-reproduction.

7.5 Community

The size and structure of bat communities are difficult to express. While there is a large number of methods of locating and collecting bats (cf. Greenhall and Paradiso, 1968), all of them are considerably selective. For a

*The construction of a stratified Lincoln index has been described by Overton (in Giles, 1971: *Wildlife management techniques*, Washington). Throughout this chapter, much information was gathered from general ecological papers and manuals as well as from the mammal literature. Due to the limited space for references only those pertaining to bats are listed and it is believed that the reader may find references to sources of more general character in other places in this book.

Table 7.5 *A numerical model of the population of* Myotis myotis *in central Bohemia (Czechoslovakia). Basic data: observation period = 10 years (1966 through 1975); size of study area = 1200 km^2; number of banded individuals = 3604; number of recoveries = 1304; population size = 2440 ± 425 individuals; population density = 0.02 individuals per ha; observed sex ratio (males:females) = 0.474:0.526; observed fraction of reproducing females = 0.442; relative natality = relative mortality = 0.422; average age (all) = 2 years, 8.4 months; average age of non-juvenile individuals = 3 years, 10.3 months; observed maximum age = 14 years, 8 months; calculated maximum age = 16 years, 6 months; probability of survival in age class 1 to 10 = $\Pi(1 - 0.0875\,k)$. Explanations: N_i = abundance of class i; $p_i\,(i, i+1)$ = sectional (annual) probability of survival; $p_i\,(0, i+1)$ = probability of attaining age i for newborns; $D_i\,(i, i+1)$ = number of died individuals of class i; $N♀_i$ = abundance of females in class i; $f♀_i$ = fraction of fertile females; $I\,(i, i+1)$ = abundance of young born by females of class i. (By kind permission of I. Horáček.)*

Age class	Abundance data			Mortality data			Natality data		
i	N_i	$N_i\,(\sum_{i=0}^{15} N_i)^{-1}$	$N_i\,(\sum_{i=1}^{15} N_i)^{-1}$	$p_i\,(i, i+1)$	$p_i(0, i+1)$	$D_i\,(i, i+1)$	$N♀_i$	$f♀_i$	$I\,(i, i+1)$
0	850.000	0.2966	–	0.5189	0.5189	408.955	450.000	0.00	0.000
1	441.065	0.1539	0.2188	0.9125	0.4734	38.594	233.505	0.10	23.351
2	402.471	0.1404	0.1997	0.8250	0.3906	70.438	213.073	0.65	138.500
3	332.033	0.1159	0.1647	0.7375	0.3222	58.132	175.782	1.00	175.782
4	273.901	0.0956	0.1350	0.6500	0.2784	37.226	145.006	1.00	145.006
5	236.675	0.0826	0.1174	0.5625	0.2168	52.399	125.298	1.00	125.298
6	184.276	0.0643	0.0914	0.4750	0.1504	56.456	97.558	1.00	97.558
7	127.820	0.0446	0.0634	0.3875	0.0916	49.528	67.670	1.00	67.670
8	77.892	0.0272	0.0386	0.3000	0.0981	37.049	41.237	1.00	41.237
9	40.843	0.0143	0.0203	0.2125	0.0212	22.848	21.623	1.00	21.623
10	17.995	0.0063	0.0089	0.1250	0.0076	11.555	9.527	1.00	9.527
11	6.440	0.0022	0.0032	<0.1000	0.0020	4.720	3.409	1.00	3.409
12	1.720	0.0006	0.0009	<0.1000	0.0004	1.411	0.911	1.00	0.911
13	0.309	0.0001	0.0002	<0.1000	0.0001	0.265	0.164	1.00	0.164
14	0.044	<0.0001	<0.0001	<0.1000	<0.0001	0.043	0.023	1.00	0.023
15	0.001	<0.0001	<0.0001	<0.1000	<0.0001	0.001	<0.001	1.00	<0.001
0–15	2865.666	1.0000	–	–	–	850.000	1584.780	–	–
1–15	2015.666	–	1.0000	–	–	441.065	1134.780	–	850.000

Fig. 7.13 Bat traps of various constructions are used to collect animals flying out from their daytime shelters. The picture shows a modified Constantine trap set at the entrance to a natural water cave. (Photo by Dr V. Hrabě, scale 1:20.)

study of communities, data from, e.g. nursery population estimates are of little use – although they are numerous – due to the bias resulting from omission of individually roosting bats and differing probabilities of locating populations of different size (Humphrey, 1975). In the temperate zone, relatively most reliable data are obtained from underground hibernating quarters; many papers on bat communities in Europe are based on such data. Even such samples, however, do not as a rule represent the whole hibernating community of a region. Recently, mist-netting at foraging sites in summer has become a valuable aid to synecological approach in chiropterology. Even this method has its own drawbacks, mainly because it does not comprise high flying individuals. The samples obtained, however, show high species diversity (Kunz, 1973; Humphrey, 1975; Gaisler, 1975) and their composition may be closer to the natural structure of the communities than in samples obtained by other methods (Fig. 7.14).

In practice, most students of bat communities combine different sampling methods in the belief that the greater the variety of methods the more relevant are the results with respect to the actual state. Contrary to population estimates, most community estimates do not involve the time factor and, hence, reveal only relative data. Below, a synopsis is given of the basic symbols and formulae used or suggested in several recent papers (Arata and Vaughn, 1970; Fleming *et al.*, 1972; Kunz, 1973; Humphrey, 1975; Gaisler, 1975).

The basic characteristic of a community is the number of species, or faunal size, s. The total number of individuals is denoted as N; the number of individuals of the i-th species, as N_i. For simplicity's sake, the symbol of estimation (^) is left out. The probability that an individual belongs to the i-th species is

$$P_i = \frac{N_i}{N}, \tag{7.6}$$

being the starting point to calculate species dominance, diversity and equitability. The dominance of the i-th species is usually expressed as

$$D_i = 100.p_i. \tag{7.7}$$

According to the Shannon – Weaver formula, species diversity is

$$H' = -\sum_{i=1}^{s} p_i \,.\, \log_2 p_i, \tag{7.8}$$

and, according to the Sheldon formula, species equitability is

$$E = \frac{H'}{H_{max}}, \text{ where } H_{max} = \log_2 s. \tag{7.9}$$

Contribution to diversity by the most abundant species n_1 can be expressed as:

$$H'n_1 = -p_1 \,.\, \log_2 p_1 \tag{7.10}$$

and the criterion of inequitable distribution due to the presence of a

Superabundant species: $H'n < H'n_2$. Instead of $\log_2$, some authors prefer $\log_e$ in the equations (7.8), (7.9) and (7.10).

The total number of collecting localities or sites can be denoted as a, and the number of localities at which the i-th species was collected as a_i. The frequency of occurrence of the i-th species indicates its constancy:

$$C_i = \frac{100 \,.\, a_i}{a} \,. \qquad (7.11)$$

By combining the above incides, the index of relative abundance of the i-th species can be expressed:

$$A_i = N_i \,.\, C_i, \qquad (7.12)$$

and the index of relative abundance of the whole community:

$$A = N \,.\, H'; \qquad (7.13)$$

the index is applicable only if $H'n_1 > H'n_2$.

From among these constants, s, N, D_l and H' are most frequently indicated in the chiropterological literature. In general, the faunal size (s) increases towards the equator. This can be demonstrated by selecting regions of comparable size, e.g. West Africa with 97 species (Rosevear, 1965) against Alaska with 4 species (McNab, 1971), or the Garamba National Park, 4°N, with 38 species (Verschuren, 1957) against the Voronezh State Reserve, 51°N, with 10 species (Panyutin, 1970). The hitherto infrequent observations tend to show that in the tropics, communities in moist lowland forest are the richest both as to number of species and individuals, followed by those in riparian forest, dry lowland forest and mountain forest (Lim, 1966; Fleming *et al.*, 1972). The diversity (H'), however, may not follow this rank. Each tropical bat fauna, whether it is a complex mainland fauna in a rain forest or a simple fauna on small islands, has a definite pattern and structure depending, among other things, on trophic relations (McNab, 1971; Heithaus *et al.*, 1975).

The structural diversity of the environment appears to be the major factor influencing the numbers and composition of bat communities in the temperate zone. In Nearctic bats both faunal size and community diversity are correlated with the presence or absence of structures used for roosting, implying little correlation with length of growing season or diversity of insect communities (Humphrey, 1975). Food may be a limiting factor only in high altitude and latitude environments where the growing season is too short. This is in accordance with the observations of Rachmatulina (in Strelkov and Kuzyakin, 1974) who found the largest number of species in the heterogeneous landscape of the foothills and mountains up to elevations of 1800 m. In lowland forests, the number of bat species was smaller but some of them were superabundant, and the least numbers of species and individuals occurred in the alpine zone (in Soviet Azerbaijan).

Findley (1976) presented an interesting comparison of the structure of several bat communities. His work is based on multivariate analysis of 18 morphological features. Three tropical bat communities examined resemble one another in faunal statistics, suggesting common evolutionary pressure in different and distant regions. Two temperate bat faunae under comparison differ from the tropical ones in the greater rarity of highly distinctive taxa but are not meaningfully different in species relationship or degree of ecological overlap.

In view of modern ecology it is essential to know the energetic balance of a particular community and in this respect, bat communities are virtually unknown. The main problem is associated with the determination of actual density per unit of area which is much more complicated than in the case of a single species population. As mentioned above, most available data concerning bat communities are based on relative values. The estimates of absolute values of, e.g. abundance are scarce and inexact. Nevertheless, they reveal some information and, thus, are considered worth mentioning. According to a number of Soviet workers, densities of bat communities in the central zone of deciduous forests vary between 0.5 and 7.0 individuals per ha, whereas in more southern areas of the U.S.S.R. they attain up to 13.0 individuals per ha in places (Strelkov and Kuzyakin, 1974). Gaisler (1975) compared a foraging bat community, studied by netting and direct observations, to a small terrestrial mammal community, assessed by removal trapping. The resulting density was 2.5 individual bats and 26.2 individual small mammals per ha, whereas H' was 2.956 in bats and 2.322 in small mammals. At the time being, bat census is carried out by the present author with the assistance of 50 previously trained volunteers, using simultaneous visual counts at fixed places and time intervals, following the shifts in sunset. Preliminary (and unpublished) data indicate that mean density of the bat community foraging in a Central European town 230 km^2 in area, is 0.6 individuals per ha.

While these rough estimates cannot be generalized, one may hypothesize that in the temperate zone the abundance of bat communities and, hence, the energy flow through them is comparatively low. Further speculation would suggest that in the tropics the situation is different.

7.6 Relations

7.6.1 The role of bats in ecosystems

The present status of knowledge does not make it possible to analyse the whole variety of relations that bats display within different ecosystems. The relation of bats to certain plants appears to be the best elucidated. Vegetarian bats not only consume parts of plants but also play an important part

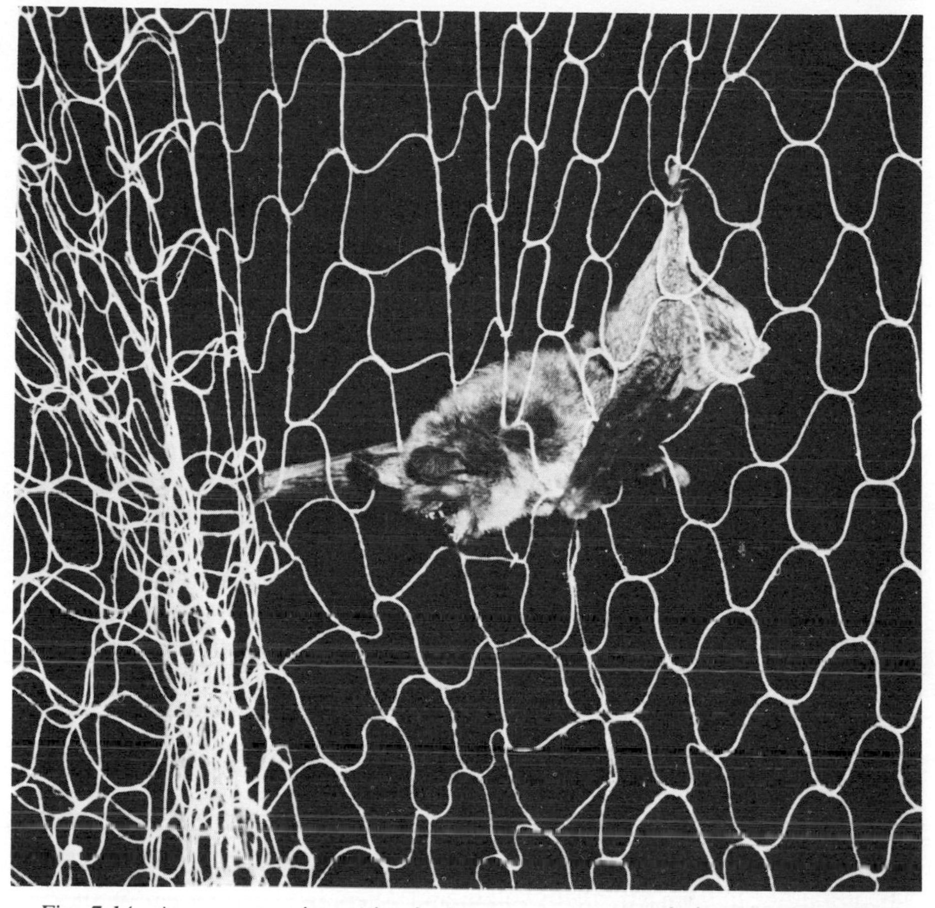

Fig. 7.14 At present, mist-netting is the most common technique to sample bats when on the wing. This bat (*Myotis mystacinus*) was captured in a white thread net made in the GDR. (Photo by Dr J. Červený, scale 1:1.2.)

in their dispersal. This phenomenon, known as chiropterochory, is widespread in the tropics and, among mammals, bats are probably the most important dispersers of seeds (Jones, in Baker *et al.*, 1976). Many bats are also pollinators of a wide variety of plants, and Howell (1976) revealed 130 genera of chiropterophilous plants. Heithaus *et al.* (1975) showed that vegetarian bats and certain plants make up a co-evolved system that illustrates the effects plants can have on animal populations, and conversely, the effects animals can have on plant populations.

Much has been written on the importance of insectivorous bats in controlling insect numbers and 'maintaining natural equilibrium'. Such papers, however, often mirrored endeavours at justifying the interest in bats rather than results of investigations. Nevertheless, it is clear that bats consume enormous numbers of arthropods and it is probable that they exert some control over the populations on which they regularly prey (Constantine in Wimsatt, 1970). Certain prey species show physiological and behavioural counteradaptations (Roeder, 1967), which also suggests that insectivorous bats and their prey form an organised system. Bats exert some other influence on arthropod populations, e.g. they enable the existence of many parasitic species (on their bodies) and commensals (in their guano).

The problems of ecological interactions between bats and nocturnal birds have been studied by Fenton and Fleming (1976). Around 230 species of birds which are nocturnal and rely to varying degrees on insects, insects and vertebrates, or fruit, present the most important potential competitors of bats. Certain birds also are the relatively most frequent predators of bats. However, bats appear effectively to dominate the aerial nocturnal adaptive zones of fruit and insect eating. There are other examples of known interactions of bats with some elements in their environment. Crespo *et al.* (1961) illustrated graphically the relations of vampire bats to some animal communities and in a similar way one could evaluate the carnivorous or piscivorous bats as well. Also, colonies of bats are known to be capable of influencing the microclimate in their shelters, particularly in tree holes and, in cases of large numbers, even in caves. In certain poorly ventilated bat caves, ammonia and carbon dioxide concentrations become very high (Constantine in Wimsatt, 1970).

7.6.2 Bats and man

Extensive literature and excellent reviews are available on this point (Constantine in Wimsatt, 1970; Sulkin and Allen, in Slaughter and Walton, 1970; Strelkov in Strelkov and Kuzyakin, 1974; Jones in Baker *et al.*, 1976). Man's interest in bats is of ancient date and these animals have been considered as objects of mystery, superstition, fear and even worship. In some regions, bats are eaten by man; on the other hand the common vampire attacks humans, and very many bat species share with man his dwelling place. In modern times a great surge of interest in bats resulted from their relation to health, economy and research activity of man.

Bats are subject to bacterial, mycotic, protozoan, rickettsial and viral diseases. They are also infested by helminths and parasitic arthropods. Of the greatest importance is their part in the transmission and dispersal of rabies virus. In America, bats are recognized as natural reservoir hosts for this agent. The major hosts of the rabies virus are probably the vampire bats, but the virus has been observed to occur in many additional, mostly insectivorous, species. Bats are able to sustain rabies infection in the absence of overt symptoms, although a small percentage of bats show abnormal behaviour or even death. The infection appears to be a closed cycle, and man, livestock and other species are infected tangentially. There are a few cases of rabies-positive bats outside America but there is no evidence of a transmission of rabies from bats to man in those cases.

At least 50 arboviruses, about half of which produce disease in man, have been observed to occur in bats (Constantine, loc. cit.). Histoplasmosis, known from bats in America and from bat guano in Africa, Malaya and Europe (Rumania), is an important mycotic disease which can seriously affect man. Of protozoal diseases, several trypanosomiases known in tropical bats and toxoplasmosis ascertained in bats even in Europe may likewise present danger for man. The importance of bats as carriers of bacterial diseases is little known; certain (namely tropical) bat species have been shown to carry even bacteria pathogenic to man.

Although there is a large number of pathogenic agents connected with bats, it would not be right to over-estimate the epidemiological importance of these mammals. With the possible exception of rabies and histoplasmosis (both mainly in tropical and subtropical America), bats seem to play only a marginal role in the ecology of human pathogens. The danger to human health, resulting from bats, is much smaller than in the case of other mammals, such as rodents or carnivores, and a general control of bats would not be reasonable.

The relation of bats to economy is both positive and negative. Leaving aside the importance of bats in consuming insects (which is obvious ecologically but problematical economically) one can see the practical contribution of bats first of all in the production of guano. Bat guano, accumulated in bat shelters in enormous quantities (up to thousands of tonnes) is an excellent fertilizer. While commercial exploitation of bat guano does not pay in most cooler regions, prosperous guano mining companies have been established in certain tropical regions. Of negative importance are fruit-eating bats, but reports of great damage to important fruit are restricted to a few regions, mainly in the Old World tropics. Beyond doubt, the vampire *Desmodus rotundus* is the most harmful species, not only because it is dangerous to man but mainly because it causes great losses among cattle. These losses are mostly not a direct result of vampire bites but they result from the infection by rabies and other diseases. As vampires tend to increase in numbers (Jones in Baker *et al.*, 1976), it is necessary to control their numbers by artificial measures.

The importance of bats as model objects of research continues to increase. They are important in medical research when studying such problems as experimental hypothermia, survival in extreme environments, and different topics of comparative anatomy, embryology, histology and cytology. Bats' ultrasonic orientation (echolocation) is subject to research by special laboratories and knowledge of it forms the basis of present day bio-acoustics. There may be still another field in which bats could be used in a more general respect, namely, as bio-indicators of environmental deterioration.

Like other animals, bats comprise species the numbers of which increase with advancing civilisation, species more or less indifferent, and species whose numbers decrease. There is no evidence of any bat species that would be totally exterminated, and no bat species has been included in the Red Data Book. It is evident, however, that many species require active protection if they are to be maintained in a sufficiently large area and in numbers affording their reproduction. It appears that the most endangered are certain species of cave origin, at least in some areas of the temperate zone. *Myotis sodalis, M. grisescens* and *M. austroriparius* in the eastern U.S.A., and *Rhinolophus hipposideros* and *R. ferrumequinum* in western and central Europe may serve as examples. The need for conservation of bats is generally agreed and an increasing number of countries (at present, about 30) protect bats by law. In those regions where bats cause appreciable damage, their control should be approached in a differentiated way and should involve only selected species and/or populations.

7.7 Acknowledgements

The author wishes to thank all his chiropterological colleagues who provided copies of their papers or photographs used in compiling this chapter. In particular, the author is obliged to Dr V. Hanák (Prague) for comments on the text, and Dr A. Krzanowski (Cracow) for pointing out certain major literary sources. Miss Z. Bauerová and Mr J. Dungel – the author's former students – and Dr I. Horáček – formerly a student of Dr Hanák – kindly submitted their own unpublished data. Of them, special thanks are due to Dr I. Horáček (Prague) who not only provided a summary of his thesis for the present chapter but also redrew all line illustrations. Last but not least, the author wishes to thank Dr R. Obrtel (Brno) for translating the chapter into English.

7.8 References

Abelentsev, V.I., Pidoplichko, I.T. & Popov, B.M. (1956) *Fauna Ukrainy,* **1**, Ssavci. Kiïv: ANURSR.

Aellen, V. (1962) Le baguement des chauves-souris au Col de Bretolet (Valais). *Archive des Sciences Physiques et Naturelles, Genève,* **14**, 365-392.

Airapetiantz, E.S. & Konstantinov, A.I. (1974) *Echolokacia v prirodě.* Leningrad: Nauka.

Allen, G.M. (1939) *Bats.* Cambridge: Harvard University.

Anciaux de Faveaux, M. (1973) Essai de synthèse sur la reproduction de chiroptères d'Afrique (region faunistique Ethiopienne). *Periodicum Biologorum, Zagreb,* **75**, 195-199.

Arata, A.A. & Vaughn, J.B. (1970) Analyses of the relative abundance and reproductive activity of bats in south-western Colombia. *Caldasia,* **10**, 517-528.

Baker, R.J., Jones, J. Knox Jr. & Carter, D.C. (1976) Biology of bats of the New World family Phyllostomatidae, 1. *Special Publications, Museum of the Texas Tech University,* **10**, 1-218.

Barbour, R.W. & Davis, W.H. (1969) *Bats of America*. Lexington: University Press.

Bateman, G.C. & Vaughan, T.A. (1974) Nightly activities of mormoopid bats. *Journal of Mammalogy,* **55**, 45-65.

Beck, A.J. & Lim, B.L. (1973) Reproductive biology of *Eonycteris spelaea,* Dobson (Megachiroptera) in West Malaysia. *Acta Tropica,* **30**, 251-260.

Belwood, J.J. & Fenton, M.B. (1976) Variation in the diet of *Myotis lucifugus. Canadian Journal of Zoology,* **54**, 1674-1678.

Bezem, J.J., Sluiter, J.W. & Heerdt, P.F. van (1960) Population statistics of five species of the bat genus *Myotis* and one of the genus *Rhinolophus* hibernating in the caves of S. Limburg. *Archives Néerlandaises de Zoologie,* **13**, 511-539.

Bezem, J.J., Sluiter, J.W. & Heerdt, P.F. van (1964) Some characteristics of the hibernating locations of various species of bats in South Limburg, 1, 2. *Proceedings of the Koninklijke Nederlandse Akademie van Wettenschappen,* C **67**, 325-336, 337-350.

Black, H.L. (1972) Differential exploitation of moths by the bat *Eptesicus fuscus* and *Lasiurus cinereus. Journal of Mammalogy,* **53**, 598-601.

Black, H.L. (1974) A north temperate bat community: structure and prey populations. *Journal of Mammalogy,* **55**, 138-157.

Bleier, W.J. (1975) Early embryology and implantation in the California leaf-nosed bat, *Macrotus californicus. Anatomical Record,* **182**, 237-254.

Böhme, W. & Natuschke, G. (1967) Untersuchungen der Jagdflugaktivität freilebender Fledermäuse in Wochenstuben mit Hilfe einer doppelseitigen Lichtschranke und einige Ergebnisse an *Myotis myotis* (Borkhausen, 1797) und *Myotis nattereri* (Kuhl, 1818). *Säugetierkundliche Mitteilungen,* **15**, 129-138.

Bonaccorso, F.J., Smythe, N. & Humphrey, S.R. (1976) Improved techniques for marking bats. *Journal of Mammalogy,* **57**, 181-182.

Brosset, A. (1966) *La biologie des chiroptères.* Paris: Masson.

Buchler, E.R. (1976) A chemiluminescent tag for tracking bats and other small nocturnal animals. *Journal of Mammalogy,* **57**, 173-176.

Christian, J.J. (1956) The natural history of a summer aggregation of the big brown bat, *Eptesicus fuscus fuscus. American Midland Naturalist,* **55**, 66-95.

Cockrum, E.L. (1969) Migration of the guano bat, *Tadarida brasiliensis. University of Kansas, Museum of Natural History, Miscellaneous Publications,* **51**, 303-336.

Constantine, D.G. (1967) Activity patterns of the Mexican free-tailed bat. *University of New Mexico Publications in Biology,* **7**, 1-79.

Cranbrook, Earl of & Barrett, H.G. (1965) Observations on noctule bats (*Nyctalus noctula*) captured while feeding. *Proceedings of the Zoological Society, London,* **144**, 1-24.

Crespo, J.A., Vanella, J.M., Blood, B.D. & DeCarlo, J.M. (1961) Observaciones ecological del vampiro, *Desmodus r. rotundus* (Geoffroy). *Revista del Museo Argentino de Ciencias Naturales,* **6**, 131-160.

Crespo, R.F., Linhart, S.B., Burns, R.J. & Mitchell, G.C. (1972) Foraging behavior of the common vampire bat related to moonlight. *Journal of Mammalogy,* **53**, 366-368.

Daan, S. (1973) Activity during natural hibernation in three species of vespertilionid bats. *Netherlands Journal of Zoology,* **23**, 1-71.
Davis, R.B. (1966) Homing performance and homing ability in bats. *Ecological Monographs,* **36**, 201-237.
Davis, R.B., Herreid, C.F. & Short, H.L. (1962) Mexican free-tailed bats in Texas. *Ecological Monographs,* **32**, 311-346.
Davis, W.B. & Dixon, J.R. (1976) Activity of bats in a small village clearing near Iquitos, Peru. *Journal of Mammalogy,* **57**, 747-749.
Davis, W.H. (1959) Disproportionate sex ratios in hibernating bats. *Journal of Mammalogy,* **40**, 16-19.
Davis, W.H. (1966) Population dynamics of the bat *Pipistrellus subflavus. Journal of Mammalogy,* **47**, 383-396.
DeCoursey, G. & DeCoursey, P.J. (1964) Adaptive aspects of activity rhythms in bats. *Biological Bulletin,* **126**, 14-27.
Dinale, G. (1968) Studi sui chirotteri italiani, 7, Sul raggiungimento della maturità sessuale nei chirotteri europei ed in particolare nei Rhinolophidae. *Archivio Zoologico Italiano,* **53**, 51-71.
Dwyer, P.D. (1966) The population pattern of *Miniopterus schreibersii* (Chiroptera) in northeastern New South Wales. *Australian Journal of Zoology,* **14**, 1073-1137.
Dwyer, P.D. (1970) Foraging behaviour of the Australian large-footed *Myotis* (Chiroptera). *Mammalia,* **34**, 76-80.
Dwyer, P.D. (1971) Temperature regulation and cave-dwelling in bats: an evolutionary perspective. *Mammalia,* **35**, 424-455.
Eisentraut, M. (1936) Beitrag zur Mechanik des Fledermaus-fluges. *Zeitschrift für Wissenschaftliche Zoologie,* **148**, 159-188.
Eisentraut, M. (1937) *Die deutschen Fledermäuse.* Leipzig: Schöps.
Eisentraut, M. (1947) Die Bedeutung von Temperatur und Klima im Leben der Chiropteren. *Biologischer Zentralblatt,* **66**, 236-251.
Eisentraut, M. (1949) Beobachtungen über Lebensdauer und jährliche Verlustziffern bei Fledermäusen, insbesondere bei *Myotis myotis. Zoologischer Jahrbücher, Systematik und Ökologie*, **78**, 193-216.
Eisentraut, M. (1951) Die Ernährung der Fledermäuse (Michrochiroptera). *Zoologischer Jahrbücher, Systematik und Ökologie,* **79**, 114-177.
Fenton, M.B. (1974) Feeding ecology of insectivorous bats. *Bios*, **45**, 3-15.
Fenton, M.B. (1975) Observations on the biology of some Rhodesian bats, including a key to the Chiroptera of Rhodesia. *Life Science Contributions,* Royal Ontario Museum, **104**, 1-27.
Fenton, M.B. & Flemming, T.H. (1976) Ecological interactions between bats and nocturnal birds. *Biotropica,* **8**, 104-110.
Fenton, M.B., Jacobson, S.L. & Stone, R.N. (1973) An automatic ultrasonic sensing system for monitoring the activity of some bats. *Canadian Journal of Zoology,* **51**, 291-299.
Fenton, M.B. & Morris, G.K. (1976) Opportunistic feeding by desert bats (*Myotis* spp.). *Canadian Journal of Zoology,* **54**, 526-530.
Findley, J.S. (1976) The structure of bat communities. *American Naturalist,* **110**, 129-139.
Findley, J.S., Studier, E.H. & Wilson, D.E. (1972) Morphologic properties of bat wings. *Journal of Mammalogy,* **53**, 429-444.
Fleming, T.H., Hooper, E.T. & Wilson, D.E. (1972) Three Central American bat communities: structure, reproductive cycles, and movement patterns. *Ecology,* **53**, 555-569.

Gaisler, J. (1963) The ecology of lesser horseshoe bat (*Rhinolophus hipposideros hipposideros* Bechstein, 1800) in Czechoslovakia, 1; 2, Ecological demands, problem of synanthropy. *Acta Societatis Zoologicae Bohemoslovacae,* **27**, 211-233, 322-327.

Gaisler, J. (1966) A tentative ecological classification of colonies of the European bats. *Lynx,* **6**, 35-39.

Gaisler, J. (1966) Reproduction in the lesser horseshoe bat (*Rhinolophus hipposideros hipposideros* Bechstein, 1800) *Bijdragen tot de Dierkunde,* **36**, 45-64.

Gaisler, J. (1973) Netting as a possible approach to study bat activity. *Periodicum Biologorum, Zagreb,* **75**, 129-134.

Gaisler, J. (1975) A quantitative study of some populations of bats in Czechoslovakia (Mammalia: Chiroptera). Acta Scientiarum Naturalium Academiae Scientiarum Bohemoslovacae, *Brno,* **9**, 1-44.

Gaisler, J. & Hanák, V. (1969) Ergebnisse der zwanzigjährigen Beringung von Fledermäusen (Chiroptera) in der Tschechoslowakei: 1948-1967. *Acta Scientiarum Naturalium Academiae Scientiarum Bohemoslovacae, Brno,* **3**, 1-33.

Gaisler, J. & Klíma, M. (1968) Das Geschlechterverhältnis bei Feten und Jungen einiger Fledermausarten. *Zeitschrift für Säugetierkunde,* **33**, 352-357.

Gilette, D.D. (1975) Evolution of feeding strategies in bats. *Tebiwa,* **18**, 39-48.

Goehring, H.H. (1972) Twenty-year study of *Eptesicus fuscus* in Minnesota. *Journal of Mammalogy,* **53**, 201-207.

Goodwin, G.G. & Greenhall, A.M. (1961) A review of the bats of Trinidad and Tobago. *Bulletin of the American Museum of Natural History,* **122**, 187-302.

Greenhall, A.M. (1972) The biting and feeding habits of the vampire bat, *Desmodus rotundus. Journal of Zoology,* London, **168**, 451-461.

Greenhall, A.M. & Paradiso, J.L. (1968) *Bats and bat banding.* Publications of the Bureau of Sport, Fisheries and Wildlife Resources, Washington, **72**, 1-48.

Greenhall, A.M., Schmidt, U. & Lopez-Forment, C.W. (1971) The attacking behavior of the vampire bat, *Desmodus rotundus*, under field conditions in Mexico. *Biotropica,* **3**, 136-141.

Griffin, D. (1958) *Listening in the dark.* New Haven: Yale University.

Haensel, J. (1974) Uber die Beziehungen zwischen verschiedenen Quartiertypen des Mausohrs, *Myotis myotis* (Borkhausen 1797), in den brandenburgischen Bezirken der DDR. *Milu,* **3**, 542-603.

Hall, J.S. & Brenner, F.J. (1968) Summer netting of bats at a cave in Pennsylvania. *Journal of Mammalogy,* **49**, 779-781.

Hall, J.S., Cloutier, R.J. & Griffin, D.R. (1957) Longevity records and notes on tooth wear of bats. *Journal of Mammalogy,* **38**, 407-409.

Hamilton, R.B. & Stalling, D.T. (1972) *Lasiurus borealis* with five young. *Journal of Mammalogy,* **53**, 190.

Hartman, F.A. (1963) Some flight mechanisms of bats. *Ohio Journal of Science,* **63**, 59-64.

Heithaus, E.R., Fleming, T.H. & Opler, P.A. (1975) Foraging patterns and resource utilization in seven species of bats in a seasonal tropical forest. *Ecology,* **56**, 841-854.

Herreid, C.F. (1964) Bat longevity and metabolic rate. *Experimental gerontology,* **1**, 1-9.

Herreid, C.F. (1967) Mortality statistics of young bats. *Ecology,* **48**, 310-312.

Hill, J.E. (1974) A new family, genus and species of bat (Mammalia: Chiroptera) from Thailand. *Bulletin of the British Museum (Natural History), Zoology,* **27**, 303-336.

Horst, R. (1972) Bats as primary producers in an ecosystem. *Bulletin of the National Speleological Society,* **34**, 49-54.
Howell, D.J. (1976) Plant-loving bats, bat-loving plants. *Natural History,* **85**, 52-59.
Humphrey, S.R. (1975) Nursery roosts and community diversity of nearctic bats. *Journal of Mammalogy,* **56**, 321-346.
Humphrey, S.R. & Cope, J.B. (1977) Survival rates of the endangered Indiana bat, *Myotis sodalis. Journal of Mammalogy,* **58**, 32-36.
Humphrey, S.R. & Kunz, T.H. (1976) Ecology of a pleistocene relict, the western big-eared bat (*Plecotus townsendii*), in the southern Great Plains. *Journal of Mammalogy,* **57**, 470-494.
Hůrka, L. (1966) Beitrag zur Bionomie, Ökologie und zur Biometrik der Zwergfledermaus (*Pipistrellus pipistrellus* Schreber, 1774) (Mammalia: Chiroptera) nach den Beobachtungen in Westböhmen. *Acta Societatis Zoologicae Bohemoslovacae,* **30**, 228-246.
Husar, S.L. (1976) Behavioral character displacement: evidence of food partitioning in insectivorous bats. *Journal of Mammalogy,* **57**, 331-338.
Jones, C. (1972) Comparative ecology of three pteropid bats in Rio Muni, West Africa. *Journal of Zoology,* London, **167**, 353-370.
Kingdon, J. (1974) *East African mammals*, **2**, A. London & New York: Academic Press.
Klevezal, G.A. & Kleinenberg, S.E. (1967) *Opredělenie vozrasta mlekopitajuščich po sloistym strukturam zubov i kosti.* Moskva: Nauka.
Kock, D. (1972) Fruit-bats and bat-flowers. *Bulletin of the East African Natural History Society,* 123-126.
Kolb, A. (1959) Über die Nahrungsaufnahme einheimischer Fledermäuse vom Boden. *Zoologischer Anzeiger,* **22**, Supplement, 162-168.
Kolb, A. (1973) Riechverhalten und Riechlaute der Mausohr-fledermaus *Myotis myotis. Zeitschrift für Säugetierkunde,* **38**, 277-284.
Krzanowski, A. (1969) The protection of bats. *Lynx,* **10**, 41-44.
Kulzer, E. (1965) Temperaturregulation bei Fledermäusen (Chiroptera) aus verschiedenen Klimazonen. *Zeitschrift für vergleichende Physiologie,* **50**, 1-34.
Kulzer, E. (1972) Der Winterschlaf der Fledermäuse – eine stammesgeschichtliche Anpassung. *Laichinger Höhlenfreund*, **7**, 9-20.
Kunz, T.H. (1973) Population studies of the cave bat (*Myotis velifer*): reproduction, growth, and development. *Occasional Papers, Museum of Natural History, University of Kansas,* **15**, 1-43.
Kunz, T.H. (1973) Resource utilization: temporal and spatial components of bat activity in central Iowa. *Journal of Mammalogy,* **54**, 14-32.
Kunz, T.H. (1974) Reproduction, growth, and mortality of the vespertilionid bat, *Eptesicus fuscus,* in Kansas. *Journal of Mammalogy,* **55**, 1-13.
Kunz, T.H. & Brock, C.F. (1975) A comparison of mist nets and ultrasonic detectors for monitoring flight activity of bats. *Journal of Mammalogy,* **56**, 907-911.
Kuzyakin, A.P. (1950) *Letuchie myshi.* Moskva: Nauka.
Laufens, G. (1972) *Freilanduntersuchungen zur Aktivitätsperiodik dunkelaktiver Säuger*. Dissertation Thesis Abstract. Köln: Universität.
LaVal, R.K. (1973) Observations on the biology of *Tadarida brasiliensis cynocephala* in southeastern Louisiana. *American Midland Naturalist,* **89**, 112-120
Lawlor, T.E. (1973) Aerodynamic characteristics of some neotropical bats. *Journal of Mammalogy,* **54**, 71-78.

Lim, B.L. (1966) Abundance and distribution of Malaysian bats in different ecological habitats. *Federation Museums Journal, Kuala Lumpur,* **11**, 61-76.

Lord, R.D., Muradali, F. & Lazaro, L. (1976) Age composition of vampire bats (*Desmodus rotundus*) in northern Argentina and southern Brazil. *Journal of Mammalogy,* **57**, 573-575.

McNab, B.K. (1971) The structure of tropical bat faunas. *Ecology,* **52**, 352-358.

McNab, B.K. (1974) The behavior of temperate cave bats in a subtropical environment. *Ecology,* **55**, 943-958.

Mills, R.S., Barrett, G.W. & Farrell, M.P. (1975) Population dynamics of the big brown bat (*Eptesicus fuscus*) in southwestern Ohio. *Journal of Mammalogy,* **56**, 591-604.

Mutere, F.A. (1967) The breeding biology of equatorial vertebrates: Reproduction in the fruit bat, *Eidolon helvum,* at latitude 0°20′ N. *Journal of Zoology,* London, **153**, 153-161.

Mutere, F.A. (1968) The breeding biology of the fruit bat *Rousettus aegyptiacus* E. Geoffroy living at 0°22′ S. *Acta Tropica,* **25**, 97-108.

Mutere, F.A. (1969) Flight activity of the tropical Microchiroptera, *Tadarida (Chaerephon) pumila* Cretzschmar and *Tadarida (Mops) condylura* A. Smith. *Lynx,* **10**, 53-59.

Mutere, F.A. (1973) On the food of the Egyptian fruit bat *Rousettus aegyptiacus,* E. Geoffroy. *Periodicum biologorum, Zagreb,* **75**, 159-162.

Nelson, J.E. (1965) Movements of Australian flying foxes (Pteropodidae: Megachiroptera). *Australian Journal of Zoology,* **13**, 53-73.

Norberg, U.M. (1970) Hovering flight of *Plecotus auritus* Linnaeus. *Bijdragen tot de Dierkunde,* **40**, 62-66.

Novick, A. (1958) Orientation in paleotropical bats, 1, Microchiroptera. *Journal of Experimental Zoology,* **138**, 81-154.

Nyholm, E.S. (1965) Zur Ökologie von *Myotis mystacinus* (Leisl.) und *M. daubentoni* (Leisl.) (Chiroptera). *Annales Zoologici Fennici,* **2**, 77-123.

O'Farrell, M.J. & Bradley, W.G. (1970) Activity patterns of bats over a desert spring. *Journal of Mammalogy,* **51**, 18-26.

Packard, R.L. & Mollhagen, T.R. (1973) Bats of Texas caves. In *Natural History of Texas Caves,* Dallas, 122-132.

Panyutin, K.K. (1970) Ekologia letuchikh myshey v lesnykh landshaftakh. Dissertation Thesis Abstract. Moskva: University.

Pearson, O.P., Koford, M.R. & Pearson, A.K. (1952) Reproduction of the lump-nosed bat (*Corynorhinus rafinesquei*) in California. *Journal of Mammalogy,* **33**, 273-320.

Pennycuick, C.J. (1973) Wing profile shape in a fruit-bat gliding in a wind tunnel, determined by photogrammetry. *Periodicum Biologorum, Zagreb,* **75**, 77-82.

Perez, G.S.A. (1973) Notes on the ecology and life history of Pteropidae on Guam. *Periodicum Biologorum,* Zagreb, **75**, 163-168.

Racey, P.A. (1975) The prolonged survival of spermatozoa in bats. *Biological Journal of the Linnean Society,* **7**, Supplement 1, 385-416.

Racey, P.A. (1977) The environmental control of reproduction in hibernating bats. *Proceedings of the Fourth International Bat Research Conference,* Nairobi, in print.

Racey, P.A. & Kleiman, D.G. (1970) Maintenance and breeding in captivity of some vespertilionid bats, with special reference to the noctule. *International Zoo Yearbook,* **10**, 65-70.

Racey, P.A., Suzuki, F. & Medway, Lord (1975) The relationship between stored spermatozoa and the oviducal epithelium in bats of the genus *Tylonycteris*. In *The Biology of spermatozoa,* ed. Hafez, E.S.E. & Thibault, C.G. Basel: Karger.

Racey, P.A. & Tam, W.H. (1974) Reproduction in male *Pipistrellus pipistrellus* (Mammalia: Chiroptera). *Journal of Zoology,* London, **172**, 101-122.

Rachmatulina, I.K. (1971) *Rukokrylye Azerbaidzhana.* Dissertation Thesis Abstract. Baku: University.

Rachmatulina, I.K. (1971) Razmnozhenie, rost i razvitie netopyrej-karlikov v Azerbaidzhane. *Ekologia,* **2**, 54-61.

Roeder, K.D. (1967) Predator and prey. *Bulletin of the Entomological Society of America,* **13**, 6-9.

Roer, H. (1960) Vorläufige Ergebnisse der Fledermaus-Beringung un Literaturübersicht. *Bonner Zoologische Beiträge,* **11**, Sonderheft, 234-263.

Roer, H. (1969) Zur Ernährungsbiologie von *Plecotus auritus* (L.) (Mam. Chiroptera). *Bonner Zoologische Beiträge,* **20**, 378-383.

Roer, H. (1970) Zur Wasserversorgung der Microchiropteren *Eptesicus zuluensis vansoni* (Vespertilionidae) und *Sauromys petrophilus erongensis* (Molossidae) in der Namibwüste. *Bijdragen tot de Dierkunde,* **40**, 71-73.

Roer, H. (1971) Weitere Ergebnisse und Aufgaben der Fledermausberingung in Europa. *Decheniana-Beihefte,* **18**, 121-144.

Roer, H. (1972) Zur Bestandsentwicklung der Kleinen Hufeisennase (Chiroptera, Mam.) im westlichen Mitteleuropa. *Bonner Zoologische Beiträge,* **23**, 325-337.

Roer, H. (1973) Über die Ursachen hoher Jugendmortalität beim Mausohr, *Myotis myotis* (Chiroptera, Mammalia). *Bonner Zoologische Beiträge,* **24**, 332-341.

Roer, H. (1974) Fledermaus-Invasion in einer rheinischen Grossstadt. *Rheinische Heimatpflege,* **2**, 98-103.

Roer, H. & Egsbaek, W. (1966) Zur Biologie einer skandinavischen Population der Wasserfledermaus (*Myotis daubentoni*) (Chiroptera). *Zeitschrift für Säugetierkunde,* **31**, 440-453.

Rosevear, D. (1965) *The bats of West Africa.* London: British Museum (Natural History).

Ross, A. (1967) Ecological aspects of the food habits of insectivorous bats. *Proceedings of the Western Foundation of Vertebrate Zoology,* **1**, 205-264.

Ryberg, O. (1947) *Studies on bats and bat parasites.* Stockholm: Svensk Natur.

Saint Girons, H., Brosset, A. & Saint Girons, M.C. (1969) Contribution à la connaissance du cycle annuel de la chauve-souris *Rhinolophus ferrumequinum* (Schreber, 1774). *Mammalia,* **33**, 357-470.

Slaughter, B.H. & Walton, D.W. (1970) *About bats.* Dallas: Southern Methodist University.

Sluiter, J.W. & Heerdt, P.F. van (1957) Distribution and decline of bat populations in S. Limburg from 1942 till 1957. *Natuurhistorisch Maandblad,* **46**, 134-143.

Sluiter, J.W. & Heerdt, P.F. van (1966) Seasonal habits of the noctule bat (*Nyctalus noctula*). *Archives Néerlandaises de Zoologie,* **16**, 423-439.

Sluiter, J.W., Heerdt, P.F. van & Gruet, M. (1971) Paramètres de population chez le grand rhinolophe fer-à-cheval (*Rhinolophus ferrum-equinum* Schreber), estimés par la méthode des reprises après baguages. *Mammalia,* **35**, 254-272.

Sluiter, J.W., Heerdt, P.F. van & Voûte, A.M. (1971) Contribution to the population biology of the pond bat, *Myotis dasycneme* (Boie, 1825). *Decheniana-Beihefte,* **18**, 1-44.

Smith, J.D. (1972) Systematics of the chiropteral family Mormoopidae.

Miscellaneous Publications, Museum of Natural History, University of Kansas, **56**, 1-132.

Sologor, E.A. & Petrusenko, A.A. (1973) K izutcheniu pitania rukokrylykh (Chiroptera) srednego Pridnjeprovia. *Vestnik Zoologii, Kiev,* 7, 40-45.

Stebbings, R.E. (1966) A population study of bats of the genus *Plecotus. Journal of Zoology, London,* **150**, 53-75.

Stebbings, R.E. (1969) Observer influence on bat behaviour. *Lynx,* **10**, 93-100.

Strelkov, P.P. (1962) The peculiarities of reproduction in bats (Vespertilionidae) near the northern border of their distribution. In *Symposium theriologicum,* ed. Kratochvíl, J. & Pelikán, J. Praha: CAS.

Strelkov, P.P. (1969) Migratory and stationary bats (Chiroptera) of the European part of the Soviet Union. *Acta Zoologica Cracoviensia,* **14**, 393-439.

Strelkov, P.P. & Kuzyakin, A.P. (1974) *Materialy Pervovo vsesojuznovo sovestchania po rukokrylym (Chiroptera).* Leningrad: ZIN AN SSSR.

Tamsitt, J.R. (1967) Niche and species diversity in neotropical bats. *Nature, London,* **213**, 784-786.

Tinkle, D.W. & Milstead, W.W. (1960) Sex ratios and population density in hibernating *Myotis. American Midland Naturalist,* **63**, 327-334.

Turner, D.C. (1975) *The vampire bat.* Baltimore & London: J. Hopkins University.

Tuttle, M.D. (1976) Population ecology of the gray bat (*Myotis grisescens*): philopatry, timing and patterns of movement, weight loss during migration, and seasonal adaptive strategies. *Occasional Papers, Museum of Natural History, University of Kansas,* **54**, 1-38.

Vaughan, T.A. (1976) Nocturnal behavior of the African false vampire bat (*Cardioderma cor*). *Journal of Mammalogy,* **57**, 227-248.

Vaughan, T.A. & O'Shea, T.J. (1976) Roosting ecology of the pallid bat, *Antrozous pallidus. Journal of Mammalogy,* **57**, 19-42.

Verschuren, J. (1957) Ecologie, biologie et systématique des cheiroptères. *Exploration du parc national de la Garamba.* Bruxelles: IPNCB.

Villa, R.B. (1966) *Los murcielagos de Mexico.* Mexico: Universidad Nacional.

Voûte, A.M., Sluiter, J.W. & Grimm, M.P. (1974) The influence of the natural light-dark cycle on the activity rhythm of pond bats (*Myotis dasycneme* Boie, 1825) during summer. *Oecologia,* **17**, 221-244.

Webster, F.A. & Griffin, D.R. (1962) The role of the flight membranes in insect capture by bats. *Animal Behaviour,* **10**, 332-340.

Whitaker, J.O. Jr. & Black, H. (1976) Food habits of cave bats from Zambia, Africa. *Journal of Mammalogy,* **57**, 199-204.

Williams, T.C., Ireland, L.C. & Williams, J.M. (1973) High altitude flights of the free-tailed bat, *Tadarida brasiliensis,* observed with radar. *Journal of Mammalogy,* **54**, 807-821.

Wilson, D.E. (1971) Ecology of *Myotis nigricans* (Mammalia: Chiroptera) on Barro Colorado Island, Panama Canal Zone. *Journal of Zoology, London,* **163**, 1-13.

Wilson, D.E. (1973) Bat faunas: a trophic comparison. *Systematic Zoology,* **33**, 14-29.

Wilson, D.E. (1973) Reproduction in neotropical bats. *Periodicum Biologorum, Zagreb,* **75**, 215-217.

Wilson, D.E. & Findley, J.S. (1970) Reproductive cycle of a neotropical insectivorous bat, *Myotis nigricans. Nature,* **225**, 1155.

Wilson, D.E. & Findley, J.S. (1972) Randomness in bat homing. *American Naturalist,* **106**, 418-424.

Wimsatt, W.A. (1969) Some interrelations of reproduction and hibernation in mammals. *Symposia of the Society for Experimental Biology,* **23**, 511-549.
Wimsatt, W.A. (1969) Transient behavior, nocturnal activity patterns and feeding efficiency of vampire bats (*Desmodus rotundus*) under natural conditions. *Journal of Mammalogy,* **50**, 233-244.
Wimsatt, W.A. (1970) *Biology of bats,* 1, 2. New York & London: Academic Press.

Ecology of small marsupials 8

C.H. TYNDALE-BISCOE

8.1 Introduction

Marsupials are almost wholly confined to the Neotropical and Australasian regions of the world, where they comprise substantial components of the mammalian faunas. In South and Central America all 60 species are small and represent 7.4 per cent of the total mammalian species (Keast, 1972), while in New Guinea and Australia small marsupials represent 27 and 38 per cent respectively of the mammalian faunas (Ziegler, 1977; Keast, 1972). There are almost as many species of indigenous rodents in Australia as there are small marsupials, and in New Guinea and South America also small rodents and bats are as abundant or more abundant than marsupials (Table 8.1). In none of these three areas do marsupials occupy similar niches to species of either of the two other orders; there are no true flying marsupials and none is strictly seed-eating. The niches that small marsupials do occupy are those of small insectivores, small to medium carnivore/omnivore, and arboreal leaf-, fruit- and nectar-eater. In other regions of the world, especially tropical Africa and Asia, these niches are filled by tupaiids and prosimian primates such as tarsiers, pottos, lorises and lemurs, which do not occur in either Australasia or South America and by species of Insectivora, such as shrews, which are absent from Australasia and only extend into the most northerly parts of South America.

This division of niches reflects the past history of mammals. Marsupials were widespread in North America during the Cretaceous and in South America throughout the Tertiary but, apart from rare occurrences in the Eocene of Europe are unknown from the Eurasian land mass and Africa. In Australia and New Guinea the fossil record extends back to the Oligocene, and the prevailing view is that marsupial ancestors entered Australia from South America via Antarctica in the late Cretaceous or early Tertiary, and adaptive radiation probably took place in the absence of eutherian competition (Clemens, 1977). Murid rodents probably entered Australia via South East Asia and New Guinea in the Miocene or later and filled many

Table 8.1 *A comparison of the numbers of species less than 5 kg of Marsupialia, Chiroptera and Rodentia in Australia, New Guinea and Neotropica*

	Total	Marsupialia	Chiroptera	Rodentia
Australia[1]	364	139	89	124
New Guinea[2]	172	46	82	44
Neotropica[1]	810	60	375	222

[1] Data from Keast (1972) Table 3.
[2] Data from Ziegler (1977) Table 7.1.

apparently unoccupied niches. It is surprising in the circumstances that marsupials had not already filled the usual rodent niches by the Miocene. However, grasses only achieved their dominance in the Miocene and it may be that the rodents, having evolved rapidly in the Old World by exploiting this abundant new food source, pre-empted the newly evolved seed-eating niches when they entered Australia.

In South America marsupials extend from sea level to the sub-alpine zone at 3500 m and from the tropics to the cool temperate climate of southern Chile (Hershkovitz, 1972). They are predominantly forest dwellers, living on the ground as omnivores or in trees as frugivores or insectivores; only one is a leaf eater, *Dromiciops* in Chile. Since the Pliocene, when the isthmus of Panama emerged, marsupials have extended north as far as Mexico, their northern limit being defined by the limit of tropical forest.

Only one species of marsupial extends into North America. Until recently *Didelphis virginiana* was thought to represent a northern extension of the lowland *Didelphis marsupialis* of South America but Gardiner (1973) has shown that *D. virginiana* is a distincly new species that evolved very recently in Central America where it is now sympatric with *D. marsupialis*. Evidence from archaeological sites in North America suggest that it did not occupy its present northern distribution earlier than 4000 years ago, so that it is a recent invader of North America. The ecology of American marsupials has been comprehensively reviewed by Hunsaker (1977b).

In Australia and New Guinea small marsupials occur in a much wider variety of habitats, ranging from tropical rain forests, through eucalyptus forests and woodland savannah to semi-arid desert (Keast, 1972). In the forests and woodland there are a considerable number of leaf eating arboreal species (Phalangeroidea) as well as nectar and insect eaters (Burramyidae, Dasyuridae). The species that occupy the grassland and deserts are predominantly insectivorous (Dasyuridae and Peramelidae), but the small hare wallabies and rat kangaroos (Macropodidae) are grass eaters.

It was customary 30 years ago to view marsupials as distinctively inferior mammals but the very large body of modern research into marsupial

Table 8.2 *Comparisons of gestation lengths and relative birth weights in marsupials and eutherian mammals.*

Species	Gestation (days)	Possible litter size	Individual birth weight (g)	Maternal weight (kg)	Birth weight/maternal weight × per cent	
Metatheria						
Antechinus stuartii	26–35	8–10	0.016	0.03	0.05	0.5 for litter
Isoodon macrourus	14	6	0.18	1.1	0.02	0.10 for litter
Didelphis virginiana	13	13	0.16	1.4	0.01	0.14 for litter
Trichosurus vulpecula	17	1	0.22	2.5	0.01	0.01 for litter
Setonix brachyurus	27	1	0.34	3.5	0.01	0.01 for litter
Eutheria						
Mesocricetus auratus	16	5	2.2	0.096	2.3	11.5 for litter
Rattus norvegicus	22	7	4.5	0.2	2.3	16.1 for litter
Oryctolagus cuniculus	32	6	57	1.9	3.0	18.0 for litter

Data for eutheria from Parker (1977).

biology does not support this view. On the contrary, there seems to be an extraordinary degree of similarity between metatherian and eutherian physiology, biochemistry and behavior (*see* Tyndale-Biscoe, 1973; Stonehouse and Gilmore, 1977; Hunsaker, 1977a for reviews). The only truly distinctive feature of marsupials is their mode of reproduction which must have influenced the ecological strategies they evolved. Elucidation of these strategies is the main contribution that marsupial studies can make to general ecological theory and this chapter will be largely devoted to considering the best understood of these.

8.1.1 Distinctive features of marsupial reproduction

Gestation

The gestation periods of small marsupials are of similar duration to those of small eutherians (Table 8.2). However, the young at birth are very small; none weighs as much as 1 g and the young of the smallest species may weigh as little as 10 mg. As a proportion of the mother's body weight it is tiny and ranges from 0.01 per cent to 0.05 per cent (Table 8.2). Even the whole litter of polytocous species, such as *Antechinus*, are only 0.5 per cent of their mother's weight. In contrast single eutherian young weigh 2.3 to 3 per cent of the mother's weight and for example, a litter of rabbits at birth may weigh 18 per cent of the doe's own weight. Thus the investment in resources for pregnancy made by a female marsupial is far less than that made by any eutherian and in ecological terms is probably negligible (Parker, 1977).

Lactation

Growth of the young takes place during the suckling phase, which is relatively much longer than in eutherian mammals. For the female marsupial the major investment in reproduction occurs during the second half of this long period of lactation while the young are still wholly dependent on her. During this time the weight of the litter may equal or exceed that of the mother, e.g. *Antechinus*. Even in a monotocous species such as the brush possum, *Trichosurus vulpecula*, the outlay is considerable as shown by the weight of the suckled mammary gland which increases 12-fold between parturition and 140 days when the young begins to leave the pouch (Fig. 8.1). The end of pouch life is also the most critical period in the survival of the young, when their physiological control is not completely developed and when they are changing from milk to an adult diet. In the brush possum Dunnet (1964) observed that mortality was most severe in the young between five months and ten months of age, and in the greater glider,

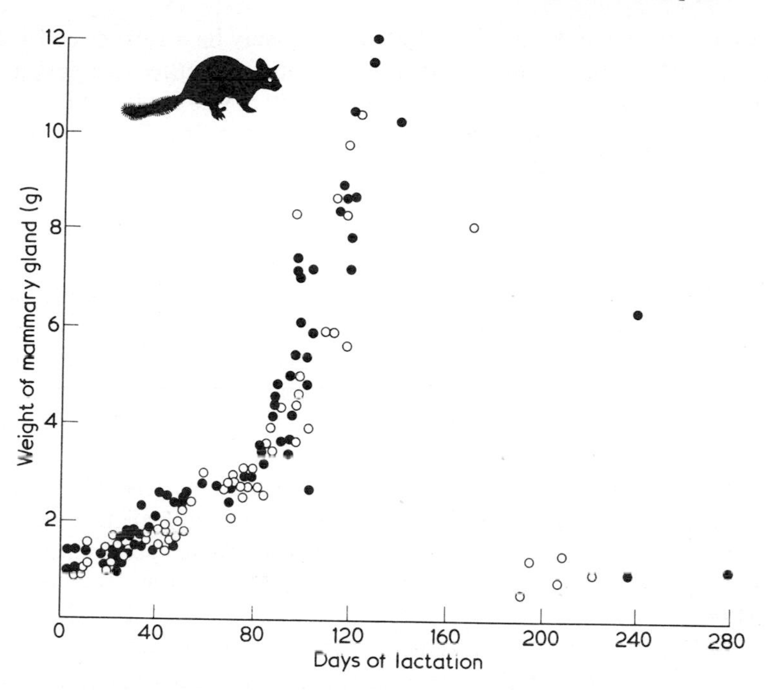

Fig. 8.1 Relationship between weight of suckled mammary gland and duration of lactation in the brush possum, *Trichosurus vulpecula*. Open circles, glands of females less than 2 year; closed circles, older females. (From Smith, Brown & Frith, 1969, *CSIRO Wildlife Research*, **14**, 187.)

Schoinobates volans (Tyndale-Biscoe and Smith, 1969a) and in the ringtail possum, *Pseudocheirus peregrinus* (Thomson and Owen, 1964) the heaviest mortality occurs after emergence from the pouch.

Seasonal breeding

In many marsupials, especially those occupying temperate climates or places with marked annual seasons, the period of weaning or pouch exit rather than the period of birth coincides with the most favourable time of the year. Thus in *Antechinus stuartii* births occur in winter and weaning in late spring or summer, while in the brush possum with a longer pouch life most births occur in autumn (April-May) and the young leaves the pouch in late spring. In the small wallaby, *Setonix brachyurus*, females come into oestrus in January or February, which is the hottest, driest time of the year when the animals are in a very poor state of nutrition but the young emerge from the pouch eight months later after new plant growth has been stimulated by winter rains.

The long nursing period limits the number of litters that a female can raise each year, while the drain on her reserves may be a further constraint which limits lactation to one part of the year only. Another strategy, available to polytocous marsupials, is to vary the litter size itself. This may be achieved at the time the young are born or part or all of the litter may be lost during the course of lactation.

Litter size

During the first half of lactation each young is attached permanently to one teat so the maximum litter size is determined by the number of teats. For the insectivorous and omnivorous species this is generally more than 4 and in some species exceeds 10, so that considerable variation in litter size is possible. Conversely, none of the herbivorous marsupials have more than 4 teats and some have only two. The smaller arboreal species and the forest dwelling rat kangaroo, *Hypsiprymnodon moschatus*, may have litters up to the maximum of 4 but the larger arboreal species and all the other rat kangaroos (Potoroinae) and wallabies (Macropodinae) are monotocous. Twins are very rare among these species and seldom survive to the end of pouch life.

The marsupial mode of reproduction requires the female to make her investment of resources in the offspring slowly over a relatively long period but allows her to terminate it at almost any stage if conditions become adverse. By comparison the female eutherian makes a large investment across the placenta during pregnancy, which once embarked upon is difficult to terminate, but she gains the advantage of completing the process faster. As a result, small eutherian mammals, such as rodents and lagomorphs, have potential rates of increase far greater than small marsupials and, in the short term this enables them to colonise new habitat more rapidly and to respond by genetic change more rapidly as well. In the longer term this advantage seems to be less significant as will be seen in the New Zealand situation.

8.2 Breeding strategies of polytocous marsupials

8.2.1 American opossums

The American opossums of the genus *Didelphis* exemplify the way litter size and breeding season are adapted to environmental factors. *Didelphis marsupialis* is restricted to the warm humid habitats of South and Central America up to an altitude of 2000 metres. It is replaced by *D. albiventris* in subtropical to cool temperate habitats up to 3500 m and as far as latitude 39°S. In North America *D. virginiana,* like *D. albiventris,* is able to occupy cool temperate habitat as far north as the Great Lakes at latitude 45°N. The three closely related species thus occupy a very wide latitudinal and altitudinal range and a great diversity of habitats.

The mean litter size of *D. virginiana* in the northern parts of its range is 9 but in Florida and Texas it is about 6. At Panama the mean litter size of *D. marsupialis* is likewise 6 but in parts of Colombia it is less than 5 and further south at Rio de Janeiro (23°S) it is 7 to 8 (Tyndale-Biscoe and Mackenzie, 1976). In both species the number of eggs shed at one ovulation greatly exceeds these litter sizes so that considerable wastage or competition must occur in the uterus and at the time of birth. In *D. virginiana* the usual number of teats is 13 and in *D. marsupialis* 9 but maximum litter sizes are rare, so that other factors must be involved in determining the actual number of young that obtain a teat. The posterior teats are more frequently occupied than the anterior, so there may be a posteroanterior gradient in the development of the mammary glands that limits litter size to the nutritional capacity of the mother at the time she gives birth.

In *D. virginiana* breeding commences in January in lower latitudes such as in Florida and Texas, but in February or even in March further north (Hamilton, 1958). In the zone of sympatry of the two species in Nicaragua breeding also commences in January as it does for *D. marsupialis* in Panama and Colombia. However, south of the Equator, in Brazil, this species commences to breed in June or July which would seem to implicate photoperiod as the proximate factor for the initiation of breeding in both species. It is not so clear what determines the close of breeding and hence the number of litters per year.

For *D. virginiana* two litters are the rule, although a third litter may be possible in Texas. Further north no females were found with pouch young after September and, in New York State even a second litter is uncommon (Hamilton, 1958). The close of breeding in North America is probably due to diminishing food resources and cooling temperatures in the autumn. Nevertheless, at lower latitudes, where low temperatures are not a factor, *D. marsupialis* displays similar patterns with two litters a year in Panama (Fleming, 1973), eastern Colombia and Brazil (Tyndale-Biscoe and MacKenzie, 1976). At each of these sites the second litters are weaned towards the end of a marked wet season and the females enter anoestrus. In Panama, Fleming (1973) considers that the breeding season of *Didelphis* may be so timed that weaning coincides with the main fruiting times from May to September. In Nicaragua and western Colombia there is no marked dry and rainy season and three peaks of births occurred in February, May and August.

For both *D. marsupialis* and *D. virginiana,* litter size and the average number of litters per year appear to be inversely related to each other and both are related to latitude and annual patterns of climate. Thus in both species the total annual productivity of the females is much the same throughout the wide latitudinal range they occupy, but is achieved by different strategies in response to local climatic conditions. Where the favourable period for rearing young is restricted to six months, the females produce either one or two large litters per year, whereas at the other extreme, as in western Colombia females produce three small litters annually. The small amount of evidence indicates that *D. albiventris* females in the highlands of Colombia and Brazil have two small litters per year and hence a lower productivity than *D. marsupialis*.

In Panama, two other species of marsupial, smaller than *Didelphis*, have similar diet and habitat preferences (Fleming, 1973). These are *Philander opossum* and *Metachirus nudicaudatus* and both have breeding patterns similar to *Didelphis*, with two litters a year. However, the much smaller mouse opossum, *Marmosa robinsoni* has only one litter a year, but the average litter size of 10 is larger than in any of the other three. As a result of this the productivity of all four species is similar.

8.2.2 Australian bandicoots

The reproductive strategies of the bandicoots (Peramelidae) differ from the opossums. The reproductive sequence is much more rapid than in any other marsupial. Gestation of *Perameles nasuta* is 12 days and suckling lasts 45 days so that production of a litter is accomplished in two months (Stodart, 1977). The female can return to oestrus within 10 days of weaning, so that a succession of litters can be produced in one year. The same pattern has been described for three other species, *P. gunnii* in Tasmania (Heinsohn, 1966), *Isoodon obesulus* in Tasmania and Victoria (Stoddart and Braithwaite, in press) and *I. macrourus* in New South Wales and Queensland (Gordon, 1974). All these species have eight teats but the mean litter size is less than 4. Although the ovulation rate is low another reason for the small litter size is that the teats used by the previous litter are too large for the newborn young to attach to and require a month or more to fully regress. While litters of 5 and 6 do occur, they are rare, as are litters of 1 or 2.

In the more southern regions of Tasmania, Victoria and New South Wales, which receive regular winter rainfalls, the breeding seasons of bandicoots are restricted to the months of July to March when insect and invertebrate food is most plentiful. However in Queensland where food supply is less variable, *I. macrourus* breeds throughout the year.

In New South Wales the breeding of *I. macrourus* appeared to start just after the females increased in weight and ceased when they lost weight. Similarly, Stoddart and Braithwaite (in press) found that litter size was positively correlated with body weight of the mother in *I. obesulus*. However, in this species the onset of breeding was so highly synchronised among the females in three successive years that they conclude that an additional signal, probably increasing photoperiod after the winter solstice, must be involved. Heinsohn (1966) came to the same conclusion in regard to the onset of breeding of *P. gunnii*.

Female *P. gunnii* produced 3 or 4 litters in a season (mean 3.8) but the earliest and latest litters were smaller (2.14 and 2.25) than the middle ones (2.9 and 2.8) which would reflect the changing body weight of the females. Productivity of females of this species was thus about 9.5 young/female/year. However, survival of the first litter was higher than that of later ones because the young were weaned at the time of most abundant food. They were able to establish themselves and breed by the end of the season of their own birth, which would contribute yet further to the potential rate of increase of the population. A similar pattern of changing litter·size was observed in *I. macrourus* in New South Wales, but the same species in Queensland breeding all the year had a mean litter size of 3.4. The bandicoots thus have a reproductive pattern that is responsive to environmental and nutritional conditions; under adverse conditions breeding is delayed or litter size reduced, while under good nutritional conditions reproductive performance may be influenced by female body weight and the potential rate of population growth increased. The primary factor that makes this strategy possible is the very rapid process of reproduction compared to other marsupials of comparable size. It is not unreasonable to suppose that the highly developed chorio-allantoic placenta of bandicoots may be in part responsible, by enabling the young to be born after a very short gestation at a more advanced stage of development than other marsupials.

8.3 Semelparity — an unique marsupial strategy

Twenty-five species of dasyurids occupy the niches of insectivores and small carnivores throughout Australia and most are members of two genera. Species of *Sminthopsis* are strictly terrestrial and all that have been studied are seasonally polyoestrous. *Antechinus* species, on the other hand, are scansorial and monoestrous. Each genus is represented by different species in a wide range of habitats from tropical forest to desert. The species are generally allopatric although there may be zones of sympatry (Fig. 8.2).

8.3.1 Life-history of *Antechinus* species

The ecology of *Antechinus stuartii* is the best known of all this group largely because of its highly unusual life-history, which has stimulated a considerable amount of research (Lee, Bradley and Braithwaite, 1977). It is a forest dwelling species that occurs in eastern Australia from 18°S near Townsville to the south coast of Victoria at 38°S (Fig. 8.2). It is sympatric in part of its range with *A. swainsonii*, a larger forest dwelling species, but with the very

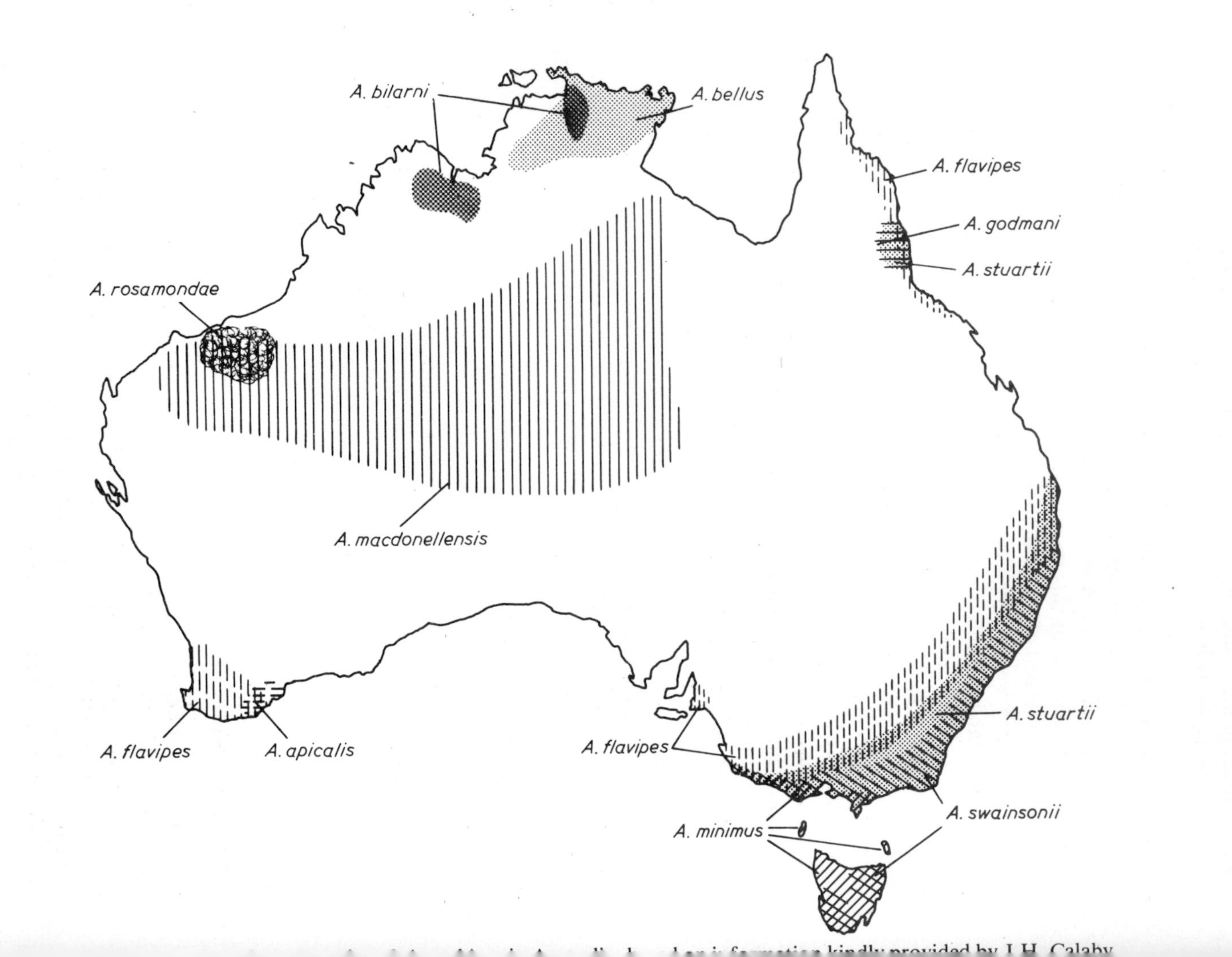
A. bilarni
A. bellus
A. flavipes
A. godmani
A. stuartii
A. rosamondae
A. macdonellensis
A. flavipes
A. apicalis
A. flavipes
A. stuartii
A. swainsonii
A. minimus

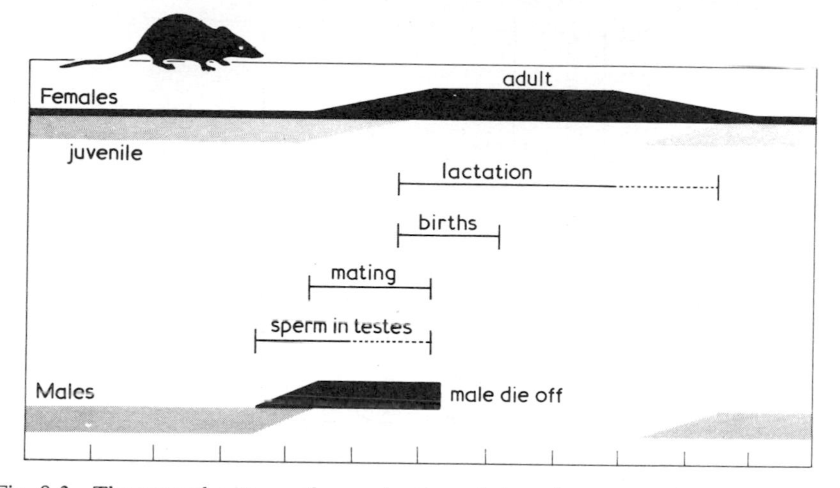

Fig. 8.3 The annual pattern of reproduction of *Antechinus stuartii*, near Canberra. (From Woolley, P., 1973.) Breeding patterns, and the breeding and laboratory maintenance of dasyurid marsupials. *Experimental Animals* **22**, 168.

similar-sized *A. flavipes*, which inhabits dry woodland, it is largely allopatric. *A. flavipes* extends to South Australia and a subspecies occurs in south western Australia. Along the south east coast of Australia and in Tasmania *A. minimus* occupies treeless sedgelands. All these species share with *A. stuartii* a similar life history, the most remarkable feature of which is the total demise of the adult male population at the close of the breeding season.

In Queensland at 27°S the young are weaned in February and become sexually mature in June (Wood, 1970). The males become increasingly aggressive towards each other and move greater distances. In test encounters they are highly territorial from then until mid-September when the females come into oestrus and mating occurs (Braithwaite, 1974). Copulation is pursued strenuously for 2 weeks, with individual acts lasting several hours each. Ovulation occurs spontaneously during this time and pregnancy lasts 25–31 days. An excess of young is delivered so that usually all 8 teats become occupied (average litter size is 7.5). Births occur in late October, and by this time none of the males is still alive. In Wood's study the last males were trapped in early October. For about 2 months the young are continuously attached to the mothers teats and for a further 2 months they are left in a nest to which the female returns to suckle them. The litter is weaned and the young disperse in February. Most of the adult females disappear from the population, so that less than 20 per cent contribute to the next breeding season.

Near Canberra, (Woolley, 1966) and near Melbourne (Lee *et al.*, 1977), the pattern is the same but occurs about one month earlier in the year

Table 8.3 *Summary of life history events in species of* Antechinus *in eastern Australia*

Species	Locality	♂ territory formation	Mating	♂ die off	Birth	Weaning	Reference
A. stuartii	Queensland	M – Jne	S–O	early Oct.	O–N	Feb.	1
A. stuartii	A.C.T.	M – Jne	Jl–S	S	A–O	Jan.	5
A. stuartii	Victoria	M – Jne	Aug	Late A	S	Dec.	1
A. flavipes	A.C.T.	May	Jn–Jl	–	Jl–A	–	5
A. flavipes	S. Aust.	June	July	–	Aug	–	2
A minimus	Victoria	–	June	July	July	Nov.	4
A. swainsonii	Victoria	–	July	–	Aug	–	3

[1] Lee, Bradley and Braithwaite (1977)
[2] Inns (1976)
[3] Leonard (1976)
[4] Wainer (1976)
[5] Woolley (1966)

(Fig. 8.3). In *A. flavipes* near Canberra the sequence begins a month earlier than in *A. stuartii,* so that males of this species have disappeared before the females of *A. stuartii* have come into oestrus (Table 8.3). Despite the overlap in range the effective reproductive isolation maintains sympatry.

Similarly in Victoria *A. swainsonii* females give birth one month (Leonard, 1976) and female *A. minimus* give birth two months earlier than female *A. stuartii* and Wainer (1976) has suggested that the difference may be related to a difference in feeding habits and available food; in *A. minimus*, which lives on the ground and hunts for larval insects, lactation coincides with the winter peak of larvae, whereas *A. stuartii*, which hunts above the ground, breeds later when adult insects are abundant.

8.3.2 Cause of synchronous death of males

Much of the recent research has been directed to understanding the phenomenon of synchronous male die off and its significance in the evolution of the species. Two hypotheses have been examined. One that the males undergo accelerated senescence and the other that they become acutely stressed by the intensive aggressive and sexual encounters at this time.

The increasing activity and aggressiveness coincides with enlargement of the testes, penis, prostate and cowpers glands (Woolley, 1966), and with the onset of sperm production so, although no measurements have been published, it is reasonable to presume that circulating testosterone levels are high at this time. The males grow heavier than females and their metabolic rate as determined by oxygen consumption and food intake increases so that both functions reach a maximum immediately before the brief mating period. Body weight then drops precipitately and they enter negative

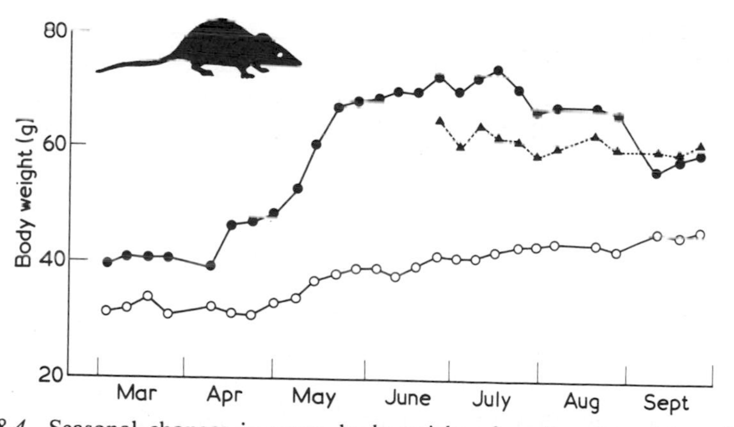

Fig. 8.4 Seasonal changes in mean body weight of captive *Antechinus flavipes*. Closed circles, intact males; closed triangles, castrated males; open circles, intact females. (From Inns, R.W., 1976, *Australian Journal of Zoology* **24**, 525.)

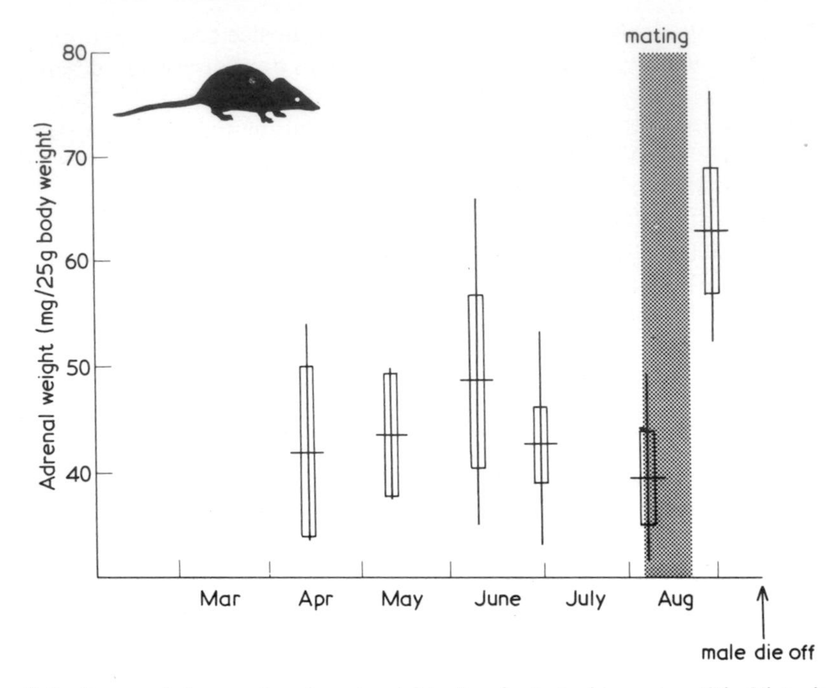

Fig. 8.5 Seasonal changes in adrenal weight of male *Antechinus stuartii* in Victoria. Adrenal weight has been adjusted to a standard 25 g body weight and is expressed as mean ± S.E. and range. (From Barnett, J.L., 1973, *Australian Journal of Zoology,* **21**, 508.)

nitrogen balance. Inns (1976) has found the same to hold for *A. flavipes,* and when he castrated males at the beginning of the season in June he abolished in them the upward rise in metabolism and body weight (Fig. 8.4), which implicates testosterone.

Barnett (1973) observed increased adrenocortical weight in males immediately before their demise (Fig. 8.5). This was correlated with increased corticosteroids in the blood and of glycogen in the liver, as well as falls in sodium and glucose. These several effects were not found in females or in captive males but could be induced in captive males by injections of cortisol acetate, when a high proportion died (Lee *et al.*, 1977). It is not clear what the final cause of death is but an important effect of corticosteroids is to lower the immune reaction to pathogens.

Field caught *A. stuartii* were examined for helminth parasites and blood protozoa through the breeding season (Beveridge and Barker, 1976). The incidence and severity of helminth infestations was low in all animals of both sexes until shortly before the male die off when both increased in males but not in females (Fig. 8.6). Nevertheless the pathological changes associated with the parasites were not considered sufficient to cause death. On the

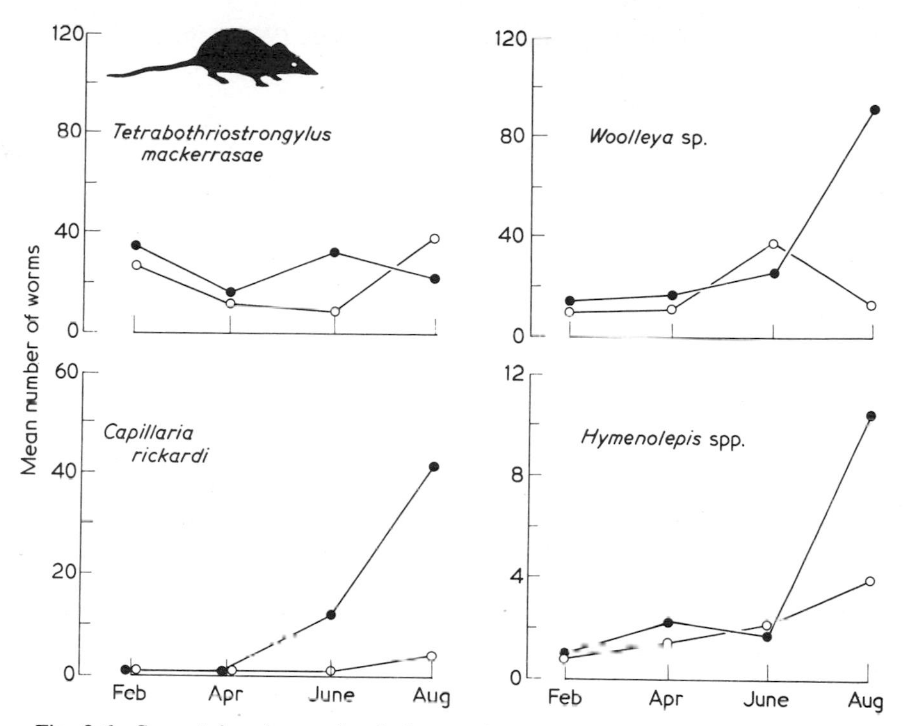

Fig. 8.6 Seasonal and sex-related changes in the numbers (mean + S.E.) of nematodes and cestodes in *Antechinus stuartii* in Victoria. Closed circles, males; open circles, females. (After Beveridge, I. & Barker, I.K., 1976, *Australian Journal of Zoology* **24**, 269.)

other hand males collected at the end of the mating season and observed continuously in the laboratory until they became moribund had several pathologic states including, anaemia, involuted splenic follicles and gastrointestinal ulceration and haemorrhage associated with bacterial infection. Death could not be ascribed to any one organism, but it seems that a number of common pathogens increase at this time, which may reflect reduced immunological resistance brought about by the high corticosteroid levels.

Of the two hypotheses first proposed the weight of evidence now favours stress rather than senescence as the cause of death of males and a variety of factors, endocrine, metabolic and pathologic contribute to the final result. Nevertheless, the quite unique synchrony of death of all the males of one population and the occurrence of the phenomenon in several species in different habitats indicates that it must have a genetic basis. Furthermore, the cells of the testis are programmed to undergo only one cycle of activity, because males that have been artificially brought through the critical period in captivity and lived into their second year had shrunken testes that lacked a seminiferous epithelium and Leydig cells (Woolley, 1966; Inns, 1976).

8.3.3 Significance of semelparity

Two important consequences for the survival of the species follow from the male die off. First, there is no second chance for females that fail to conceive, except for the small number that survive to the next season; the long single oestrous period of the females, the greatly increased activity of the males at the mating period and the vigorous copulation are all presumably adaptations that ensure a high rate of conception. Second, there is no opportunity for rapid growth of the population, since no female can conceive more than one litter in a year and the litter size is almost invariably the maximum. Since these two consequences would appear to put the species at a serious disadvantage what compensating advantage does this pattern of breeding afford? Braithwaite and Lee (1979) have examined this, pointing out that semelparity (or single breeding), which is common among bacteria, plants and invertebrates, is very rare among vertebrates. It is of selective advantage to species with a short life span that live in a habitat where conditions favourable to reproduction occur briefly once a year. In these circumstances selection will favour those individuals that maximise reproductive effort at that time, even to the exclusion of surviving to a second breeding season. While many small mammals and birds live for little more than a year they can usually produce several litters in that time, so that semelparity is not an advantage to them. On the other hand most small marsupials take longer to raise a litter to independence and the total investment by the female parent is greater. Thus in *A. stuartii* the time from conception to weaning is 5 months and the litter of young by then weigh 3 times as much as the mother. As mentioned earlier this period coincides for the different species with the time of maximum abundance of their main insect food. If females were to breed again in February the main burden would fall on them in winter when insect food is scarce.

Semelparity occurs in species of *Antechinus* that live in the forests of eastern and south Australia with winter rainfall and a marked seasonal abundance of insects, and probably occurs in *A. bellus* and *A. bilarni* which live in a monsoonal climate with marked changes in abundance of food (J.H. Calaby, pers. comm.). On the other hand the New Guinea species, *A. melanurus*, and *A. godmani* of north Queensland, which live in tropical forests do not exhibit semelparity (Dwyer, 1977). The desert living species of Western Australia, *A. rosamondae* and *A. macdonnellensis*, and *A. apicalis* from the south-west are polyoestrous and the males do not die synchronously after their first breeding season either. Similarly, none of the many species of *Sminthopsis* so far investigated are monoestrous or exhibit synchronous male die off, even though some species live in the same habitat as species of *Antechinus* that do show the phenomenon. Semelparity is therefore not to be explained solely in terms of the peculiar features of marsupial reproduction.

Semelparity can be an effective isolating mechanism for sympatric species. *A. stuartii* and *A. flavipes* are sympatric in dry sclerophyll forest near Canberra but are reproductively isolated because *A. flavipes* males die in July–August before *A. stuartii* females come into oestrus at the end of September. Similarly *A. stuartii* and *A. swainsonii* breed out of phase and in northern Australia *A. bellus* and *A. bilarni* do too.

8.3.4 Response to fire

Fire is a natural feature of the forests in which *Antechinus* species live so it is pertinent to ask how they react to fire. Leonard (1976) examined the response of three species of small mammal to experimental control burning, which involved relatively cool fuel reduction burns of the forest litter. He found that all three species survived the burn in dry sclerophyll forest but some individuals succumbed to the hotter fires of wet sclerophyll forest. However, the subsequent responses were related to the consequent loss of cover. Breeding of *A. swainsonii* was reduced but *A. stuartii* bred normally. Newsome, McIlroy and Catling (1975) have described the effects on populations of 5 small mammals to a devastating forest fire that occurred on December 16–18, 1972 on the Nadgee Fauna Reserve in New South Wales, and their recovery in the aftermath and for the next two years. In general the fire annihilated the populations of the two native rodents, *Rattus fuscipes* and *R. lutreolus* and the two dasyurids *A. swainsonii* and *A. stuartii*. The two species of *Rattus* disappeared from some areas immediately but the *Antechinus* species disappeared only after the breeding season of 1973. The alien species *Mus musculus*, which had never been trapped before the fire, appeared in the first year and was very abundant in 1974. In the subsequent years up to 1977 the two species of *Rattus* became re-established at their former levels and *Mus musculus* reverted to being an uncommon species. *Antechinus stuartii* slowly recovered but not to its former level and *A. swainsonii* was still absent five years after the fire. This is a rare but important example of the consequences of the eutherian and the extreme marsupial reproductive strategies after very severe population depletion.

8.4 Contrasting strategies of arboreal leaf eating marsupials

The eucalypt forests of eastern Australia sustain 12 species of arboreal marsupial that range in size from 20 g pigmy possums, *Cercartetus* spp. to the 7 kg koala, *Phascolarctos cinereus*. Four of these species have been studied in some detail and their several strategies for forest life can be usefully compared.

8.4.1 Ringtail possum and greater glider

Within two families, the Petauridae and Burramyidae, there are three pairs of closely related arboreal species in which one species possesses gliding membranes and the other species is non-volant, so that evolution of the gliding mode has evolved independently several times. The largest pair at 1 kg, are the common ringtail possum *Pseudocheirus peregrinus* and the greater glider, *Schoinobates volans*. They belong to the same family, Petauridae, yet their patterns of life, utilization of the forest and reproductive strategies are very different indeed.

Habitat requirements

The ringtail possum is socially gregarious and lives in the lower storey of the forest, being particularly abundant in woodland, low scrub, along water courses and in areas of forest regenerating after partial clearance. It travels across the ground but Thomson and Owen (1964) could find no evidence that ground plants or insects were eaten, and the diet seemed to consist exclusively of foliage of arboreal species, particularly *Eucalyptus, Acacia* and *Leptospermum* and other minor components. In pine plantations they will eat pine needles and pollen, and they eat orchard and garden shrubs in suburban areas. For shelter they build nests, or dreys, in closely branched trees and especially favour clumps of mistletoe that parasitise eucalypts. They are socially gregarious and several ringtails may occupy one drey and feed together in the same tree. Because the dreys are also essential for the survival of young after they cease to ride their mother's back, requirements for drey sites, such as mistletoe, correlate closely with abundance of ringtails.

Ringtail possums are rare inhabitants of wet sclerophyll forest and do not live in the large trees, but this is the prime habitat of the greater glider in which it is the most abundant arboreal marsupial. Gliders feed exclusively on the canopy foliage of the largest *Eucalyptus* trees and retire by day to spouts and hollows formed by broken boughs in these same large trees. They seldom descend to the ground or even to the lower storey of the forest. Unlike ringtail possums, greater gliders are solitary for most of the year. A nearest neighbour analysis of their daytime distribution, obtained when forest was being clear-felled (Tyndale-Biscoe and Smith, 1969a) showed that each sex was more uniformly spaced than they would be if randomly distributed and that they were not clustered as that of a gregarious species

Table 8.4 *Some parameters of populations of four arboreal marsupials*

Species	Productivity, young/female/year	Pouch life, days	Mother-young association, days	Sexual maturity of females, years	Mortality by first year: Males	Mortality by first year: Females	Adult sex ratio, per cent male	References
Brush possum *Trichosurus vulpecula*	1.3	135	175	1	0.83	0.50	40	Dunnet (1964)
Bobuck *Trichosurus caninus*	0.5	175	>240	2–3	0.57	0.64	50	How (1976)
Ringtail possum *Pseudocheirus peregrinus*	1.7	110	180	1	0.68	0.68	48	Thomson and Owen (1964)
Greater glider *Schoinobates volans*	<0.7	120	240	2	0.40	0.20	39	Tyndale-Biscoe and Smith (1969a)

would be. This pattern may not reflect a circumscribed home range, as gliders have been observed to travel considerable distances at night, but it does suggest that there is some active repulsion between members of the population. Owing to the inaccessible nature of their habitat we do not know how this is achieved; the animals are generally silent, but they have well developed paracloacal glands which produce a secretion with a pungent odour, so an olfactory basis is likely.

Reproductive strategies

The reproductive strategies of the two species are very different: ringtail possums first breed in their second year; gliders, not until their third. The females of both species are polyoestrous and the main breeding season begins in May. It is highly restricted in the glider and less than 70 per cent of females give birth to a single offspring. If the young is lost it is not replaced because the males are aspermatogenic after June. Conversely, ringtails give birth to 1–3 offspring (average 1.9) and lost young are probably replaced as births continue to occur until November. It is even possible for females 2 years and older to rear two litters in a year although, in Victoria where the study was done, this was very rare and the overall success rate was 90 per cent. Thus the productivity of the glider is less than 0.7 and of the ringtail about 1.7 young per female per year (Table 8.4), while the difference in the potential rate of increase is further increased by the earlier maturation of the ringtail.

Thomson and Owen (1964) calculated that mortality of ringtails to the end of the first year was 0.68 and was then much lower until the fourth year when it again increased, so that no animals survived their fifth year. The sex ratio remained constant at all ages at 48 per cent males which indicates that there was no sex specific mortality. The main predator of both species in sclerophyll forests is the powerful owl *Ninox strenua*.

Mortality during the first year is low in the glider but is distributed very unevenly between the sexes. Until weaning the sex ratio is 50 per cent male but it changes to less than 40 per cent male at the time the young begins to dissociate from the mother and the same unbalanced sex ratio prevails in all subsequent age classes. We concluded that a male specific mortality factor operating at weaning imposes an imbalanced sex ratio upon the whole adult population because the same pattern was found in all samples over five years and in two separate localities in New South Wales and Queensland. The second unusual feature of the glider population was that the number of adult females that bred equalled the number of adult males. We postulated that the size of the glider population may be regulated by the interplay of these two functions. If the male specific mortality is density dependent and the number of adult males directly determines on a 1:1 basis the number of females that will breed, population size may be regulated by controlling fecundity through a behavioural response to density, instead of by severe mortality of a reproductive surplus. In the terminology of MacArthur and Wilson (1967) the glider is a resource conserver or *K*-selected species and the ringtail an *r*-selected or opportunistic species.

8.4.2 Brush possum and bobuck

The brush possum, *Trichosurus vulpecula*, and the bobuck, *T. caninus*, are the other pair of closely-related arboreal species that have been well studied in recent years and they similarly display different ends of the *r-K* continuum.

Habitat requirements

The bobuck is restricted to the eucalypt forests of eastern Australia and rain forest in the northern part of its range, and does not occur in Tasmania or Western Australia, whereas the brush possum, *Trichosurus vulpecula*, is widely distributed in all wooded habitats of Australia and Tasmania. Where the two species are sympatric, *T. vulpecula* is the less abundant.

Within the forest *T. caninus* selects a wide range of plant species for food, many of which are found on the forest floor or the lower storey, whereas *T. vulpecula* in the same forest subsists on foliage of eucalypts, often of a single species (Owen and Thomson, 1965). In other habitats the brush possum will use other food species in a similar manner. It is probably this ability to exploit the most abundant plant species available that has enabled the brush possum to colonise so many habitats in Australia. Another factor, however, is its high fecundity and potential rate of increase.

Reproductive strategies

The breeding biology of the brush possum has been studied in several habitats in Australia (Dunnet, 1964; How, 1976; Smith, Brown and Frith, 1969). Females reach sexual maturity at one year and 80 per cent produce one young each in autumn. Pouch life is 135 days and the young becomes independent at 6 months so it is possible for females to produce a second young in spring. The number that do so varies between localities and between seasons. Assuming that half the females do, the overall fecundity is about 1.3 young per female per year (Table 8.4). The survival of young to six months is high but during the subsequent dispersal phase there is a considerable mortality, especially of males (Dunnet, 1964). Thus the adult sex ratio is 40 per cent males but, since the species is polygamous, all females breed. There is very little social organization after the mother-young association ceases and the high mortality of newly independent young is thought to be caused by their failure to find den sites of their own and to establish a home range in habitat already occupied by adults. These attributes of high fecundity and early dispersal of the young provide the means for this species to exploit new habitats rapidly, or to re-establish in an old habitat after the population has been reduced by natural factors or by man.

It is not clear from studies so far, what factors restrict the bobuck, but its social organization and population structure are clearly adapted to survival in a stable ecosystem. Females reach sexual maturity at 2 years and there is only one period of births in early autumn (How, 1976; Owen and Thomson, 1965). The pouch life is 175 days and the young does not become independent until 240 days. Even after this time the young one may remain in close association with its mother for another year. How (1976) showed that when this occurs the female either does not breed while her young of the previous year is with her or, if she does, the new young seldom survives beyond early pouch life. Thus the productivity of the bobuck is about 0.5. For those young that do survive, life expectancy is higher than for young brush possums. The young may remain with its mother for two years and there is no selective mortality of young males so that the adult sex ratio is 50 per cent. There is also evidence from How's study of very close overlap of home ranges of particular male and female pairs, which suggests that the species may be monogamous and that the adults have a high site attachment and they hold territory exclusively. These several features of the mountain possum's biology appear to be adaptations for a stable environment where it is advantageous to the species to maintain a relatively constant population size by regulating fecundity.

The bobuck is better adapted than the brush possum to forest and displaces it in this habitat, but its very adaptations of low fecundity and high site attachment make it less resilient and adaptable to other habitats, or to alterations of its own habitat. In Tasmania, where the bobuck does not occur, the island subspecies of the brush possum, *T. vulpecula fuliginosus* inhabits the forests and resembles the mountain possum in being more terrestrial in habit than the southeastern mainland species, *T. v. vulpecula* (Owen and Thomson, 1965). The absence of the bobuck from Tasmania despite the presence of suitable forest suggests that it failed to penetrate a barrier of unsuitable habitat at the time that the island was joined to the mainland. The more versatile brush possum did do so and, in the absence of competition from the bobuck, occupied the forest.

8.4.3 Concept of forest dependent species

In both the pairs of species considered, the *r*-selected species are the more versatile and adaptable and occur in a wider range of habitats, whereas the

Table 8.5 *Distribution of 32 species of forest dwelling marsupials in relation to their dependence on wet sclerophyll forest.*

Forest dependence	South-eastern Australia	Tasmania	South-western Australia
Marginal	18	10	6
Non dependent residents	7	4	2
Dependent residents	7	1*	0
Total	32	15	8

**Petaurus breviceps.*

Data from Tyndale-Biscoe and Calaby (1975) Tables 1 and 2.

two *K*-selected species are wholly restricted to the moist habitats of wet sclerophyll eucalyptus forest and tropical rain forest. Furthermore, while the ringtail and brush possum are represented by several subspecies or closely related species throughout the Australian mainland and Tasmania, the bobuck and greater glider are monotypic and are restricted to the forests of eastern Australia. They are not even known from the forests of Tasmania or from South Western Australia, and their absence from these two isolated areas of forest is a further reflection upon their total dependence on wet sclerophyll forest (Tyndale-Biscoe and Calaby, 1975). Evidence from pollen analysis from the Bass Strait islands suggests that even when Tasmania was joined to southern Australia 10 000 years ago the vegetation of the isthmus was a heathland type and there would have been a discontinuity between the forests of Victoria and Tasmania as there is between western and eastern Australia today. When the distributions of the other forest living marsupials are examined (Table 8.5) we find that all seven species that live exclusively in the wet sclerophyll forest association of eastern Australia are absent from south western Australia and Tasmania, while the non-dependent forest residents like the brush possum and ringtail occur in these other forested areas as well. Only one species, the sugar glider, *Petaurus breviceps,* appears to be anomalous because although it is considered to be a dependent species, it is abundant in Tasmanian forests. However, it is unknown in Tasmania as a fossil from pre-European times and there is evidence from early colonial records that it was introduced in about 1840. Its successful colonisation since then is further evidence that the absence of dependent species from Tasmanian and south western forests really does reflect a long standing barrier to the movement of these species.

8.4.4 Response to forestry practices

The concept of dependent species is very pertinent to assessing the effects on forest wildlife of the various commercial practices now being used to manage or exploit the forests and in relation to the liberation of the brush possum to New Zealand.

Low intensity control burning, designed to reduce the fuel bed, does not appear to affect even the ground living species, as we have noted earlier, and even fairly intense fires, which burn the larger trees appear to have little effect on arboreal species. They are protected from flash burn in their daytime refuges, and foliage regeneration by eucalypts is rapid after fire. However, management practices that involve the cutting of overmature stems to increase timber production reduced the available homesites for arboreal species and so may cause a reduction in population. There is insufficient information to know what the precise requirements for den sites are for different species and in our study there were considerably more trees with suitable hollows than there were greater gliders but How considered that competition between gliders, brush possums and bobucks for dens was a factor in limiting populations of these three species in the wet sclerophyll forest in his study area.

Clear felling of indigenous forest for woodchip or for replacement with selected species of eucalypt or pine has an immediate and final effect on the glider population (Tyndale-Biscoe and Smith, 1969b). Greater gliders are usually unharmed at tree fall but most disappear from the site within one week and the few that do survive are generally those whose home tree stood close to the boundary with uncleared forest. Marked animals did not become established in adjacent forest, even when the glider population there had been experimentally reduced to make homesites available. Similarly How considered that the disappearance of bobucks after clear felling was due to the high site attachment of the species.

On the other hand, although ringtail possums are more liable to be killed at tree fall the species survives clear felling operations and becomes re-established in secondary growth along water courses and fire trails and in the pine plantations. So does the brush possum. Thus the two species that have been classified as dependent residents of wet sclerophyll forest, with special adaptations for survival in a stable environment, are the ones that become extinct when the forest is felled, whereas the non-dependent species are resilient and become re-established in regenerated forest.

8.5 The ecology of invasion – the brush possum in New Zealand

New Zealand separated from the land mass of Antarctica-Australia about 80 million years ago and, so far as is known from early Tertiary fossiliferous rocks, was not colonised by land mammals of any kind. Its forests evolved without leaf eating mammals, although the ground dwelling moas probably browsed seedlings and shrubs and leaf eating insects take a heavy toll of the plant growth.

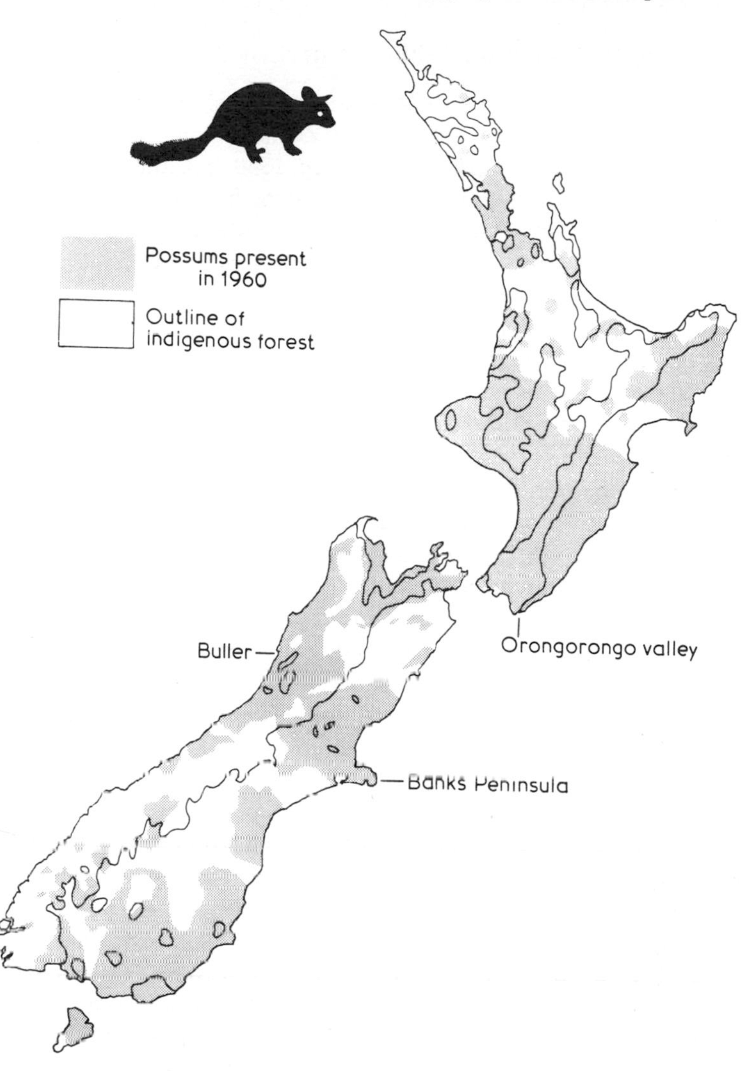

Fig. 8.7 Distribution at 1960 of the brush possum, *Trichosurus vulpecula* in New Zealand. No correlation exists between its distribution and the distribution of indigenous forest, outlined. (After Wodzicki, 1965.)

Polynesian settlers brought rats and dogs and they are thought to have induced the grasslands of the eastern side of the South Island by extensive firing of the podocarp forests. However, the advent of European settlement, as in Australia, caused the greatest changes to the forests of New Zealand. Not only were large tracts cleared for farming but many species of browsing and grazing mammals were introduced (Wodzicki, 1965).

8.5.1 Introduction and establishment

The Tasmanian brush possum (*Trichosurus vulpecula*) was highly regarded as a fur bearer and was imported to New Zealand soon after settlement. The earliest liberation may have been before 1840 but the period of major import began in 1890 mainly from Tasmania, but also from Victoria and New South Wales. Since the latter animals would have been collected near Sydney and Melbourne it is probable that they were all brush possums and not bobucks (*T. caninus*). However, bobucks may have been imported subsequently as introduction of this species was recommended as late as 1936 (Pracy, 1974). The brush possum was firmly established by the turn of the century and the success of its colonisation surpassed all expectations. The few places where initial introductions failed could be attributed to insufficient numbers released and to the lack of suitable dry shelter at the site of liberation (Gilmore, 1977). No other species of mammal has become so widespread and well established in New Zealand (Fig. 8.7). While the rabbit became exceedingly abundant and caused great damage in agricultural land until it was brought under control in the 1960s, it never penetrated more than a few kilometres into forested land. The possum on the other hand is firmly established throughout the indigenous forests as well as in exotic pine plantations, in farmland and in suburbia.

The first serious study of the possum in New Zealand was made by Kirk in 1919 who concluded that the damage to forest vegetation was insignificant, and the revenue from possum skins far outweighed it anyway. This view prevailed until the 1940s when the effects of possum browsing became more evident and by the end of the decade government opinion, especially that of the Forest Service concerned with catchment forests, regarded the possum as a pest equally as important as the rabbit and the red deer.

A bounty scheme was introduced in 1951, which lasted for 8 years. In that time the bounty was paid on 8 million possums, with no noticeable effect on the possum population and the scheme was abandoned (Gilmore, 1977). Up to this time very little work had been done on the biology of the possum to understand how it had achieved its prodigious success in New Zealand. A small beginning was made in the Orongorongo valley, near Wellington, in 1946 but serious research did not begin until 1966, when the Ecology Division of DSIR began to study the long term interactions between possums, other mammals and the indigenous forest (Gibb and Flux, 1973).

8.5.2 Effects on indigenous forest

Possums were first liberated near the Orongorongo valley in 1893 and trapping was carried out from 1921. By 1946 the density of possums was estimated to be about 6/ha, which is five or six times the density recorded from forests in Australia. The density has remained the same to the present time and it is generally thought to be a stable population. We can review this work by looking at the effect that possums have had on the forest and the effect the forest has had on the possums. Ruth Mason (1958) examined the stomach contents of 135 possums trapped over a one year period in 1946–47 and at the same time observed the vegetation near to the trap lines. She analysed the relationship between incidence of species eaten and the proximity and abundance of species around the trap sites and concluded that only species near the trap lines were eaten, that there was a marked preference in the diet for a few species of plants and that the mean number of species in any one stomach was 3.1. The four most favoured species in descending order were *Fuschia excortica, Metrosideros robusta, Alectryon excelsus* and *Weinmannia racemosa* (Fig. 8.8).

Twenty-five years later, Alice Fitzgerald (1976) began a comprehensive

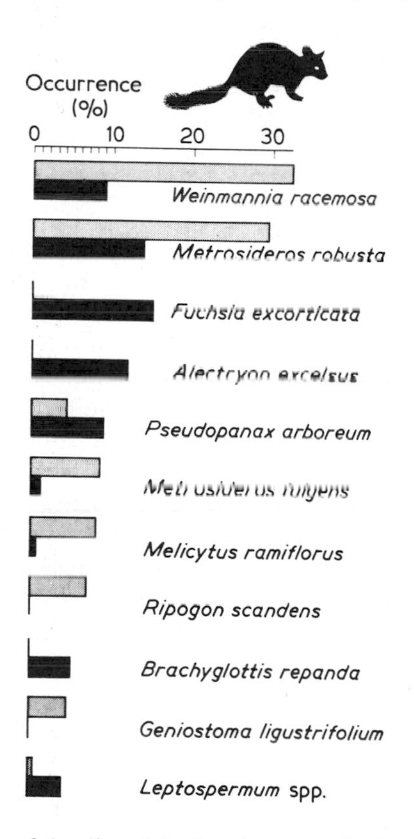

Fig. 8.8 Comparison of the diet of the brush possum in the Orongorongo valley as determined by Fitzgerald in 1969–1973 (light stipple) and by Mason in 1946–1947 (dark stipple). (From Fitzgerald, A.E., 1976, *New Zealand Journal of Zoology*, **3**, 416.)

study in the same forest of the plant species' abundance, their phenology (times of growth, flowering, fruiting and setting seed) and the seasonal pattern of possum use, derived from analysis of faecal pellets. Her results agree with Mason's so far as the possum's preference for selecting a few species to eat, but the order of preference has changed in the intervening years. *Fuchsia* and *Alectryon* are now entirely absent from the diet, whereas *Metrosideros* and *Weinmannia* together comprise over 50 per cent and several other species, not found in the earlier study now comprise up to 10 per cent of the diet (Fig. 8.8). The two species which previously had been highly preferred by the possums are no longer abundant anywhere in the study area and, outside the study area, the few extant plants are heavily browsed, so the conclusion is that selective browsing by possums has removed these two shrubby species from the forest and the possums have turned to other less favoured species.

On the study area *Metrosideros robusta*, the northern tree rata, is now of prime importance in the diet and the number of trees is about 10/ha, similar to the number of possums. Between 1970 and 1975, 24 of the trees were studied regularly by Meads (1976) from the ground and from platforms in the crown and 26 other trees were observed less intensively. Two points in Meads' study are of special interest. Although the forest contained about one large rata tree for each adult possum, only a few trees were being heavily browsed while adjacent trees remained untouched or only lightly browsed. Several possums would be observed together in a heavily browsed tree, so that it would appear that these trees are especially attractive to the possums. Similar observations made before on other species of tree from other localities had led to the conclusion that browsing stimulates the production of fresh shoots which are especially attractive to possums. However, Meads does not agree that browsing stimulates new growth on rata or that possums prefer new shoots.

Rata flowers in December–January and vegetative growth occurs from January–March. The leaves stay on the tree for 3–4 years with most being shed in December. Various species of phasmid insects eat the blade of the leaves but generally leave the petiole intact and the damaged leaves survive. On the other hand possums tear away blade and petiole and the leaves so damaged are shed prematurely and this leads to the defoliated appearance of the possum-browsed tree. Four of the ratas that were examined every two months provide the evidence of how this happened. At the start of each year the trees put on new growth but much of the new foliage was taken before it was fully grown. The damaged leaves would fall prematurely and the 3–4 year old leaves fall in December. Each year some more branches would fail to put out new shoots so that the tree became progressively defoliated as it lost its old leaves. Meads prevented possums from getting to the crowns of five heavily browsed trees by sheathing the trunks with iron sheeting and over the next five years four of them recovered fully. In the sheathed trees the numbers of phasmid insects increased, but their predation did not induce premature leaf fall and the trees recovered. Of the 50 unprotected trees, ten were dead by the end of five years. Clearly 20 per cent mortality in five years cannot be sustained for long by a species that usually lives for over 200 years.

What seems to emerge from these studies is that the possum population at Orongorongo is successively removing from the forest its preferred food plants. First *Fuchsia* and *Alectryon*, both woody shrubs, have been reduced to rare species, and now the other preferred species, which are both trees, are dying out as a consequence of the behaviour of possums to browse intensively on a few plants. Similarly, in the Westland forests of the South Island there is widespread death of the Southern rata (*Metrosideros umbellata*) which has presumably resulted from the same cause, though here aggravated by wild ungulates in the same forest. On the other hand Gilmore (1967) reported a different selection of food plants by possums or Banks Peninsula, where grasses and clover were important components and in South Westland *Fuchsia* was not so extensively used as at Orongorongo.

8.5.3 Forest influence on the possum population

How does the forest at Orongorongo affect the possum population? As mentioned above the density has remained at about 6–10/ha for the past 30 years, so that the population in the short term at least is stable. We noted before that the possum in Australia has a considerable potential for increase and is an *r* selected species. Crawley's (1973) study at Orongorongo and the continuation of it by Brockie and Bell (1979) to the present suggests that the population is behaving more like the bobuck, a *K* selected species, does in New South Wales. Crawley (1973) states that in the periods 1953–62 and 1966–68, 73 per cent of the adult females produced only one young a year in autumn and none bred in spring, so that productivity was 0.7 (compare Table 8.4). Furthermore, although females in other regions of New Zealand and Australia can breed at one year at Orongorongo few females bred before their third year. The annual recruitment of subadults to the population was 28 per cent, which indicates that one third to one quarter of the young must have died in their first year. The annual disappearance rate of adults was 26 per cent so that the mean life expectancy in this population is about 3 years. Nevertheless, some males marked when young lived for 9 years and some females for 13 years. With a longer period of study Brockie and Bell have observed that breeding is significantly affected by the season. The autumn of 1968 was unusually wet and little fruit was produced whereas 1971 was drier and an exceptionally heavy crop of fruit and seed was set. In 1971 the young were born earlier in the year than in 1968 and yearling possums were 36 per cent heavier, whereas more possums died in

1968 than in 1971. So although the possums seem to have an abundant supply of food, as leaves, throughout the year their breeding and survival are directly affected by seasonal factors.

The results from the Orongorongo studies can be compared with less precise data from other parts of New Zealand. It is possible to determine the age of a wild caught possum by counting annual cementum rings in the molar teeth (Pekelharing, 1970). Using this technique Brockie and Bell (1979) have analysed samples from regions in the North and South Islands. Despite differences in habitat and past history the age structures and inferred mortalities were very similar to the Orongorongo population. Life expectancy ranged from 2.5–3.9 years and about one third of the adult population die each year, mostly in the winter.

This finding has an important bearing on attempts to reduce the possum populations by poisoning or other direct eradication methods. Such methods to be really effective must destroy more than one third of the resident adult population before they achieve any effect that is not already taking place naturally each winter. Furthermore, we know that the possum has a potential fecundity of 1.7 which is not being fully realised in the Orongorongo forest with a stable population. If a stable population such as this one could be reduced further the reduction would be made good at the next breeding season and the only overall effect of the control measures would be to alter the ratio of young to adults.

8.5.4 Possums, cattle and tuberculosis

The need for effective control of possum populations has become very urgent, since the discovery in 1970 that possums were becoming infected with bovine tuberculosis, and may in turn be reinfecting cattle. The ecological aspects of this are interesting although the resolution of the problem is most intractable.

Possums living in forest such as that at the Orongorongo valley move short distances and have a home range of about 2 ha (Crawley, 1973), but radio tracking has disclosed that animals living in marginal country between forest and farmland will travel many kilometres from the forest to feed on pastures (Green, 1979), Clover is especially attractive and densities of 25 possums/ha have been recorded on such land. Cattle, on the other hand, if free to do so will penetrate one or two kilometres from farmland into the forest to browse and shelter. There is thus a zone of overlap between the two species where they share the same food plants and are liable to come into close contact at night. The bovine tubercle bacillus is viable for several hours on the ground at night but is rapidly destroyed in sunlight, so that the potential for infection of possums has long existed. There is no definite evidence that possums contracted bovine tuberculosis before 1970, but since then infected possums have been found in 23 localities of the North and South Island, the most severely affected region being the Buller district of the South Island. Cattle are run on the cleared river flats there which extend into the valleys. Forest covers the steep and mountainous country on each side, which makes it particularly difficult to carry out effective control operations on possums. Large sums are being spent on aerial poisoning with sodium fluoroacetate (1080) but it is very difficult to judge how effective this has been in reducing the possum population. Furthermore, there is serious concern that many other species, especially indigenous birds, may be affected by the poison. The cost of operations in the Buller district has been estimated at $12 to $14 per head of cattle at risk (Coleman, 1979), which is a large proportion of the current price of beef cattle. Indeed, it may be that the cost of attempted control exceeds the profitability of the stock at risk. When one recalls how rapidly a possum population can recover from reduction to even half its size, alternative solutions must be seriously considered. One of these is to return to the original motive for introducing possums to New Zealand, and abandon cattle in favour of exploiting possums for their

Table 8.6 *The status of medium-sized and large marsupials in Central Australia.*

Species	Cattle country	Unstocked desert
Red kangaroo (*Megaleia rufa*)	Abundant	Present
Euro (*Macropus robustus*)	Present	Present
Rock wallaby (*Petrogale lateralis*)	Present	Present
Spectacled hare-wallaby (*Lagorchestes conspicillatus*)	Probably extinct	Present
Desert hare-wallaby (*Lagorchestes hirsutus*)	Extinct	Only present at Tanami
Rat-kangaroo (*Bettongia lesueur*)	Extinct	Known still to Aborigines
Nail-tailed wallaby (*Onychogalea lunata*)	Extinct	Present
Rabbit-eared bandicoot or bilbie (*Macrotis lagotis*)	Very rare	Common
Desert bandicoot (*Perameles eremiana*)	Extinct	Known still to Aborigines
Pig-footed bandicoot (*Chaeropus ecaudatus*)	Extinct	Probably extinct

From Newsome (1971) Table 5.

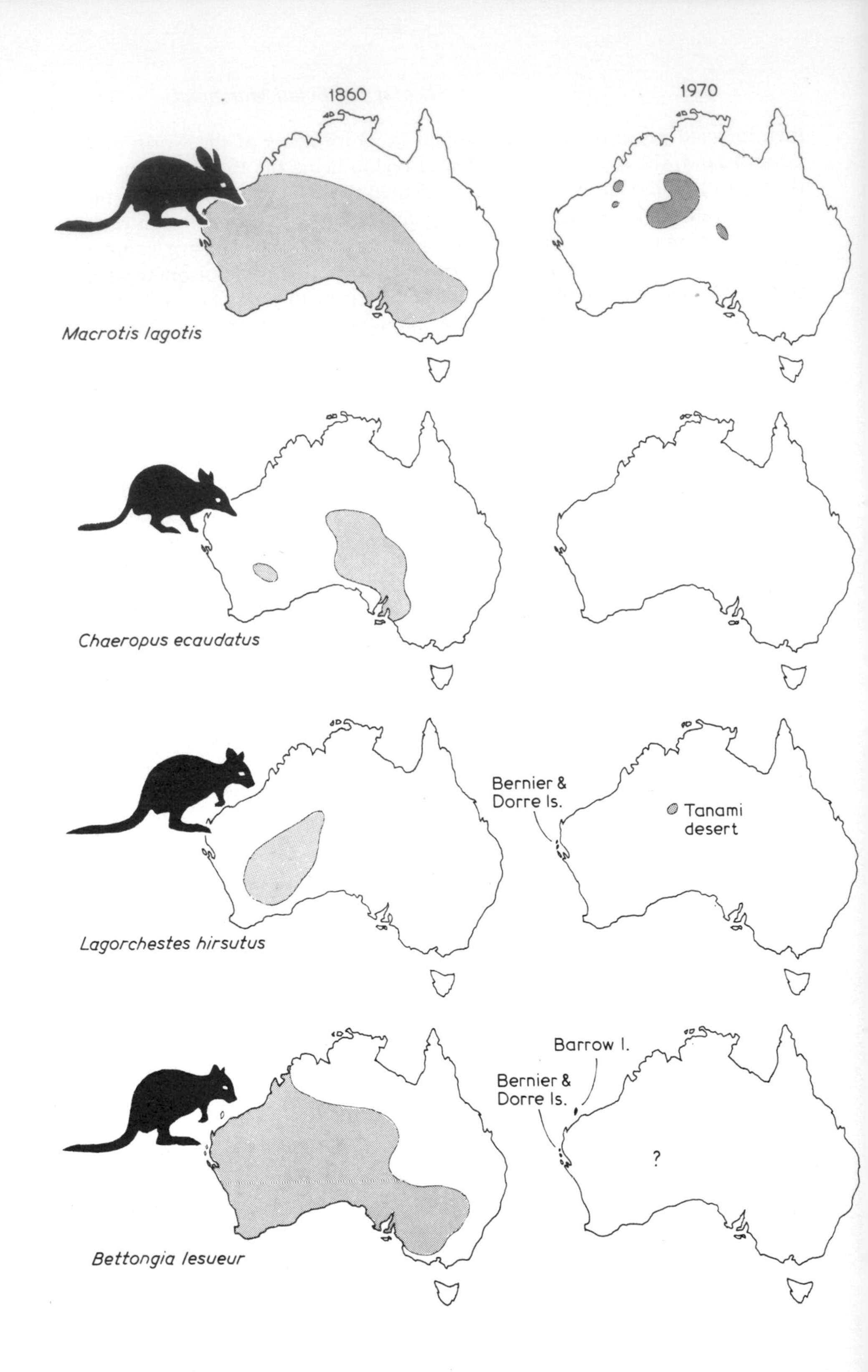

1860
1970
Macrotis lagotis
Chaeropus ecaudatus
Bernier &
Dorre Is.
Tanami
desert
Lagorchestes hirsutus
Barrow I.
Bernier &
Dorre Is.
?
Bettongia lesueur

pelts. In 1976 there were 6000 licensed possum trappers in New Zealand and 1.5 million skins were exported, which earned $4.5 million.

The history of the rabbit and the possum in New Zealand, provide an interesting comparison of how the marsupial and eutherian mode of reproduction affects the pattern of colonisation by invading species. Both species were introduced to New Zealand in 1850. The rabbit had become a devastating plague 20 years later, whereas the possum was not recognized as a serious pest until 90 years after its first liberation. The build up of the possum population was much slower than that of the rabbit, presumably because of its much lower reproductive potential. However, its ability to use a great variety of plants as food and to occupy a greater diversity of habitats has enabled it to occupy New Zealand far more extensively than the rabbit, so that in the long term the marsupial mode of reproduction has proved to be no disadvantage to it at all and it is now a far more intractable problem than the rabbit.

8.6 The ecology of extinction — small marsupials of the inland plains of Australia

The marsupials of the inland plains of Australia have been the group most affected by European settlement. Of 21 species originally present 9 are extinct and all but 3 are rare. The three that are not rare are the large kanga roos which greatly increased in abundance, whereas the small macropods and bandicoots declined. Newsome (1975) has reviewed the factors that have caused the large kangaroos to increase and has attempted to discern the causes of the decline of the smaller species (Newsome, 1971). Unfortunately their decline occurred in an undramatic way before ecological research had begun in Australia, so little is known of their biology. What is clear is that most of the species were abundant when the country was first explored and the decline in abundance is closely associated with cattle raising (Table 8.6). According to Newsome (1971, 1975) three main types of ecological change appear to have been involved; competition for food, removal of shelter and increased predation. Four species, whose past and present status are shown in Fig. 8.9 illustrate these three aspects.

Bettongia lesueur is a small (1 kg) rat kangaroo which makes extensive burrows for daytime shelter and feeds on a varied herbivorous diet, which includes grass. Once it was abundant and widespread from the Western Australian coast to New South Wales but is now restricted to three islands and doubtfully a few mainland localities. Its demise seems to have resulted

Fig. 8.9 Past and present distributions of four marsupials that have declined since stock were introduced to the inland plains of Australia. (After Newsome, 1971.)

from direct competition for food with stock and rabbits. Rabbits may have displaced *Bettongia* from their burrows, but their final extinction was associated with the later spread of the fox.

A more subtle competition may have been responsible for the decline of the rabbit eared bandicoot *Macrotis lagotis*. It also makes burrows for daytime shelter but its diet is almost exclusively termites (Smyth and Philpott, 1968). The termites, *Drepanotermes rubriceps* and *D. perniger* live in large colonies and feed on grasses for which they compete with stock. Overstocking and drought combined to reduce the grass so that the termite colonies declined and so did their predators, the bandicoots.

The small hare wallabies, *Lagorchestes hirsutus* and *L. conspicillatus* used to be so abundant that cattlemen would course them for sport; now the former is almost extinct and the latter is rare. Both species are grazers which found daytime refuge in tall tussocks. Newsome (1971) suggests that the first effects of cattle may have led to an increase of these species because the grazing induced green shoots to grow from the tussocks. However, as grazing pressure increased the tussocks were progressively reduced and the hare wallabies' shelters disappeared. Similarly the pigfooted bandicoot *Chaeropus ecaudatus* made daytime nests in tussocks and became acutely vulnerable to heat and natural predators, such as wedgetail eagles and dingoes when this was eaten from over them. The nail-tailed wallabies, *Onychogalea lunata* and *O. fraenata* may have declined for the same reason.

8.7 Acknowledgements

In preparing this chapter several people let me see manuscript papers or otherwise provided insights and help from their own experience and I wish to thank especially John Barnett, Richard Braithwaite, Ben Bell, Bob Brockie, John Calaby, Alice Fitzgerald, John Gibb, Tony Lee, Alan Newsome and Kasimierz Wodzicki.

8.8 References

Barnett, J.L. (1973) A stress response in *Antechinus stuartii* (Macleay). *Australian Journal of Zoology,* **21**, 501-513.

Beveridge, I. & Barker, I.K. (1976) The parasites of *Antechinus stuartii* Macleay from Powelltown, Victoria with observations on seasonal and sex-related variations in numbers of helminths. *Australian Journal of Zoology,* **24**, 265-272.

Braithwaite, R.W. (1974) Behavioural changes associated with the population cycle of *Antechinus stuartii* (Marsupialia). *Australian Journal of Zoology,* **22**, 45-62.

Braithwaite, R.W. & Lee, A.K. (1979) A mammalian example of semelparity. *American Naturalist*, **113**, 151-155.

Brockie, R.E. & Bell, B.D. (1979) Age structure and mortality of possum populations. *Proceedings of Symposium on Marsupials in New Zealand,* Victoria University of Wellington.

Clemens, W.A. (1977) Phylogeny of the marsupials. In: *The Biology of Marsupials.* Eds: B. Stonehouse and D.P. Gilmore. Macmillan, London, pp. 51-68.

Coleman, J.D. (1979) Tuberculosis/possum control – an expensive business. *Proceedings of Symposium on Marsupials in New Zealand,* Victoria University of Wellington.

Crawley, M.C. (1973) A live-trapping study of Australian brush-tailed possums, *Trichosurus vulpecula* (Kerr), in the Orongorongo valley, Wellington, New Zealand. *Australian Journal of Zoology,* **21**, 75-90.

Dunnet, G.M. (1964) A field study of local populations of the brush-tailed possum, *Trichosurus vulpecula* in eastern Australia. *Proceedings of the Zoological Society of London,* **142**, 665-695.

Dwyer, P.D. (1977) Notes on *Antechinus* and *Cercartetus* (Marsupialia) in the New Guinea Highlands. *Proceedings of the Royal Society of Queensland,* **88**, 69-73.

Fitzgerald, A.E. (1976) Diet of the opossum *Trichosurus vulpecula* (Kerr) in the Orongorongo valley, Wellington, New Zealand, in relation to food-plant availability. *New Zealand Journal of Zoology,* **3**, 399-419.

Fleming, T.H. (1973) The reproductive cycles of three species of opossums and other mammals in the Panama Canal Zone. *Journal of Mammalogy,* **54**, 439-455.

Gardiner, A.L. (1973) The systematics of the genus *Didelphis* (Marsupialia: Didelphidae) in north and middle America. *Special publications of the Museum Texas Tech University,* **4**, 1-81.

Gibb, J.A. & Flux, J.E.C. (1973) Mammals. In: *The natural history of New Zealand.* Ed: G.R. Williams. Reed, Wellington.pp. 334-371.

Gilmore, D.P. (1967) Foods of the Australian Opossum (*Trichosurus vulpecula* Kerr) on Banks Peninsula, Canterbury, and a comparison with other selected areas. *New Zealand Journal of Science,* **10**, 235-279.

Gilmore, D.P. (1977) The success of marsupials as introduced species. In: *The Biology of Marsupials.* Eds: Stonehouse, B. & Gilmore, D.P. MacMillan, London, pp. 169-178.

Gordon, G. (1974) Movements and activity of the short-nosed bandicoot, *Isoodon macrourus* Gould (Marsupialia). *Mammalia,* **38**, 405-431.

Green, W.Q. (1979) A progress report on movements of possums between native forest and pasture. *Proceedings of Symposium on Marsupials in New Zealand,* Victoria University of Wellington.

Hamilton, W.J. (1958) Life history and economic relations of the opossum (*Didelphis marsupialis virginiana*) in New York State. *Memoirs of Cornell University Agricultural Experimental Station,* **354**, 1-48.

Heinsohn, G.E. (1966) Ecology and reproduction of the Tasmanian bandicoots (*Perameles gunni* and *Isoodon obesulus*). *University of California Publications in Zoology,* **80**, 1-96.

Hershkovitz, P. (1972) The recent mammals of the Neotropical region: a zoogeographic and ecological review. In: *Evolution, mammals, and southern continents.* Eds: Keast, A., Erk, F.C. & Glass, B. State University of New York, Albany. pp. 311-432.

How, R.A. (1976) Reproduction, growth and survival of young in the mountain possum, *Trichosurus caninus* (Marsupialia). *Australian Journal of Zoology,* **24**, 189-199.

Hunsaker II, D. (1977a) *The Biology of Marsupials*. Academic Press, N.Y.

Hunsaker II, D. (1977b) Ecology of new world marsupials. In: *The Biology of Marsupials*. Ed: Hunsaker II, D. Academic Press, N.Y. pp. 95-158.

Inns, R.W. (1976) Some seasonal changes in *Antechinus flavipes* (Marsupialia: Dasyuridae). *Australian Journal of Zoology*, **24**, 523-531.

Keast, A. (1972) Comparisons of contemporary mammal faunas of southern continents. In: *Evolution, Mammals, and Southern Continents*. Eds: Keast, A., Erk, F.C. & Glass, B. State University of New York, Albany. pp. 433-490.

Lee, A.K., Bradley, A.J. & Braithwaite, R.W. (1977) Corticosteroid levels and male mortality in *Antechinus stuartii*. In: *The Biology of Marsupials*. Eds: Stonehouse, B. & Gilmore, D.P. Macmillan, London. pp. 209-220.

Leonard, B.V. (1976) The effect of fire on small mammals in south eastern Australia. *Australian Mammal Society Bulletin*, **3**, 16.

Mason, R. (1958) Foods of the Australian opossum (*Trichosurus vulpecula*, Kerr) in New Zealand indigenous forest in the Orongorongo valley, Wellington. *New Zealand Journal of Science*, **1**, 590-613.

McArthur, R.H. & Wilson, E.O. (1967) *The theory of island biogeography*. Princeton University Press, Princeton.

Meads, M.J. (1976) Effects of opossum browsing on northern rata trees in the Orongorongo Valley, Wellington, New Zealand. *New Zealand Journal of Zoology*, **3**, 127-139.

Newsome, A.E. (1971) Competition between wildlife and domestic stock. *The Australian Veterinary Journal*, **47**, 577-586.

Newsome, A.E. (1975) An ecological comparison of the two arid zone kangaroos of Australia, and their anomalous prosperity since the introduction of ruminant stock to their environment. *The Quarterly Journal of Biology*, **50**, 389-424.

Newsome, A.E., McIlroy, J. & Catling, P. (1975) The effects of an extensive wildfire on populations of twenty ground vertebrates in south-east Australia. *Proceedings of the Ecological Society of Australia*, **9**, 107-123.

Owen, W.H. & Thomson, J.A. (1965) Notes on the comparative ecology of the common brushtail and mountain possums in eastern Australia. *Victorian Naturalist*, **82**, 216-217.

Parker, P. (1977) An ecological comparison of marsupial and placental patterns of reproduction. In: *The Biology of Marsupials*. Eds: Stonehouse, B. & Gilmore, D.P. MacMillan, London. pp. 273-286.

Pekelharing, C.J. (1970) Cementum deposition as an age indicator in the brush-tailed possum, *Trichosurus vulpecula* Kerr (Marsupialia). *Australian Journal of Zoology*, **18**, 71-76.

Pracy, L.T. (1974) Introduction and liberation of the opossum into New Zealand. *New Zealand Forest Service Information Series* No. 45.

Smith, M.J., Brown, B.K. & Frith, H.J. (1969) Breeding of the brush-tailed possum, *Trichosurus vulpecula* (Kerr), in New South Wales. *CSIRO Wildlife Research*, **14**, 181-193.

Smyth, D.R. & Philpott, C.M. (1968) Field notes on rabbit bandicoots, *Macrotis lagotis* Reid (Marsupialia), from central Western Australia. *Transactions of the Royal Society of South Australia*, **92**, 3-14.

Stodart, E. (1977) Breeding and behaviour of Australian bandicoots. In: *The Biology of Marsupials*. Eds: Stonehouse, B. & Gilmore, D.P. MacMillan, London, pp. 179-192.

Stoddart, D.M. & Braithwaite, R.W. (in press) A strategy for utilization of regenerating heathland habitat in the brown bandicoot (*Isoodon obesulus*; Marsupialia, Peramelidae) *Journal of Animal Ecology*.

Stonehouse, B. & Gilmore, D.P. (1977) *The Biology of Marsupials*. Macmillan, London.

Thomson, J.A. & Owen, W.H. (1964) A field study of the Australian ringtail possum, *Pseudocheirus peregrinus* (Marsupialia: Phalangeridae). *Ecological Monographs,* **34**, 27-52.

Tyndale-Biscoe, H. (1973) *Life of Marsupials*. Edward Arnold, London.

Tyndale-Biscoe, C.H. & Calaby, J.H. (1975) Eucalypt forests as refuge for wildlife. *Australian Forestry,* **38**, 117-133.

Tyndale-Biscoe, C.H. & Mackenzie, R.B. (1976) Reproduction in *Didelphis marsupialis* and *D. albiventris* in Colombia. *Journal of Mammalogy,* **57**, 249-265.

Tyndale-Biscoe, C.H. & Smith, R.F.C. (1969a) Studies on the marsupial glider, *Schoinobates volans* (Kerr). II. Population structure and regulatory mechanisms. *Journal of Animal Ecology,* **38**, 637-650.

Tyndale-Biscoe, C.H. & Smith, R.F.C. (1969b) Studies on the marsupial glider, *Schoinbates volans* (Kerr). III. Response to habitat destruction. *Journal of Animal Ecology,* **38**, 651-659.

Wainer, J.W. (1976) Studies of an island population of *Antechinus minimus* (Marsupialia, Dasyuridae). *The Australian Zoologist,* **19**, 1-7.

Wodzicki, K. (1965) The status of some exotic vertebrates in the ecology of New Zealand. In: *The Genetics of Colonizing Species*. Eds: Baker, H.G. & Stebbins, G.L. Academic Press, N.Y. pp. 425-458.

Wood, D.H. (1970) An ecological study of *Antechinus stuartii* (Marsupialia) in a south-east Queensland rain forest. *Australian Journal of Zoology,* **18**, 185-207.

Woolley, P. (1966) Reproduction in *Antechinus* spp. and other Dasyurid marsupials. *Symposia of the Zoological Society of London,* **15**, 281-294.

Ziegler, A.C. (1977) Evolution of New Guinea's marsupial fauna in response to a forested environment. In: *The Biology of Marsupials*. Eds: Stonehouse, B. & Gilmore, D.P. MacMillan, London. pp. 117-140.

Index